Real Analysis and Foundations

Steven G. Krantz is a professor of mathematics at Washington University in St. Louis. He has previously taught at UCLA, Princeton University, and Pennsylvania State University. He has written more than 130 books and more than 250 scholarly papers and is the founding editor of the *Journal of Geometric Analysis*. An AMS Fellow, Dr. Krantz has been a recipient of the Chauvenet Prize, Beckenbach Book Award, and Kemper Prize. He received a Ph.D. from Princeton University.

Textbooks in Mathematics

Series editors:
Al Boggess, Kenneth H. Rosen

The Elements of Advanced Mathematics, Fifth Edition
Steven G. Krantz

Differential Equations
Theory, Technique, and Practice, Third Edition
Steven G. Krantz

Geometry and Its Applications, Third Edition
Walter J. Meyer

Transition to Advanced Mathematics
Danilo R. Diedrichs and Stephen Lovett

Modeling Change and Uncertainty
Machine Learning and Other Techniques
William P. Fox and Robert E. Burks

Abstract Algebra
A First Course, Second Edition
Stephen Lovett

Multiplicative Differential Calculus
Svetlin Georgiev, Khaled Zennir

Applied Differential Equations
The Primary Course
Vladimir A. Dobrushkin

Introduction to Computational Mathematics: An Outline
William C. Bauldry

Mathematical Modeling the Life Sciences
Numerical Recipes in Python and MATLAB™
N. G. Cogan

https://www.routledge.com/Textbooks-in-Mathematics/book-series/
CANDHTEXBOOMTH

Real Analysis and Foundations

Fifth Edition

Steven G. Krantz

CRC Press
Taylor & Francis Group
Boca Raton London New York

CRC Press is an imprint of the
Taylor & Francis Group, an **informa** business

A CHAPMAN & HALL BOOK

Fifth edition published 2022
by CRC Press
6000 Broken Sound Parkway NW, Suite 300, Boca Raton, FL 33487-2742

and by CRC Press
2 Park Square, Milton Park, Abingdon, Oxon, OX14 4RN

© 2022 Steven G. Krantz

First edition published by CRC Press 1991
Second edition published by CRC Press 2004
Third edition published by CRC Press 2013
Fourth edition published by CRC Press 2016

CRC Press is an imprint of Taylor & Francis Group, LLC

ISBN: 978-1-032-10272-6 (hbk)
ISBN: 978-1-032-12026-3 (pbk)
ISBN: 978-1-003-22268-2 (ebk)

DOI: 10.1201/9781003222682

Typeset in Palatino
by MPS Limited, Dehradun

To Stan Philipp, who taught me real analysis.

And to Walter Rudin, who wrote the books from which I learned.

Contents

Preface

Overview

The subject of real analysis, or "advanced calculus," has a central position in undergraduate mathematics education. Yet, because of changes in the preparedness of students, and because of their early exposure to calculus (and therefore lack of exposure to certain other topics) in high school, this position has eroded. Students unfamiliar with the value of rigorous, axiomatic mathematics are ill-prepared for a traditional course in mathematical analysis.

Thus, there is a need for a book that simultaneously introduces students to rigor, to the *need* for rigor, and to the subject of mathematical analysis. The correct approach, in my view, is not to omit important classical topics like the Weierstrass Approximation theorem and the Ascoli-Arzela theorem, but rather to find the simplest and most direct path to each. While mathematics should be written "for the record" in a deductive fashion, proceeding from axioms to special cases, this is *not* how it is learned. Therefore, (for example) I *do* treat metric spaces (a topic that has lately been abandoned by many of the current crop of analysis texts). I do so not at first but rather at the end of the book as a method for unifying what has gone before. And, I do treat Riemann–Stieltjes integrals, but only after first doing Riemann integrals. I develop real analysis gradually, beginning with treating sentential logic, set theory, and constructing the integers.

The approach taken here results, in a technical sense, in some repetition of ideas. But, again, this is how one learns. Every generation of students comes to the university, and to mathematics, with its own viewpoint and background. Thus, I have found that the classic texts from which we learned mathematical analysis are often no longer suitable, or appear to be inaccessible, to the present crop of students. It is my hope that my text will be a suitable source for modern students to learn mathematical analysis. Unlike other authors, I do not believe that the subject has changed; therefore I have not altered the fundamental content of the course. But, the point of view of the audience has changed, and I have written my book accordingly.

The current crop of real analysis texts might lead one to believe that real analysis is simply a rehash of calculus. Nothing could be further from the truth. But, many of the texts written 30 years ago are simply too dry and austere for today's audience. My purpose here is to teach today's students the mathematics that I grew to love in a language that speaks to them.

Prerequisites

A student with a standard preparation in lower division mathematics—calculus and differential equations—has adequate preparation for a course based on this text. Many colleges and universities now have a "transitions" course that helps students develop the necessary mathematical maturity for an upper divi- sion course such as real analysis. I have taken the extra precaution of providing a mini-transitions course in my Chapter 0. Here, I treat logic, basic set theory, methods of proof, and constructions of the number systems. Along the way, students learn about mathematical induction, equivalence classes, completeness, and many other basic constructs. In the process of reading these chapters, written in a rigorous but inviting fashion, the student should gain both a taste for and an appreciation for the use of rigor. While many instructors will want to spend some class time with this chapter, others will make it assigned reading and begin the course proper with Chapter 1.

How to Build a Course from This Text

Chapters 2 through 8 present a first course in real analysis. I begin with the simplest ideas—sequences of numbers—and proceed to series, topology (on the real line only), limits and continuity of functions, and differentiation of functions. These are followed of course by the integral and sequences and series of functions.

The order of topics is similar to that in traditional books like *Principles of Mathematical Analysis* by Walter Rudin, but the treatment is more gentle. There are many more examples, and much more explanation. I do not short-change the really interesting topics like compactness and connectedness. The exercise sets provide plenty of drill, in addition to the more traditional "Prove this, Prove that." If it is possible to obtain a simpler presentation by giving up some generality, I always opt for simplicity.

Today, many engineers and physicists are required to take a term of real analysis. Chapters 2 through 8 are designed for that purpose. For the more mathematically inclined, this first course serves as an introduction to the more advanced topics treated in the second part of the book.

In Chapter 7, I give a rather traditional treatment of the integral. First, the Riemann integral is covered, then the Riemann–Stieltjes integral. I am careful to establish the latter integral as the natural setting for the integration by parts theorem. I establish explicitly that series are a special case of the Riemann– Stieltjes integral. Functions of bounded variation are treated briefly and their utility in integration theory is explained.

The usual material on sequences and series of functions in Chapter 8 (*including* uniform convergence) is followed by a somewhat novel chapter on "Special Functions." Here, I give a rigorous treatment of the elementary transcendental functions as well as an introduction to the gamma function and its application to Stirling's formula.

I feel strongly, based in part on my own experience as a student, that analysis of several variables is a tough nut the first time around. In particular, college juniors and seniors are not (except perhaps at the very best schools) ready for differential forms. Therefore, my treatment of functions of several variables in Chapter 10 is brief, it is only in $\mathbb{R}^3$, and it excludes any reference to differential forms. The main interests of this chapter, from the student's point of view, are (i) that derivatives are best understood using linear algebra and matrices and (ii) that the inverse function theorem and implicit function theorem are exciting new ideas. There are many fine texts that cover differential forms and related material and the instructor who wishes to treat that material in depth should supplement my text with one of those.

Chapter 11 is dessert. For I have waited until now to introduce the language of metric spaces. But now comes the power, for I prove and apply both the Baire category theorem and the Ascoli-Arzela theorem. This is a suitable finish to a year-long course on the elegance and depth of rigorous reasoning.

Chapters 12 and 13 cover differential equations and harmonic analysis. These are obviously topics that the instructor will want to dip into as interest dictates. They serve to show the student what real analysis is for, and why it is important. It is here for color and for texture.

I would teach my second course in real analysis by covering all of Chapters 9 through 13. Material in Chapters 12 and 13 is easily omitted if time is short.

Audience

This book is intended for college juniors and seniors and some beginning graduate students. It addresses the same niche as the classic books of Apostol, Royden, and Rudin. However, the book is written for today's audience in today's style. All the topics which excited my sense of wonder as a student—the Cantor set, the Weierstrass nowhere differentiable function, the Weierstrass approximation theorem, the Baire category theorem, the Ascoli-Arzela theorem—are covered. They can be skipped by those teaching a course for which these topics are deemed inappropriate. But they give the subject real texture.

What Is New in This Edition

A fifth edition of a longstanding textbook should offer the user something new. We have many new exercises in every section. These include both drill exercises and thought exercises.

We have historical passages to introduce the student to some of the main characters in this drama. Many of these are colorful personages, and it is well to become acquainted with them.

As noted, we have a new chapter on metric spaces and other advanced topics. This is a nice payoff for the student who has worked hard through the first ten chapters.

We have brief passages that introduce the student to modern topics and open questions in real analysis.

And, we have a Chapter 0 that reviews "transition" material about logic, set theory, number systems, and axiomatics. This will be background for many but preparation for some. It should be a useful reference for all concerned.

Our goal here is to make this text as useful and accessible as possible, both for the student and for the instructor.

Steven G. Krantz

Acknowledgments

It is a pleasure to thank Ken Rosen for contributing a number of ideas and insights for improvement of the text.

Of course, I am always grateful to my editor Bob Ross for his encouragement and support.

— Steven G. Krantz
St. Louis, Missouri

Background Material

0.1 Number Systems

In this section, we treat the elementary number systems that you first learned about in grade school. These include the natural or counting numbers, the integers, and the rational numbers. You should think of this as a high-level review.

Our treatment of the more-sophisticated real and complex numbers comes in Chapter 1.

0.1.1 The Natural Numbers

Mathematics deals with a variety of number systems. The simplest number system is $\mathbb{N}$, the *natural numbers*. This is just the set of positive integers $\{1, 2, 3, \ldots\}$. In a rigorous course of logic, the set $\mathbb{N}$ is constructed from the axioms of set theory. However, in this book, we shall assume that you are familiar with the positive integers and their elementary properties.

The principal properties of $\mathbb{N}$ are as follows:

1. The number 1 is a natural number.
2. If n is a natural number, then there is another natural number $\hat{n}$, which is called the *successor* of n. (We think of the successor of n as the number that comes after n.)
3. $1 \neq \hat{n}$ for every natural number n.
4. If $\hat{m} = \hat{n}$, then $m = n$.
5. *(Principle of Induction)* If $Q(n)$ is a property of the natural number n and if
   a. The property $Q(1)$ holds;
   b. Whenever $Q(j)$ holds, then it follows that $Q(\hat{j})$ holds;
   then all natural numbers have the property Q.

These rules, or *axioms*, are known as the *Peano Axioms* for the natural numbers (named after Giuseppe Peano (1858–1932) who developed them).

DOI: 10.1201/9781003222682-1

We take it for granted that the usual set of natural numbers satisfies these rules. We see that 1 is in the set $\mathbb{N}$ of natural numbers. Each positive integer has a "successor"—after 1 comes 2, after 2 comes 3, and so forth. The number 1 is not the successor of any other positive integer. Two positive integers with the same successor must be the same. The last axiom is more subtle but makes good sense: if some property $Q(n)$ holds for $n = 1$ and if whenever it holds for n then it also holds for $n + 1$, then we may conclude that Q holds for all positive integers.

We will spend the remainder of this section exploring Axiom **(5)**, the Principle of Induction.

Example 0.1: Let us prove that, for each positive integer n, it holds that

$$1 + 2 + \cdots + n = \frac{n \cdot (n + 1)}{2}.$$

We denote this equation by $Q(n)$, and follow the scheme of the Principle of Induction.

First, $Q(1)$ is true, because both the left and the right side of the equation equal 1. Now, assume that $Q(n)$ is true for some natural number n. Our job is to show that it follows that $Q(n + 1)$ is true.

Since $Q(n)$ is true, we know that

$$1 + 2 + \cdots + n = \frac{n \cdot (n + 1)}{2}.$$

Let us add the quantity $n + 1$ to both sides. Thus,

$$1 + 2 + \cdots + n + (n + 1) = \frac{n \cdot (n + 1)}{2} + (n + 1).$$

The right side of this new equality simplifies and we obtain

$$1 + 2 + \cdots + (n + 1) = \frac{(n + 1) \cdot ((n + 1) + 1)}{2}.$$

However, this is just $Q(n + 1)$ or $Q(\hat{n})$! *We have assumed $Q(n)$ and have proved $Q(\hat{n})$*, just as the Principle of Induction requires.

Thus, we may conclude that property Q holds for all positive integers, as desired.   □

The formula that we derived in Example 0.1 was probably known to the ancient Greeks. However, a celebrated anecdote credits Carl Friedrich

Gauss (1777–1855) with discovering the formula when he was nine years old. Gauss went on to become (along with Isaac Newton and Archimedes) one of the three greatest mathematicians of all time.

The formula from Example 0.1 gives a neat way to add up the integers from 1 to n, for any n, without doing any work. Any time that we discover a new mathematical fact, there are generally several others hidden within it. The next example illustrates this point.

Example 0.2: The sum of the first m positive even integers is $m \cdot (m + 1)$. To see this, note that the sum in question is

$$2 + 4 + 6 + \cdots + 2m = 2(1 + 2 + 3 + \cdots + m).$$

However, by the first example, the sum in parentheses on the right is equal to $m \cdot (m + 1)/2$. It follows that

$$2 + 4 + 6 + \cdots + 2m = 2 \cdot \frac{m \cdot (m + 1)}{2} = m \cdot (m + 1). \qquad \square$$

The second example could also be performed by mathematical induction (without using the result of the first example).

Example 0.3: Now, we will use mathematical induction incorrectly to prove a statement that is completely preposterous:

All horses are the same color.

There are finitely many horses in existence, so it is convenient for us to prove the slightly more technical statement

Any collection of k horses consists of horses that are all the same color.

Our statement $Q(k)$ is this last displayed statement.

Now, $Q(1)$ is true: *one horse is the same color*. (Note: this is not a joke, and the error has not occurred yet.)

Suppose next that $Q(k)$ is true: we assume that any collection of k horses has the same color. Now, consider a collection of $\hat{k} = k + 1$ horses. Remove one horse from that collection. By our hypothesis, the remaining k horses have the same color.

Now, replace the horse that we removed and remove a different horse. Again, the remaining k horses have the same color.

We keep repeating this process: remove each of the $k + 1$ horses one by one and conclude that the remaining k horses have the same color.

Therefore, every horse in the collection is the same color as every other. So, all $k + 1$ horses have the same color. The statement $Q(k + 1)$ is thus proved (assuming the truth of $Q(k)$) and the mathematical induction is complete.

Where is our error? It is nothing deep—just an oversight. The argument we have given is wrong when $\hat{k} = k + 1 = 2$ for removing one horse from a set of two and the remaining (*one*) horse is the same color. Now, replace the removed horse and remove the other horse. The remaining (*one*) horse is the same color. *So what?* We cannot conclude that the two horses are colored the same. Thus, the mathematical induction breaks down at the outset; the reasoning is incorrect. □

Proposition 0.4: *Let a and b be real numbers and n a natural number. Then*

$$(a + b)^n = a^n + \frac{n}{1}a^{n-1}b + \frac{n(n-1)}{2 \cdot 1}a^{n-2}b^2$$
$$+ \frac{(n(n-1)(n-2)}{3 \cdot 2 \cdot 1}a^{n-3}b^3$$
$$+ \cdots + \frac{n(n-1)\cdots 2}{(n-1)(n-2)\cdots 2 \cdot 1}ab^{n-1} + b^n.$$

Proof: The case $n = 1$ being obvious, proceed by mathematical induction. □

Example 0.5: The expression

$$\frac{n(n-1)\cdots(n-k+1)}{k(k-1)\cdots 1}$$

is often called the kth *binomial coefficient* and is denoted by the symbol

$$\binom{n}{k}.$$

Using the notation $m! = m \cdot (m-1) \cdot (m-2) \ldots 2 \cdot 1$, for m, a natural number, we may write the kth binomial coefficient as

$$\binom{n}{k} = \frac{n!}{(n-k)! \cdot k!}.$$
□

0.1.2 The Integers

Now, we will apply the notion of an equivalence class to *construct* the integers (both positive and negative). There is an important point of knowledge to be noted here. For the sake of having a reasonable place to begin our work, we considered the natural numbers $\mathbb{N} = \{1, 2, 3, ...\}$ as given. Since the natural numbers have been used for thousands of years to keep track of objects for barter, this is a plausible thing to do. Even people who has no knowledge of mathematics accept the positive integers. However, the number zero and the negative numbers are a different matter. It was not until the fifteenth century that the concepts of zero and negative numbers started to take hold—for they do not correspond to explicit collections of objects (five fingers or ten shoes) but rather to *concepts* (zero books is the lack of books; minus four pens means that we owe someone four pens). After some practice, we get used to negative numbers, but explaining in words what they mean is always a bit clumsy.

It is much more satisfying, from the point of view of logic, to *construct* the integers (including the negative whole numbers and zero) from what we already have, that is, from the natural numbers. We proceed as follows. Let $A = \mathbb{N} \times \mathbb{N}$, the set of ordered pairs of natural numbers. We define a relation (see Section 0.2.6) $\mathcal{R}$ on A and A as follows:

$$(a, b) \text{ is related to } (a', b') \text{ if } a + b' = a' + b$$

See also Section 0.2.6 for the concept of equivalence relation.

Theorem 0.6: *The relation $\mathcal{R}$ is an equivalence relation.*

Proof: That (a, b) is related to (a, b) follows from the trivial identity $a + b = a + b$. Hence, $\mathcal{R}$ is reflexive. Second, if (a, b) is related to (a', b'), then $a + b' = a' + b$; hence, $a' + b = a + b'$ (just reverse the equality), hence (a', b'), is related to (a, b). So, $\mathcal{R}$ is symmetric.

Finally, if (a, b) is related to (a', b') and (a', b') is related to (a'', b''), then we have

$$a + b' = a' + b \text{ and } a' + b'' = a'' + b'.$$

Adding these equations gives

$$(a + b') + (a' + b'') = (a' + b) + (a'' + b').$$

Cancelling a' and b' from each side finally yields

$$a + b'' = a'' + b.$$

Thus, (a, b) is related to (a'', b''). Therefore, $\mathcal{R}$ is transitive. We conclude that $\mathcal{R}$ is an equivalence relation. □

Now, our job is to understand the equivalence classes, which are induced by $\mathcal{R}$. (We will ultimately call this number system the integers $\mathbb{Z}$.) Let $(a, b) \in A$ and let $[(a, b)]$ be the corresponding equivalence class. If $b > a$, then we will denote this equivalence class by the integer $b - a$. For instance, the equivalence class $[(2, 7)]$ will be denoted by 5. Notice that if $(a', b') \in [(a, b)]$, then $a + b' = a' + b$; hence, $b' - a' = b - a$. Therefore, the integer symbol that we choose to represent our equivalence class is *independent of that element of the equivalence class that is used to compute it*.

If $(a, b) \in A$ and $b = a$, then we let the symbol 0 denote the equivalence class $[(a, b)]$. Notice that if (a', b') is any other element of $[(a, b)]$, then it must be that $a + b' = a' + b$; hence, $b' = a'$; therefore, this definition is unambiguous.

If $(a, b) \in A$ and $a > b$, then we will denote the equivalence class $[(a, b)]$ by the symbol $-(a - b)$. For instance, we will denote the equivalence class $[(7, 5)]$ by the symbol -2. Once again, if (a', b') is related to (a, b), then the equation $a + b' = a' + b$ guarantees that our choice of symbol to represent $[(a, b)]$ is unambiguous.

Thus, we have given our equivalence classes names, and these names *look just like* the names that we usually give to integers: there are positive and negative integers, and zero. However, we want to see that these objects *behave* like integers. (As you read on, use the intuitive, nonrigorous mnemonic that the equivalence class $[(a, b)]$ stands for the integer $b - a$.)

First, do these new objects that we have constructed *add* correctly? Well, let $X = [(a, b)]$ and $Y = [(c, d)]$ be two equivalence classes. *Define* their sum to be $X + Y = [(a + c, b + d)]$. We must check that this is unambiguous. If $(\tilde{a}, \tilde{b})$ is related to (a, b) and $(\tilde{c}, \tilde{d})$ is related to (c, d), then, of course, we know that

$$a + \tilde{b} = \tilde{a} + b$$

and

$$c + \tilde{d} = \tilde{c} + d.$$

Adding these two equations gives

$$(a + c) + (\tilde{b} + \tilde{d}) = (\tilde{a} + \tilde{c}) + (b + d)$$

hence, $(a + c, b + d)$ is related to $(\tilde{a} + \tilde{c}, \tilde{b} + \tilde{d})$. Thus, adding two of our equivalence classes gives another equivalence class, as it should.

Example 0.7: To add 5 and 3, we first note that 5 is the equivalence class $[(2, 7)]$ and 3 is the equivalence class $[(2, 5)]$. We add them componentwise and find that the sum is $[(2 + 2, 7 + 5)] = [(4, 12)]$. Which equivalence class is this answer? Looking back at our prescription for giving names to the equivalence classes, we see that this is the equivalence class that we called 12 − 4 or 8. So, we have rediscovered the fact that 5 + 3 = 8. Check for yourself that if we were to choose a different representative for 5—say (6, 11)—and a different representative for 3—say (24, 27)—then the same answer would result.

Now, let us add 4 and −9. The first of these is the equivalence class [(3, 7)] and the second is the equivalence class [(13, 4)]. The sum is, therefore, [(16, 11)], and this is the equivalence class that we call −(16 − 11) or −5. That is the answer that we would expect when we add 4 to −9.

Next, we add −12 and −5. Previous experience causes us to expect the answer to be −17. Now, −12 is the equivalence class [(19, 7)] and −5 is the equivalence class [(7, 2)]. The sum is [(26, 9)], which is the equivalence class that we call −17.

Finally, we can see in practice that our method of addition is unambiguous. Let us redo the second example using [(6, 10)] as the equivalence class represented by 4 and [(15, 6)] as the equivalence class represented by −9. Then, the sum is [(21, 16)], and this is still the equivalence class −5 as it should be. □

The assertion that the result of calculating a sum—no matter which representatives we choose for the equivalence classes—will give only one answer is called the "fact that addition is *well defined*." For our definitions to make sense, it is essential that we check this property of well-definedness.

Remark 0.8: What is the point of this section? Everyone knows about negative numbers, so why go through this abstract construction? The reason is that, until one sees this construction, negative numbers are just imaginary objects—placeholders if you will—which are a useful notation but which do not exist. Now, they *do* exist. They are a collection of equivalence classes of pairs of natural numbers. This collection is equipped with certain arithmetic operations, such as addition, subtraction, and multiplication. We now discuss these last two. □

If $x = [(a, b)]$ and $y = [(c, d)]$ are integers, we define their *difference* to be the equivalence class $[(a + d, b + c)]$; we denote this difference by $x - y$.

Example 0.9: We calculate 8 − 14. Now, $8 = [(1, 9)]$ and $14 = [(3, 17)]$. Therefore,

$$8 - 14 = [(1 + 17, 9 + 3)] = [(18, 12)] = -6,$$

as expected.

As a second example, we compute $(-4) - (-8)$. Now,

$$-4 - (-8) = [(6, 2)] - [(13, 5)] = [(6 + 5, 2 + 13)] = [(11, 15)] = 4.$$

Of course, that is the answer that we expect when we subtract −8 from −4.

Remark 0.10: When we first learn that $(-4) - (-8) = (-4) + 8 = 4$, the explanation is a bit mysterious: why is "minus a minus equal to a plus"? Now, there is no longer any mystery: this property follows *from our construction* of the number system $\mathbb{Z}$.   □

Finally, we turn to multiplication. If $x = [(a, b)]$ and $y = [(c, d)]$ are integers, then we define their product by the formula

$$x \cdot y = [(a \cdot d + b \cdot c, a \cdot c + b \cdot d)].$$

This definition may be a surprise. Why did we not define $x \cdot y$ to be $[(a \cdot c, b \cdot d)]$? There are several reasons: first of all, the latter definition would give the wrong answer; moreover, it is not unambiguous (different representatives of x and y would give a different answer). If you recall that we think of $[(a, b)]$ as representing $b - a$ and $[(c, d)]$ as representing $d - c$, then the product should be the equivalence class that represents $(b - a) \cdot (d - c)$. That is the motivation behind our definition.

We proceed now to an example.

Example 0.11: We compute the product of −3 and −6. Now,

$$(-3) \cdot (-6) = [(5, 2)] \cdot [(9, 3)] = [(5 \cdot 3 + 2 \cdot 9, 5 \cdot 9 + 2 \cdot 3)] = [(33, 51)]$$
$$= 18,$$

which is the expected answer.

As a second example, we multiply −5 and 12. We have

$$- 5 \cdot 12 = [(7, 2)] \cdot [(1, 13)] = [(7 \cdot 13 + 2 \cdot 1, 7 \cdot 1 + 2 \cdot 13)] = [(93, 33)]$$
$$= -60.$$

Finally, we show that 0 times any integer A equals zero. Let $A = [(a, b)]$. Then,

$$0 \cdot A = [(1, 1)] \cdot [(a, b)] = [(1 \cdot b + 1 \cdot a, 1 \cdot a + 1 \cdot b)]$$
$$= [(a + b, a + b)]$$
$$= 0.$$
                                                                         □

Remark 0.12: Notice that one of the pleasant by-products of our construction of the integers is that we no longer have to give artificial explanations for why the product of two negative numbers is a positive number or why the product of a negative number and a positive number is negative. These properties instead follow automatically from our construction.   □

Of course, we will not discuss division for integers; in general, division of one integer by another makes no sense *in the universe of the integers*.

In the rest of this book, we will follow the standard mathematical custom of denoting the set of all integers by the symbol $\mathbb{Z}$. We will write the integers not as equivalence classes, but in the usual way as $\cdots -3, -2, -1, 0, 1, 2, 3, \ldots$. The equivalence classes are a device that we used to *construct* the integers in hand; we may as well write them in the simple, familiar fashion.

In an exhaustive treatment of the construction of $\mathbb{Z}$, we would prove that addition and multiplication are commutative and associative, prove the distributive law, and so forth. However, the purpose of this section is to demonstrate modes of logical thought rather than to be thorough.

0.1.3 The Rational Numbers

In this section, we use the integers, together with a construction using equivalence classes, to build the rational number system. Let A be the set $\mathbb{Z} \times (\mathbb{Z} \setminus \{0\})$. Here, the symbol "\" stands for "subtraction of sets": $\mathbb{Z} \setminus \{0\}$ denotes the set of all elements of $\mathbb{Z}$ *except* 0. In other words, A is the set of ordered pairs (a, b) of integers subject to the condition that $b \neq 0$. (*Think, intuitively and nonrigorously, of this ordered pair as "representing" the fraction* a/b.) We definitely want it to be the case that certain ordered pairs represent the same number. For instance,

The number $\frac{1}{2}$ should be the same number as $\frac{3}{6}$.

This example motivates our equivalence relation. Declare (a, b) to be related to (a', b') if $a \cdot b' = a' \cdot b$. (*Here, we are thinking, intuitively and nonrigorously, that the fraction* a/b *should equal the fraction* a'/b' *precisely when* a · b' = a' · b.)

Is this an equivalence relation? Obviously, the pair (a, b) is related to itself, since $a \cdot b = a \cdot b$. In addition, the relation is symmetric: if (a, b) and (a', b') are pairs and $a \cdot b' = a' \cdot b$, then $a' \cdot b = a \cdot b'$. Finally, if (a, b) is related to (a', b') and (a', b') is related to (a'', b''), then we have both

$$a \cdot b' = a' \cdot b \text{ and } a' \cdot b'' = a'' \cdot b'.$$

Multiplying the left sides of these two equations together and the right sides together gives

$$(a \cdot b') \cdot (a' \cdot b'') = (a' \cdot b) \cdot (a'' \cdot b').$$

If $a' = 0$, then it follows immediately that both a and a'' must be zero. So, the three pairs (a, b), (a', b'), and (a'', b'') are equivalent and there is nothing to prove. Thus, we may assume that $a' \neq 0$. We know *a priori* that $b' \neq 0$; therefore, we may cancel common terms in the last equation to obtain

$$a \cdot b'' = b \cdot a''.$$

Thus, (a, b) is related to (a'', b''), and our relation is transitive.

The resulting collection of equivalence classes will be called the set of *rational numbers*, and we shall denote this set with the symbol $\mathbb{Q}$.

Example 0.13: The equivalence class $[(4, 12)]$ in the rational numbers contains all of the pairs $(4, 12)$, $(1, 3)$, and $(-2, -6)$. (Of course, it contains infinitely many other pairs as well.) This equivalence class represents the fraction $4/12$, which we sometimes also write as $1/3$ or $-2/(-6)$.   □

If $[(a, b)]$ and $[(c, d)]$ are rational numbers, then we define their *product* to be the rational number

$$[(a \cdot c, b \cdot d)].$$

This is well defined, because if (a, b) is related to $(\tilde{a}, \tilde{b})$ and (c, d) is related to $(\tilde{c}, \tilde{d})$, then we have the equations

$$a \cdot \tilde{b} = \tilde{a} \cdot b \text{ and } c \cdot \tilde{d} = \tilde{c} \cdot d.$$

Multiplying together the left sides and the right sides, we obtain

$$(a \cdot \tilde{b}) \cdot (c \cdot \tilde{d}) = (\tilde{a} \cdot b) \cdot (\tilde{c} \cdot d).$$

Rearranging, we have

$$(a \cdot c) \cdot (\tilde{b} \cdot \tilde{d}) = (\tilde{a} \cdot \tilde{c}) \cdot (b \cdot d).$$

However, this says that the product of $[(a, b)]$ and $[(c, d)]$ is related to the product of $[(\tilde{a}, \tilde{b})]$ and $[(\tilde{c}, \tilde{d})]$. So, multiplication is unambiguous (i.e., well defined).

Example 0.14: The product of the two rational numbers $[(3, 8)]$ and $[(-2, 5)]$ is

$$[(3 \cdot (-2), 8 \cdot 5)] = [(-6, 40)] = [(-3, 20)].$$

This is what we expect: the product of $3/8$ and $-2/5$ is $-3/20$.   □

If $q = [(a, b)]$ and $r = [(c, d)]$ are rational numbers and if r is not zero (that is, $[(c, d)]$ is not the equivalence class zero—in other words, $c \neq 0$), then we define the quotient q/r to be the equivalence class

$$[(ad, bc)].$$

We leave it to you to check that this operation is well defined.

Example 0.15: The quotient of the rational number $[(4, 7)]$ by the rational number $[(3, -2)]$ is, by definition, the rational number

$$[(4 \cdot (-2), 7 \cdot 3)] = [(-8, 21)].$$

This is what we expect: the quotient of $4/7$ by $-3/2$ is $-8/(21)$. □

How should we add two rational numbers? We could try declaring $[(a, b)] + [(c, d)]$ to be $[(a + c, b + d)]$, but this will not work (think about the way that we usually add fractions). Instead, we define

$$[(a, b)] + [(c, d)] = [(a \cdot d + c \cdot b, b \cdot d)].$$

We turn now to an example.

Example 0.16: The sum of the rational numbers $[(3, -14)]$ and $[(9, 4)]$ is given by

$$[(3 \cdot 4 + 9 \cdot (-14), (-14) \cdot 4)] = [(-114, -56)] = [(57, 28)].$$

This is consistent with the usual way that we add fractions:

$$-\frac{3}{14} + \frac{9}{4} = \frac{57}{28}. \qquad □$$

Notice that the equivalence class $[(0, 1)]$ is the rational number that we usually denote by 0. It is the additive identity, for if $[(a, b)]$ is another rational number then

$$[(0, 1)] + [(a, b)] = [(0 \cdot b + a \cdot 1, 1 \cdot b)] = [(a, b)].$$

A similar argument shows that $[(0, 1)]$ times any rational number gives $[(0, 1)]$ or 0.

Of course, the concept of subtraction is really just a special case of addition (that is $x - y$ is the same thing as $x + (-y)$). So, we shall say nothing further about subtraction.

In practice, we will write rational numbers in the traditional fashion:

$$\frac{2}{5}, \frac{-19}{3}, \frac{22}{2}, \frac{24}{4}, \dots.$$

In mathematics, it is generally not wise to write rational numbers in mixed form, such as $2\frac{3}{5}$, because the juxtaposition of two numbers could easily be mistaken for multiplication. Instead, we would write this quantity as the improper fraction 13/5.

Definition 0.17: A set S is called a *field* if it is equipped with a binary operation (usually called addition and denoted "+") and a second binary operation (called multiplication and denoted "·") such that the following axioms are satisfied:

A1. S is closed under addition: if $x, y \in S$, then $x + y \in S$.

A2. Addition is commutative: if $x, y \in S$, then $x + y = y + x$.

A3. Addition is associative: if $x, y, z \in S$, then $x + (y + z) = (x + y) + z$.

A4. There exists an element, called 0, in S, which is an additive identity: if $x \in S$, then $0 + x = x$.

A5. Each element of S has an additive inverse: if $x \in S$, then there is an element $-x \in S$ such that $x + (-x) = 0$.

M1. S is closed under multiplication: if $x, y \in S$, then $x \cdot y \in S$.

M2. Multiplication is commutative: if $x, y \in S$, then $x \cdot y = y \cdot x$.

M3. Multiplication is associative: if $x, y, z \in S$, then $x \cdot (y \cdot z) = (x \cdot y) \cdot z$.

M4. There exists an element, called 1, which is a multiplicative identity: if $x \in S$, then $1 \cdot x = x$.

M5. Each nonzero element of S has a multiplicative inverse: if $0 \neq x \in S$, then there is an element $x^{-1} \in S$ such that $(x^{-1}) \cdot x = 1$. The element x^{-1} is sometimes denoted $1/x$.

D1. Multiplication distributes over addition: if $x, y, z \in S$, then

$$x \cdot (y + z) = x \cdot y + x \cdot z.$$

Eleven axioms is a lot to digest all at once, but in fact these are all familiar properties of addition and multiplication of rational numbers that we use every day: the set $\mathbb{Q}$, with the usual notions of addition and multiplication, forms a field. The integers, by contrast, do not: nonzero elements of $\mathbb{Z}$ (except 1 and -1) do not have multiplicative inverses *in the integers*.

Let us now consider some consequence of the field axioms.

Theorem 0.18: *Any field has the following properties:*

1. *If $z + x = z + y$, then $x = y$.*
2. *If $x + z = 0$, then $z = -x$ (the additive inverse is unique).*
3. $-(-y) = y$.
4. *If $y \neq 0$ and $y \cdot x = y \cdot z$, then $x = z$.*
5. *If $y \neq 0$ and $y \cdot z = 1$, then $z = y^{-1}$(the multiplicative inverse is unique).*
6. $(x^{-1})^{-1} = x$.
7. $0 \cdot x = 0$.
8. *If $x \cdot y = 0$, then either $x = 0$ or $y = 0$.*
9. $(-x) \cdot y = -(x \cdot y) = x \cdot (-y)$.
10. $(-x) \cdot (-y) = x \cdot y$.

Proof: These are all familiar properties of the rationals, but now we are considering them for an arbitrary field. We prove just a few to illustrate the logic.

To prove **(1)** we write

$$z + x = z + y \Rightarrow (-z) + (z + x) = (-z) + (z + y)$$

and now Axiom **A3** yields that this implies

$$((-z) + z) + x = ((-z) + z) + y.$$

Next, Axiom **A5** yields that

$$0 + x = 0 + y$$

and hence, by Axiom **A4**,

$$x = y.$$

To prove **(7)**, we observe that

$$0 \cdot x = (0 + 0) \cdot x,$$

which by Axiom **M2** equals

$$x \cdot (0 + 0).$$

By Axiom **D1** the last expression equals

$$x \cdot 0 + x \cdot 0,$$

which by Axiom **M2** equals $0 \cdot x + 0 \cdot x$. Thus, we have derived the equation

$$0 \cdot x = 0 \cdot x + 0 \cdot x.$$

For Axioms **A4** and **A2,** let us rewrite the left side as

$$0 \cdot x + 0 = 0 \cdot x + 0 \cdot x.$$

Finally, part **(1)** of the present theorem (which we have already proved) yields that

$$0 = 0 \cdot x,$$

which is the desired result.

To prove **(8)**, we suppose that $x \neq 0$. In this case, x has a multiplicative inverse x^{-1} and we multiply both sides of our equation by this element:

$$x^{-1} \cdot (x \cdot y) = x^{-1} \cdot 0.$$

By Axiom **M3**, the left side can be rewritten and we have

$$(x \cdot x^{-1}) \cdot y = x^{-1} \cdot 0.$$

Next, we rewrite the right side using Axiom **M2**:

$$(x \cdot x^{-1}) \cdot y = 0 \cdot x^{-1}.$$

Now, Axiom **M5** allows us to simplify the left side:

$$1 \cdot y = 0 \cdot x^{-1}.$$

We further simplify the left side using Axiom **M4** and the right side using Part **(7)** of the present theorem (which we just proved) to obtain:

$$y = 0.$$

Thus, we see that if $x \neq 0$ then $y = 0$. But this is logically equivalent with $x = 0$ or $y = 0$, as we wished to prove. (If you have forgotten why these statements are logically equivalent, write a truth table. Or refer to the next section.) $\square$

Definition 0.19: Let A be a set. We shall say that A is *ordered* if there is a relation $\mathcal{R}$ on A and A satisfying the following properties:

1. If $a \in A$ and $b \in A$ then one and only one of the following holds: $(a, b) \in \mathcal{R}$ or $(b, a) \in \mathcal{R}$ or $a = b$.
2. If a, b, c are elements of A and $(a, b) \in \mathcal{R}$ and $(b, c) \in \mathcal{R}$ then $(a, c) \in \mathcal{R}$.

We call the relation $\mathcal{R}$ an *order* on A.

Rather than write an ordering relation as $(a, b) \in \mathcal{R}$, it is usually more convenient to write it as $a < b$. The notation $b > a$ means the same thing as $a < b$.

Example 0.20: The integers $\mathbb{Z}$ form an ordered set with the usual ordering $<$. We can make this ordering precise by saying that $x < y$ if $y - x$ is a positive integer. For instance,

$$6 < 8 \text{ because } 8 - 6 = 2 > 0.$$

Likewise,

$$-5 < -1 \text{ because } -1 - (-5) = 4 > 0.$$

Observe that the same ordering works on the rational numbers. □

If A is an ordered set and a, b are elements, then we often write $a \leq b$ to mean that *either $a = b$ or $a < b$*.

When a field has an ordering that is compatible with the field operations, then a richer structure results.

Definition 0.21: A field F is called an *ordered field* if F has an ordering $<$ that satisfies the following addition properties:

1. If $x, y, z \in F$ and $y < z$, then $x + y < x + z$.
2. If $x, y \in F$, $x > 0$, and $y > 0$, then $x \cdot y > 0$.

Again, these are familiar properties of the rational numbers: $\mathbb{Q}$ forms an ordered field. However, there are many other ordered fields as well (for instance, the real numbers $\mathbb{R}$ form an ordered field).

Theorem 0.22: *Any ordered field has the following properties:*

1. *If $x > 0$ and $z < y$, then $x \cdot z < x \cdot y$.*
2. *If $x < 0$ and $z < y$, then $x \cdot z > x \cdot y$.*
3. *If $x > 0$, then $-x < 0$. If $x < 0$, then $-x > 0$.*

4. *If $0 < y < x$, then $0 < 1/x < 1/y$.*
5. *If $x \neq 0$, then $x^2 > 0$.*
6. *If $0 < x < y$, then $x^2 < y^2$.*

Proof: Again, we prove just a few of these statements.

To prove **(1)**, observe that the property **(1)** of ordered fields together with our hypothesis implies that

$$(-z) + z < (-z) + y.$$

Thus, using **(A2)**, we see that $y - z > 0$. Since $x > 0$, property **(2)** of ordered fields gives

$$x \cdot (y - z) > 0.$$

Finally,

$$x \cdot y = x \cdot [(y - z) + z] = x \cdot (y - z) + x \cdot z > 0 + x \cdot z$$

(by property **(1)** again). In conclusion,

$$x \cdot y > x \cdot z.$$

To prove **(3)**, begin with the equation

$$0 = -x + x.$$

Since $x > 0$, the right side is greater than $-x$. Thus, $0 > -x$ as claimed. The proof of the other statement of **(3)** is similar.

To prove **(5)**, we consider two cases. If $x > 0$, then $x^2 \equiv x \cdot x$ is positive by property **(2)** of ordered fields. If $x < 0$, then $-x > 0$ (by part **(3)** of the present theorem, which we just proved); hence, $(-x) \cdot (-x) > 0$. However, part **(10)** of the last theorem guarantees that $(-x) \cdot (-x) = x \cdot x$; hence, we see that $x \cdot x > 0$.   □

We conclude this discussion by recording an inadequacy of the field of rational numbers; this will serve in part as motivation for learning about the real numbers in Chapter 1.

Theorem 0.23: *There is no positive rational number q such that $q^2 = q \cdot q = 2$.*

Proof: Seeking a contradiction, suppose that there is such a q. Write q in lowest terms as

$$q = \frac{a}{b},$$

with a and b greater than zero. This means that the numbers a and b have no common divisors except 1. The equation $q^2 = 2$ can then be written as

$$a^2 = 2 \cdot b^2.$$

Since 2 divides the right side of this last equation, it follows that 2 divides the left side. However, 2 can divide a^2 only if 2 divides a (because 2 is prime). We write $a = 2 \cdot \alpha$ for some positive integer α. But, then, the last equation becomes

$$4 \cdot \alpha^2 = 2 \cdot b^2.$$

Simplifying yields that

$$2 \cdot \alpha^2 = b^2.$$

Since 2 divides the left side, we conclude that 2 must divide the right side. But 2 can divide b^2 only if 2 divides b.

This is our contradiction: we have argued that 2 divides a *and* that 2 divides b. But a and b were assumed to *have no common divisors*. We conclude that the rational number q cannot exist. □

In fact, it turns out that a positive integer can be the square of a rational number if and only if it is the square of a positive integer. This assertion is a special case of a more general phenomenon in number theory known as Gauss's lemma.

Exercises

1. Construct truth tables for each of the following sentences:
   a. $(S \wedge T) \vee \sim(S \vee T)$
   b. $(S \vee T) \Rightarrow (S \wedge T)$
   c. $(\sim S \vee T) \Rightarrow \sim(S \wedge \sim T)$
   d. $S \Rightarrow (S \Rightarrow (S \Rightarrow (S \Rightarrow T)))$

2. Let

   S = All fish have eyelids.
   T = There is no justice in the world.
   U = I believe everything that I read.
   V = The moon's a balloon.

Express each of the following sentences using the letters S, T, U, and V and the connectives $\vee$, $\wedge$, $\sim$, $\Rightarrow$, and $\Leftrightarrow$. *Do not use quantifiers.*

a. If fish have eyelids, then there is at least some justice in the world.

b. If I believe everything that I read, then either the moon's a balloon or at least some fish have no eyelids.

c. If either the moon is not a balloon or if there is some justice in the world, then I doubt some of the things that I read.

d. For fish to have eyelids, it is necessary for the moon to be a balloon.

e. If fish have eyelids, then there is at least some justice in the world.

3. Let

$$S = \text{All politicians are honest.}$$
$$T = \text{Some men are fools.}$$
$$U = \text{I don't have two brain cells to rub together.}$$
$$W = \text{The pie is in the sky.}$$

Translate each of the following into English sentences:

a. $(S \wedge \sim T) \Rightarrow \sim U$

b. $W \vee (T \wedge \sim U)$

c. $W \Rightarrow (S \Rightarrow T)$

d. $S \Rightarrow (S \vee U)$

4. State the converse and the contrapositive of each of the following sentences. Be sure to label each.

a. In order for it to rain, it is necessary that there be clouds.

b. In order for it to rain, it is sufficient that there be clouds.

c. If life is a bowl of cherries, then I am not in the pits.

d. If I am not a fool, then mares eat oats.

5. Assume that the universe is the ordinary system R of real numbers. Which of the following sentences is true? Which is false? Give reasons for your answers.

a. If π is rational, then the area of a circle is $E = mc^2$.

b. If $2 + 2 = 4$, then $3/5$ is a rational number.

c. If $2 + 2 = 5$, then $2 + 3 = 6$.

d. If both $2 + 3 = 5$ and $2 \cdot 3 = 5$, then the world is flat.

6. For each of the following statements, formulate a logically equivalent one using only S, T, $\sim$, and $\vee$. (Of course, you may use as many parentheses as you need.) *Use a truth table or other means to explain why the statements are logically equivalent.*

   a. $S \Rightarrow \sim T$

   b. $\sim S \wedge \sim T$

   c. $S \Rightarrow \sim T$

   d. $S \wedge (T \vee \sim S)$

7. Translate each of the following statements into symbols, connectives, and quantifiers. Your answers should contain no words. State carefully what each of your symbols stands for. (Note: Each statement is true, but you are not required to verify the truth of the statements.)

   a. The number 5 has a positive square root.

   b. There is a quadratic polynomial equation with real coefficients that has no real root.

   c. The sum of two perfect cubes is never itself a perfect cube.

   d. If $x \cdot y \neq 0$, then $x^2 + y^2 > 0$.

8. In each of the following statements, you should treat the real number system $\mathbb{R}$ as your universe. Translate each statement into an English sentence. Your answers should contain no symbols—only words. (Note: Each statement is true, but you are not required to verify the truth of the statements.)

   a. $\exists x, (x \in \mathbb{R} \wedge x > 0 \wedge \sim \exists y, y > 0 \wedge y^2 = x)$

   b. $\exists x \forall y, (y > x) \Rightarrow (y > 5)$

   c. $\exists x \in \mathbb{R} \exists y \in \mathbb{R}, x^2 + y^2 < 2xy$

   d. $\exists x, x > 0 \wedge x^3 < x^2$

9. For each of the following statements, formulate an English sentence that is its negation:

   a. The set S contains at least two integers.

   b. Mares eat oats and does eat oats.

   c. I'm rough and I'm tough and I breathe fire.

   d. This town is not big enough for both of us.

10. Which of these pairs of statements is logically equivalent? Why?

    a. $A \vee \sim B \quad \sim A \Rightarrow B$

    b. $A \wedge \sim B \quad \sim A \Rightarrow \sim B$

    c. $A \vee (\sim A \wedge B) \quad \sim[\sim A \wedge (A \vee \sim B)]$

    d. $B \Rightarrow \sim A \quad A \Rightarrow (A \vee B)$

11. Explain why ∀ is logically equivalent to ~∃ ~.

12. Explain why ∃ is logically equivalent to ~∀ ~.

13. It is not known whether $\pi + e$ or $\pi - e$ is rational or irrational. But one of them must be irrational, even though we cannot say which one. Explain.

0.2 Logic and Set Theory

Everyday language is imprecise. Because we are imprecise by *convention*, we can make statements like

All automobiles are not alike.

and feel confident that the listener knows that we actually *mean*

Not all automobiles are alike.

We can also use spurious reasoning like

If it's raining, then it's cloudy.
It is not raining.
Therefore, there are no clouds.

and not expect to be challenged, because virtually everyone is careless when communicating informally. (Examples of this type will be considered in more detail later.)

Mathematics cannot tolerate this lack of rigor and precision. To achieve any depth beyond the most elementary level, we must adhere to strict rules of logic. The purpose of this section is to discuss the foundations of formal reasoning.

In this chapter, we will often use numbers to illustrate logical concepts. The number systems we will encounter are

- The natural numbers $\mathbb{N} = \{1, 2, 3, \ldots\}$
- The integers $\mathbb{Z} = \{\ldots, -3, -2, -1, 0, 1, 2, 3, \ldots\}$
- The rational numbers $\mathbb{Q} = \{p/q : p \text{ is an integer, } q \text{ is an integer, } q \neq 0\}$
- The real numbers $\mathbb{R}$, consisting of all terminating and non-terminating decimal expansions.

Chapter 1 will review the real and complex numbers. If you need to review the other number systems, then refer to Section 0.1 or look at Ref. [KRA1]. For now, we assume that you have seen these number systems before. They are convenient for illustrating the logical principles we are discussing.

0.2.1 "And" and "Or"

The statement

"A and B"

means that both **A** is true *and* **B** is true. For instance,

George is tall and George is intelligent.

means both that George is tall *and* George is intelligent. If we meet George and he turns out to be short and intelligent, then the statement is false. If he is tall and stupid, then the statement is false. Finally, if George is *both* short and stupid, then the statement is false. The statement is *true* precisely when both properties—intelligence and tallness—hold. We may summarize these assertions with a *truth table*. We let

A = George is tall.

and

B = George is intelligent.

The expression

A ∧ B

will denote the phrase "**A** and **B**." In particular, the symbol ∧ is used to denote "and." The letters "T" and "F" denote "True" and "False," respectively. Then, we have

A	B	A ∧ B
T	T	T
T	F	F
F	T	F
F	F	F

Notice that we have listed all possible truth values of **A** and **B** and the corresponding values of the *conjunction* **A ∧ B**.

It is a good idea to always list the truth values of **A** and **B** in a truth table in the same way. This will facilitate the comparison and contrast of truth tables.

In a restaurant, the menu often contains phrases like

soup or salad

This means that we may select soup *or* select salad, but we may not select both. This use of "or" is called the *exclusive* "or"; it is not the meaning of "or" that we use in mathematics and logic. In mathematics, we instead say that "**A or B**" is true provided that **A** is true or **B** is true or *both* are true. This is the *inclusive* meaning of the word "or." If we let **A ∨ B** denote "**A or B**" (the symbol ∨ denotes "or"), then the truth table is

A	B	A ∨ B
T	T	T
T	F	T
F	T	T
F	F	F

□

The only way that "**A or B**" can be false is if *both* **A** is false and **B** is false. For instance, the statement

Gary is handsome or Gary is rich.

means that Gary is either handsome or rich or both. In particular, he will not be both ugly and poor. Another way of saying this is that if he is poor, he will compensate by being handsome; if he is ugly, he will compensate by being rich. *But he could be both handsome and rich.*

We use the inclusive meaning of the word "or" because it gives rise to useful logical equivalences. We treat these later.

Example 0.24: The statement

$$x > 5 \text{ and } x < 7$$

is true for the number $x = 11/2$ because this value of x is both greater than 5 *and* less than 7. It is false for $x = 8$ because this x is greater than 5 but not less than 7. It is false for $x = 3$ because this x is less than 7 but not greater than 5.   □

Example 0.25: The statement

x is even and x is a perfect square

is true for $x = 4$ because both assertions hold. It is false for $x = 2$ because this x, while even, is not a square. It is false for $x = 9$ because this x, while a square, is not even. It is false for $x = 5$ because this x is neither a square nor an even number. □

Example 0.26: The statement

$x > 5$ or $x \le 2$

is true for $x = 1$ since this x is ≤ 2 (even though it is not > 5). It holds for $x = 6$ because this x is > 5 (even though it is not ≤ 2). The statement fails for $x = 3$, since this x is neither > 5 nor ≤ 2. There is no x, which is both > 5 and ≤ 2. □

Example 0.27: The statement

$x > 5$ or $x < 7$

is true for every real x. For $x = 6$, both statements are true. For $x = 2$, just the second statement is true. For $x = 8$, just the first statement is true. But, you can in fact verify the assertion for *every* real x. □

Example 0.28: The statement $(A \lor B) \land B$ has the following truth table:

A	B	A ∨ B	(A ∨ B) ∧ B
T	T	T	T
T	F	T	F
F	T	T	T
F	F	F	F

□

The words "and" and "or" are called *connectives*: their role in sentential logic is to enable us to build up (or connect together) pairs of statements. In the next subsection, we will become acquainted with the other two basic connectives "not" and "if–then."

0.2.2 "Not" and "If Then"

The statement "not A," written $\sim A$, is true whenever A is false. For example, the statement

Gene is not tall.

is true provided the statement "Gene is tall" is false. The truth table for ~**A** is as follows:

A	~ A
T	F
F	T

Although "not" is a simple idea, it can be a powerful tool when used in proofs by contradiction. To prove that a statement **A** is true using proof by contradiction, we instead assume ~**A**. We, then, show that this hypothesis leads to a contradiction. Thus, ~**A** must be false; according to the truth table, we see that the only remaining possibility is that **A** is true.

Greater understanding is obtained by combining connectives:

Example 0.29: Here is the truth table for ~ (**A** ∨ **B**):

A	B	A ∨ B	~ (A ∨ B)
T	T	T	F
T	F	T	F
F	T	T	F
F	F	F	T

Example 0.30: Now, we look at the truth table for (~**A**) ∧ (~**B**):

A	B	~ A	~ B	(~ A) ∧ (~ B)
T	T	F	F	F
T	F	F	T	F
F	T	T	F	F
F	F	T	T	T

Notice that the statements ~ (**A** ∨ **B**) and (~**A**) ∧ (~**B**) have the *same truth table* (look at the last column in each table). We call such pairs of statements *logically equivalent.*

The logical equivalence of ~ (**A** ∨ **B**) with (~**A**) ∧ (~**B**) makes good intuitive sense: the statement **A** ∨ **B** fails if and only if **A** is false *and* **B** is false. Since in mathematics we cannot rely on our intuition to establish facts, it is important to have the truth table technique for establishing logical equivalence.

A statement of the form "If **A** then **B**" asserts that whenever **A** is true then **B** is also true. This assertion (or "promise") is tested when **A** is true, because it is then claimed that something else (namely, **B**) is true as well. *However,* when **A** is false, then the statement "If **A** then **B**" *claims nothing.* Using the symbols **A** ⇒ **B** to denote "If **A** then **B**," we obtain the following truth table:

A	B	A ⇒ B
T	T	T
T	F	F
F	T	T
F	F	T

Notice that we use here an important principle of Aristotelian logic: every sensible statement is either true or false. There is no "in between" status. Thus, when **A** is false, then the statement **A** ⇒ **B** is not tested. It, therefore, cannot be false. So, it must be true. In fact, the only way that **A** ⇒ **B** can be false is if **A** is true and **B** is false.

Example 0.31: The statement **A** ⇒ **B** is logically equivalent with ∼ (**A** ∧ ∼ **B**). The truth table for the latter is

A	B	∼ B	A ∧ ∼ B	∼ (A ∧ ∼ B)
T	T	F	F	T
T	F	T	T	F
F	T	F	F	T
F	F	T	F	T

which is the same as the truth table for **A** ⇒ **B**.   □

There are in fact infinitely many pairs of logically equivalent statements. But just a few of these equivalences are really important in practice—most others are built up from these few basic ones.

Example 0.32: The statement

   If x is negative, then $-5 \cdot x$ is positive.

is true. For if $x < 0$, then $-5 \cdot x$ is indeed > 0; if $x \geq 0$, then the statement is unchallenged.   □

Example 0.33: The statement

> If $\{x > 0$ and $x^2 < 0\}$, then $x \geq 10$.

is true since the hypothesis "$x > 0$ and $x^2 < 0$" is never true.   □

Example 0.34: The statement

> If $x > 0$, then $\{x^2 < 0$ or $2x < 0\}$.

is false since the conclusion "$x^2 < 0$ or $2x < 0$" is false whenever the hypothesis $x > 0$ is true.   □

0.2.3 Contrapositive, Converse, and "If"

The statement

> **If A, then B.** or **A $\Rightarrow$ B.**

is the same as saying

> **A suffices for B.**

or as saying

> **A only if B.**

All these forms are encountered in practice, and you should think about them long enough to realize that they all say the same thing.

On the other hand,

> **If B, then A.** or **B $\Rightarrow$ A.**

is the same as saying

> **A is necessary for B.**

or as saying

> **A if B.**

We call the statement **B $\Rightarrow$ A** the *converse of* **A $\Rightarrow$ B**.

Example 0.35: The converse of the statement

> If x is a healthy horse, then x has four legs.

is the statement

> If x has four legs, then x is a healthy horse.

Notice that these statements have very different meanings: the first statement is true while the second (its converse) is false. For example, my desk has four legs but it is not a healthy horse. □

The statement

> A if and only if B.

is a brief way of saying

> If A, then B *and* if B, then A.

We abbreviate **A if and only if B** as $A \Leftrightarrow B$ or as **A if B**. Here is a truth table for $A \Leftrightarrow B$.

A	B	$A \Rightarrow B$	$B \Rightarrow A$	$A \Leftrightarrow B$
T	T	T	T	T
T	F	F	T	F
F	T	T	F	F
F	F	T	T	T

Notice that we can say that $A \Leftrightarrow B$ is true only when both $A \Rightarrow B$ and $B \Rightarrow A$ are true. An examination of the truth table reveals that $A \Leftrightarrow B$ is true precisely when A and B are either both true or both false. Thus, $A \Leftrightarrow B$ means precisely that A and B are logically equivalent. One is true *when and only when* the other is true. □

Example 0.36: The statement

$$x > 0 \Leftrightarrow 2x > 0$$

is true. For if $x > 0$, then $2x > 0$; and if $2x > 0$, then $x > 0$. □

Example 0.37: The statement

$$x > 0 \Leftrightarrow x^2 > 0$$

is false. For $x > 0 \Rightarrow x^2 > 0$ is certainly true while $x^2 > 0 \Rightarrow x > 0$ is false $((-3)^2 > 0$ but $-3 \not> 0)$.   □

Example 0.38: The statement

$$\{\sim(A \vee B)\} \Leftrightarrow \{(\sim A) \wedge (\sim B)\} \tag{0.38.1}$$

is true because the truth table for $\sim(A \vee B)$ and that for $(\sim A) \wedge (\sim B)$ are the same (we noted this fact in the last section). Thus, they are logically equivalent: one statement is true precisely when the other is. Another way to see the truth of (0.38.1) is to examine the following full truth table:

A	B	$\sim(A \vee B)$	$(\sim A) \wedge (\sim B)$	$\sim(A \vee B) \Leftrightarrow \{(\sim A) \wedge (\sim B)\}$
T	T	F	F	T
T	F	F	F	T
F	T	F	F	T
F	F	T	T	T

Given an implication

$$A \Rightarrow B,$$

the *contrapositive* statement is defined to be the implication

$$\sim B \Rightarrow \sim A.$$

The contrapositive is logically equivalent to the original implication, as we see by examining their truth tables:

A	B	$A \Rightarrow B$
T	T	T
T	F	F
F	T	T
F	F	T

and

A	B	~A	~B	(~B) ⇒ (~A)
T	T	F	F	T
T	F	F	T	F
F	T	T	F	T
F	F	T	T	T

Example 0.39: The statement

If it is raining, then it is cloudy.

has, as its contrapositive, the statement

If there are no clouds, then it is not raining.

A moment's thought convinces us that these two statements say the same thing: if there are no clouds, then it could not be raining; for the presence of rain implies the presence of clouds.   □

Example 0.40: The statement

If X is a healthy horse, then X has four legs.

has, as its contrapositive, the statement

If X does not have four legs, then X is not a healthy horse.

A moment's thought reveals that these two statements say precisely the same thing. They are logically equivalent.   □

The main point to keep in mind is that, given an implication $\mathbf{A} \Rightarrow \mathbf{B}$, its *converse* $\mathbf{B} \Rightarrow \mathbf{A}$ and its *contrapositive* $(\sim\mathbf{B}) \Rightarrow (\sim\mathbf{A})$ are two different statements. The converse is distinct from, and *logically independent from*, the original statement. The contrapositive is distinct from, but *logically equivalent to*, the original statement.

0.2.4 Quantifiers

The mathematical statements that we will encounter in practice will use the *connectives* "and," "or," "not," "if-then," and "if." They will also use *quantifiers*. The two basic quantifiers are "for all" and "there exists."

Example 0.41: Consider the statement

All automobiles have wheels.

This statement makes an assertion about *all* automobiles. It is true, just because every automobile does have wheels.

Compare this statement with the next one:

There exists a woman who is blonde.

This statement is of a different nature. It does not claim that all women have blonde hair—merely that there exists at *least one* woman who does. Since that is true, the statement is true.   □

Example 0.42: Consider the statement

All positive real numbers are integers.

This sentence asserts that something is true for all positive real numbers. It is indeed true for *some* positive real numbers, such as 1 and 2 and 193. However, it is false for at least one positive number (such as π), so the entire statement is false.

Here, is a more interesting example:

The square of any real number is positive.

This assertion is *almost* true—the only exception is the real number 0: we see that $0^2 = 0$ is not positive. But, it only takes one exception to falsify a "for all" statement. So, the assertion is false.   □

Example 0.43: Look at the statement

There exists a real number which is greater than 4.

In fact, there are lots of real numbers, which are greater than 4; some examples are 7, 8π, and 97/3. Since there is *at least one* number satisfying the assertion, the assertion is true.

A somewhat different example is the sentence

There exists a real number that satisfies the equation
$$x^3 + x^2 + x + 1 = 0.$$

There is in fact only one real number that satisfies the equation, and that is $x = -1$. Yet that information is sufficient to make the statement true.   □

We often use the symbol ∀ to denote "for all" and the symbol ∃ to denote "there exists." The assertion

$$\forall x, \; x + 1 < x$$

claims that, for every x, the number $x + 1$ is less than x. If we take our universe to be the standard real number system, this statement is false (for example, $5 + 1$ is not less than 5). The assertion

$$\exists x, \; x^2 = x$$

claims that there is a number whose square equals itself. If we take our universe to be the real numbers, then the assertion is satisfied by $x = 0$ and by $x = 1$. Therefore, the assertion is true.

Quite often we will encounter ∀ and ∃ used together. The following examples are typical:

Example 0.44: The statement

$$\forall x \; \exists y, \; y > x$$

claims that for any number x there is a number y, which is greater than it. In the realm of the real numbers, this is true. In fact, $y = x + 1$ will always do the trick.

The statement

$$\exists y \; \forall x, \; y > x$$

has quite a different meaning from the first one. It claims that there is an y, which is greater than *every* x. This is absurd. For instance, y is *not* greater than $x = y + 1$.

One sees that these two examples show that ∀ and ∃ do not commute. □

Example 0.45: The statement

$$\forall x \; \forall y, \; x^2 + y^2 \geq 0$$

is true in the realm of the real numbers: it claims that the sum of two squares is always greater than or equal to zero.

The statement

$$\exists x \; \exists y, \; x + 2y = 7$$

is true in the realm of the real numbers: it claims that there exist x and y such that $x + 2y = 7$. The numbers $x = 3$, $y = 2$ will do the job (although there are many other choices that work as well). □

We conclude by noting that ∀ and ∃ are closely related. The statements

$$\forall\, x,\, B(x) \text{ and } \sim \exists\, x,\, \sim B(x)$$

are logically equivalent. The first asserts that the statement $B(x)$ is true for all values of x. The second asserts that there exists no value of x for which $B(x)$ fails, which is the same thing.

Likewise, the statements

$$\exists\, x,\, A(x) \text{ and } \sim \forall\, x,\, \sim A(x)$$

are logically equivalent. The first asserts that there is some x for which $A(x)$ is true. The second claims that it is not the case that $A(x)$ fails for every x, which is the same thing.

Remark 0.46: Most of the statements that we encounter in mathematics are formulated using "for all" and "there exists." For example,

Through every point P not on a line ℓ there is a line parallel to ℓ.

Each continuous function on a closed, bounded interval has an absolute maximum.

Each of these statements uses (implicitly) both a "for all" and a "there exists."   □

A "for all" statement is like an *infinite conjunction*. The statement $\forall x,\, P(x)$ (when x is a natural number, let us say) says $P(1) \wedge P(2) \wedge P(3) \wedge \cdots$. A "there exists" statement is like an *infinite disjunction*. The statement $\exists x,\, Q(x)$ (when x is a natural number, let us say) says $Q(1) \vee Q(2) \vee Q(3) \vee \cdots$. Thus, it is neither practical nor sensible to endeavor to verify statements such as these using truth tables. This is one of the chief reasons that we learn to produce mathematical proofs. One of the main themes of this text is to gain new insights and to establish facts about the real number system using mathematical proofs.

0.2.5 Set Theory and Venn Diagrams

The two most basic objects in all of mathematics are sets and functions. In this section, we discuss the first of these two concepts.

A *set* is a collection of objects. For example, "the set of all blue shirts" and "the set of all lonely whales" are two examples of sets. In mathematics, we often write sets with the following "set-builder" notation:

$$\{x : x + 5 > 0\}.$$

This is read "the set of all x such that $x + 5$ is greater than 0." The universe from which x is chosen (for us, this will usually be the real numbers) is understood from context, though sometimes we may be more explicit and write

$$\{x \in \mathbb{R} : x + 5 > 0\}.$$

Here, $\in$ is a symbol that means "is an element of."

Notice that the role of x in the set-builder notation is as a *dummy variable*; the set we have just described could also be written as

$$\{s : s + 5 > 0\}$$

or

$$\{\alpha \in \mathbb{R} : \alpha + 5 > 0\}.$$

To repeat, the symbol $\in$ is used to express membership in a set; for example, the statement

$$4 \in \{x : x > 0\}$$

says that 4 is a member of (or *an element of*) the set of all numbers x which are greater than 0. In other words, 4 is a positive number.

If A and B are sets, then the statement

$$A \subset B$$

is read "A is a subset of B." It means that each element of A is also an element of B (but not vice versa!). In other words $x \in A \Rightarrow x \in B$.

Example 0.47: Let

$$A = \{x \in \mathbb{R} : \exists\, y \text{ such that } x = y^2\}$$

and

$$B = \{t \in \mathbb{R} : t + 3 > -5\}.$$

Then $A \subset B$. Why? The set A consists of those numbers that are squares— that is, A is just the nonnegative real numbers. The set B contains all numbers which are greater than -8. Since every nonnegative number (element of A) is also greater than -8 (element of B), it is correct to say that $A \subset B$.

However, it is not correct to say that $B \subset A$, because -2 is an element of B but is not an element of A.   □

We write $A = B$ to indicate that both $A \subset B$ and $B \subset A$. In these circumstances, we say that the two sets are equal: every element of A is an element of B and every element of B is an element of A.

We use a slash through the symbols $\in$ or $\subset$ to indicate negation:

$$-4 \notin \{x : x \geq -2\}$$

and

$$\{x : x = x^2\} \not\subset \{y : y > 1/2\}.$$

It is often useful to combine sets. The set $A \cup B$, called the *union* of A and B, is the set consisting of all objects which are either elements of A *or* elements of B (or both). The set $A \cap B$, called the *intersection* of A and B, is the set consisting of all objects, which are elements of *both* A and B.

Example 0.48: Let

$$A = \{x: -4 < x \leq 3\}, \quad B = \{x: -1 \leq x < 7\},$$
$$C = \{x: -9 \leq x \leq 12\}.$$

Then,

$$A \cup B = \{x: -4 < x < 7\} \quad A \cap B = \{x: -1 \leq x \leq 3\},$$
$$B \cup C = \{x: -9 \leq x \leq 12\}, \quad B \cap C = \{x: -1 \leq x < 7\}.$$

Notice that $B \cup C = C$ and $B \cap C = B$ because $B \subset C$.   □

Example 0.49: Let

$$A = \{\alpha \in \mathbb{Z} : \alpha \geq 9\}$$
$$B = \{\beta \in \mathbb{R} : -4 < \beta \leq 24\},$$
$$C = \{\gamma \in \mathbb{R} : 13 < \gamma \leq 30\}.$$

Then,

$$(A \cap B) \cap C = \{x \in \mathbb{Z} : 9 \leq x \leq 24\} \cap C = \{t \in \mathbb{Z} : 13 < t \leq 24\}.$$

Also,

$$A \cap (B \cup C) = A \cap \{x \in \mathbb{R} : -4 < x \leq 30\} = \{y \in \mathbb{Z} : 9 \leq x \leq 30\}.$$

Try your hand at calculating $A \cup (B \cup C)$. □

The symbol $\emptyset$ is used to denote the set with no elements. We call this set the *empty set*. For instance,

$$A = \{x \in \mathbb{R}: x^2 < 0\}$$

is a perfectly good set. However, there are no real numbers which satisfy the given condition. Thus A is empty, and we write $A = \emptyset$.

Example 0.50: Let

$$A = \{x : x > 8\} \text{ and } B = \{x : x^2 < 4\}.$$

Then, $A \cup B = \{x : x > 8 \text{ or } -2 < x < 2\}$ while $A \cap B = \emptyset$. □

We sometimes use a *Venn diagram* to aid our understanding of set-theoretic relationships. In a Venn diagram, a set is represented as a domain in the plane. The intersection $A \cap B$ of two sets A and B is the region common to the two domains—see Figure 0.1.

Now, let A, B, and C be three sets. The Venn diagram in Figure 0.2 makes it easy to see that $A \cap (B \cup C) = (A \cap B) \cup (A \cap C)$.

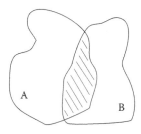

FIGURE 0.1
The intersection of two sets.

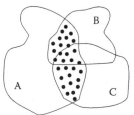

FIGURE 0.2
$A \cap (B \cup C) = (A \cap B) \cup (A \cap C)$.

If A and B are sets then $A \setminus B$ denotes those elements that are *in A but not in B*. This operation is sometimes called *subtraction of sets* or *set-theoretic difference*.

Example 0.51: Let

$$A - \{x : 4 < x\}$$

and

$$B = \{x : 6 \le x \le 8\}.$$

Then,

$$A \setminus B = \{x : 4 < x < 6\} \cup \{x : 8 < x\}$$

while

$$B \setminus A = \varnothing.$$

Notice that $A \setminus A = \varnothing$; this fact is true for any set.   □

Example 0.52: Let

$$S = \{x : 5 \le x\}$$

and

$$T = \{x : 4 < x < 6\}.$$

Then,

$$S \setminus T = \{x : 6 \le x\} \text{ and } T \setminus S = \{x: 4 < x < 5\}.$$

The Venn diagram in Figure 0.3 illustrates the fact that

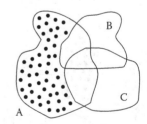

FIGURE 0.3
$A \setminus (B \cup C) = (A \setminus B) \cap (A \setminus C).$

$$A \setminus (B \cup C) = (A \setminus B) \cap (A \setminus C).$$

A Venn diagram is not a proper substitute for a rigorous mathematical proof. However, it can go a long way toward guiding our intuition.

We conclude this section by mentioning a useful set-theoretic operation and an application. Suppose that we are studying subsets of a fixed set X. We sometimes call X the "universal set." If $S \subset X$, then we use the notation ^cS to denote the set $X \setminus S$ or $\{x \in X : x \notin S\}$. The set ^cS is called *the complement of S* (in the set X).

Example 0.53: When we study real analysis, most sets that we consider are subsets of the real line $\mathbb{R}$. If $S = \{x \in \mathbb{R} : 0 \leq x \leq 5\}$, then $^cS = \{x \in \mathbb{R} : x < 0\} \cup \{x \in \mathbb{R} : x > 5\}$. If T is the set of rational numbers, then ^cT is the set of irrational numbers. □

If A, B are sets, then it is straightforward to verify that $^c(A \cup B) = {}^cA \cap {}^cB$ and $^c(A \cap B) = {}^cA \cup {}^cB$. These are known as *de Morgan's laws*. Let us prove the first of these.

If $x \in {}^c(A \cup B)$, then x is not an element of $A \cup B$. Hence, x is not an element of A and x is not an element of B. So, $x \in {}^cA$ and $x \in {}^cB$. Therefore, $x \in {}^cA \cap {}^cB$. That shows that $^c(A \cup B) \subset {}^cA \cap {}^cB$. For the reverse direction, assume that $x \in {}^cA \cap {}^cB$. Then, $x \in {}^cA$ and $x \in {}^cB$. As a result, $x \notin A$ and $x \notin B$. So, $x \notin A \cup B$. So, $x \in {}^c(A \cup B)$. This shows that $^cA \cap {}^cB \subset {}^c(A \cup B)$.

The two inclusions that we have proved establish that $^c(A \cup B) = {}^cA \cap {}^cB$.

0.2.6 Relations and Functions

In more elementary mathematics courses, we learn that a "relation" is a rule for associating elements of two sets; and a "function" is a rule that associates to each element of one set a unique element of another set. The trouble with these definitions is that they are imprecise. For example, suppose we define the function $f(x)$ to be identically equal to 1 if there is life as we know it on Mars and to be identically equal to 0 if there is no life as we know it on Mars. Is this a good definition? It certainly is not a very practical one!

More important is the fact that using the word "rule" suggests that functions are given by formulas. Indeed, some functions are; but most are not. Look at any graph in the newspaper—of unemployment, or the value of the Japanese Yen (Figure 0.4), or the gross national product. The graphs represent values of these parameters as a function of time. And, it is clear that the functions are not given by elementary formulas.

To summarize, we need a notion of function, and of relation, which is precise and flexible and which does not tie us to formulas. We begin with relations, and then specialize down to functions.

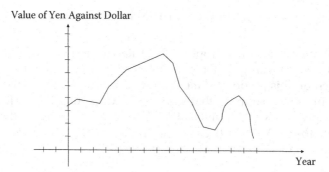

FIGURE 0.4
Value of the Yen against the Dollar.

Definition 0.54: Let A and B be sets. A *relation* $\mathcal{R}$ on A and B is a collection of ordered pairs (a, b) such that $a \in A$ and $b \in B$. (Notice that we did not say "*the* collection of all ordered pairs"—that is, a relation consists of some of the ordered pairs, but not necessarily all of them.) If a is related to b then we sometimes write $a\mathcal{R}b$ or $(a, b) \in \mathcal{R}$.

Example 0.55: Let A be the real numbers and B the integers. The set

$$\mathcal{R} = \{(\pi, 2), (3.4, -2), (\sqrt{2}, 94), (\pi, 50), (2 + \sqrt{17}, -2)\}$$

is a relation on A and B. It associates certain elements of A to certain elements of B. Observe that repetitions are allowed: $\pi \in A$ is associated to both 2 and 50 in B; also $-2 \in B$ is associated to both 3.4 and $2 + \sqrt{17}$ in A. This relation is not given by any formula or rule.

Now, let

$$A = \{3, 17, 28, 42\} \text{ and } B = \{10, 20, 30, 40\}.$$

Then,

$$\mathcal{R} = \{(3, 10), (3, 20), (3, 30), (3, 40), (17, 20), (17, 30),$$
$$(17, 40), (28, 30), (28, 40)\}$$

is a relation on A and B. In fact $a \in A$ is related to $b \in B$ precisely when $a < b$. This second relation is given by a rule.   □

Example 0.56: Let

$$A = B = \{\text{meter, pound, foot, ton, yard, ounce}\}.$$

Then,

$$\mathcal{R} = \{(\text{foot, meter}), (\text{foot, yard}), (\text{meter, yard}), (\text{pound, ton}),$$
$$(\text{pound, ounce}), (\text{ton, ounce}), (\text{meter, foot}), (\text{yard, foot}),$$
$$(\text{yard, meter}), (\text{ton, pound}), (\text{ounce, pound}), (\text{ounce, ton})\}$$

is a relation on A and B. In fact, two words are related by $\mathcal{R}$ if and only if they measure the same thing: foot, meter, and yard measure length while pound, ton, and ounce measure weight.

Notice that the pairs in $\mathcal{R}$, and in any relation, are *ordered* pairs: the pair (foot, yard) is different from the pair (yard, foot). □

Example 0.57: Let

$$A = \{25, 37, 428, 695\} \text{ and } B = \{14, 7, 234, 999\}$$

Then,

$$\mathcal{R} = \{(25, 234), (37, 7), (37, 234), (428, 14), (428, 234), (695, 999)\}$$

is a relation on A and B. In fact, two elements are related by $\mathcal{R}$ if and only if they have at least one digit in common. □

Definition 0.58: A relation $\mathcal{R}$ on a set A is said to be an *equivalence relation* if it has these three properties:

Reflexive: For any $a \in A$, it holds that $a \, \mathcal{R} \, a$.
Symmetric: If $a \, \mathcal{R} \, b$, then $b \, \mathcal{R} \, a$.
Transitive: If $a \, \mathcal{R} \, b$ and $b \, \mathcal{R} \, c$, then $a \, \mathcal{R} \, c$.

It can be proved that, if $\mathcal{R}$ is an equivalence relation, then it partitions A into pairwise disjoint equivalence classes. That is to say, if $a \in A$, then let

$$E_a = \{x \in A : a\mathcal{R}x\}.$$

We call E_a the *equivalence class* of a. Then it is the case that if $E_a \cap E_b \neq \varnothing$, then $E_a = E_b$. So the union of the E_a is all of A, and the E_a are pairwise disjoint. For all the details of the theory of equivalence classes, consult Ref. [KRA1].

A function is a special type of relation, as we shall now learn.

Definition 0.59: Let A and B be sets. A *function* from A to B is a relation $\mathcal{R}$ on A and B such that for each $a \in A$ there is one and only one pair $(a, b) \in \mathcal{R}$. We call A the *domain* of the function and we call B the *range*.[1]

Example 0.60: Let

$$A = \{1, 2, 3, 4\} \text{ and } B = \{\alpha, \beta, \gamma, \delta\}.$$

Then,

$$\mathcal{R} = \{(1, \gamma), (2, \delta), (3, \gamma), (4, \alpha)\}$$

is a function from A to B. Notice that there is precisely one pair in $\mathcal{R}$ for each element of A. However, notice that repetition of elements of B is allowed. Notice also that there is no apparent "pattern" or "rule" that determines $\mathcal{R}$. Finally, observe that not all the elements of B are used.

With the same sets A and B, consider the relations

$$S = \{(1, \alpha), (2, \beta), (3, \gamma)\}$$

and

$$\mathcal{T} = \{(1, \alpha), (2, \beta), (3, \gamma), (4, \delta), (2, \gamma)\}.$$

Then, S is not a function because it violates the rule that there be a pair for *each* element of A. Also, $\mathcal{T}$ is not a function because it violates the rule that there be *just one* pair for each element of A.   □

The relations and function described in the last example were so simple that you may be wondering what happened to the kinds of functions that we usually look at in mathematics. Now, we consider some of those.

Example 0.61: Let $A = \mathbb{R}$ and $B = \mathbb{R}$, where $\mathbb{R}$ denotes the real numbers. The relation

$$\mathcal{R} = \{(x, \, \sin x) : x \in A\}$$

is a function from A to B. For each $a \in A = \mathbb{R}$, there is one and only one ordered pair with first element a.

Now, let $S = \mathbb{R}$ and $T = \{x \in \mathbb{R} : -2 \le x \le 2\}$. Then,

$$\mathcal{U} = \{(x, \, \sin x) : x \in A\}$$

is also a function from S to T. Technically speaking, it is a different function from $\mathcal{R}$ because it has a different range. However, this distinction often has no practical importance and we shall not mention the difference. It is frequently convenient to write functions like $\mathcal{R}$ or $\mathcal{U}$ as

$$\mathcal{R}(x) = \sin x$$

and

$$\mathcal{U}(x) = \sin x.$$

□

The last example suggests that we distinguish between the set B where a function takes its values and the set of values that the function *actually assumes*.

Definition 0.62: Let A and B be sets and let f be a function from A to B. Define the *image* of f to be

$$\text{Image } f = \{b \in B : \exists\, a \in A \text{ such that } f(a) = b\}.$$

The set Image f is a subset of the range B. In general, the image *will not* equal the range.

Example 0.63: Both the functions $\mathcal{R}$ and $\mathcal{U}$ from the last example have the set $\{x \in \mathbb{R} : -1 \le x \le 1\}$ as image. In neither instance does the image equal the range. □

If a function f has domain A and range B and if S is a subset of A, then we define

$$f(S) = \{b \in B : b = f(s) \text{ for some } s \in S\}.$$

The set $f(A)$ equals the image of f.

Example 0.64: Let $A = \mathbb{R}$ and $B = \{0, 1\}$. Consider the function

$$f = \{(x, y) : y = 0 \text{ if } x \text{ is rational and }$$
$$y = 1 \text{ if } x \text{ is irrational}\}.$$

The function f is called the *Dirichlet function* (P. G. Lejeune-Dirichlet, 1805–1859). It is given by a rule, but not by a formula.
  Notice that $f(\mathbb{Q}) = \{0\}$ and $f(\mathbb{R}) = \{0, 1\}$. □

Definition 0.65: Let A and B be sets and f a function from A to B.

We say that f is *one-to-one* if whenever $(a_1, b) \in f$ and $(a_2, b) \in f$ then $a_1 = a_2$.

We say that f is *onto* if whenever $b \in B$, then there exists an $a \in A$ such that $(a, b) \in f$.

Example 0.66: Let $A = \mathbb{R}$ and $B = \mathbb{R}$. Consider the functions

$$f(x) = 2x + 5, \quad g(x) = \arctan x$$
$$h(x) = \sin x, \quad j(x) = 2x^3 + 9x^2 + 12x + 4.$$

Then, f is both one-to-one and onto, g is one-to-one but not onto, j is onto but not one-to-one, and h is neither.

Refer to Figure 0.5 to convince yourself of these assertions.   □

When a function f is both one-to-one and onto then it is called a *bijection* of its domain to its range. Sometimes we call such a function a *set-theoretic isomorphism*. In the last example, the function f is a bijection of $\mathbb{R}$ to $\mathbb{R}$.

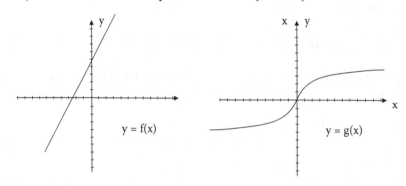

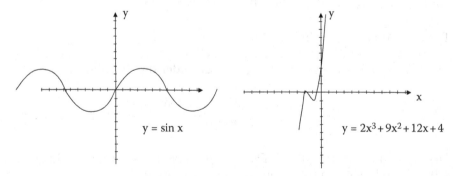

FIGURE 0.5
One-to-one and onto functions.

If f and g are functions, and if the image of g is contained in the domain of f, then we define the *composition* $f \circ g$ to be

$$\{(a, c): \exists\, b \text{ such that } g(a) = b \text{ and } f(b) = c\}.$$

This may be written more simply as

$$f \circ g(a) = f(g(a)) = f(b) = c.$$

Let f have domain A and range B. Assume for simplicity that the image of f is all of B. If there exists a function g with domain B and range A such that

$$f \circ g(b) = b \quad \forall\, b \in B$$

and

$$g \circ f(a) = a \quad \forall\, a \in A,$$

then g is called the *inverse* of f.

Clearly, if the function f is to have an inverse, then f must be one-to-one. For if $f(a) = f(a') = b$, then it cannot be that both $g(b) = a$ and $g(b) = a'$. Also, f must be onto. For if some $b \in B$ is not in the image of f, then it cannot hold that $f \circ g(b) = b$. It turns out that these two conditions are also sufficient for the function f to have an inverse: If f has domain A and range B and if f is both one-to-one and onto, then f has an inverse.

Example 0.67: Define a function f, with domain $\mathbb{R}$ and range $\{x \in \mathbb{R} : x \geq 0\}$, by the formula $f(x) = x^2$. Then f is onto but is not one-to-one (because $f(-1) = f(1)$), hence it cannot have an inverse. This is another way of saying that a positive real number has two square roots—not one.

However, the function g, with domain $\{x \in \mathbb{R} : x \geq 0\}$ and range $\{x \in \mathbb{R} : x \geq 0\}$, given by the formula $g(x) = x^2$, *does* have an inverse. In fact, the inverse function is $h(x) = +\sqrt{x}$.

The function $k(x) = x^3$, with domain $\mathbb{R}$ and range $\mathbb{R}$, is both one-to-one and onto. It, therefore, has an inverse: the function $m(x) = x^{1/3}$ satisfies $k \circ m(x) = x$, and $m \circ k(x) = x$ for all x. □

0.2.7 Countable and Uncountable Sets

One of the most profound ideas of modern mathematics is Georg Cantor's theory of the infinite (George Cantor, 1845–1918). Cantor's insight was that infinite sets can be compared by size, just as finite sets can. For instance, we think of the number 2 as *less* than the number 3; so a set with two elements is "smaller" than a set with three elements. We would like to have a similar notion of comparison for infinite sets. In this section, we will present Cantor's ideas; we will also give precise definitions of the terms "finite" and "infinite."

Definition 0.68: Let A and B be sets. We say that A and B have the *same cardinality* if there is a function f from A to B, which is both one-to-one and onto (that is, f is a bijection from A to B). We write card(A) = card(B). Some books write $|A| = |B|$.

Example 0.69: Let $A = \{1, 2, 3, 4, 5\}$, $B = \{\alpha, \beta, \gamma, \delta, \epsilon\}$, $C = \{a, b, c, d, e, f\}$. Then, A and B have the same cardinality because the function

$$f = \{(1, \alpha), (2, \beta), (3, \gamma), (4, \delta), (5, \epsilon)\}$$

is a bijection of A to B. This function is not the *only* bijection of A to B (can you find another?), but we are only required to produce one.

On the other hand, A and C do not have the same cardinality; neither do B and C.   □

Notice that if card(A) = card(B) via a function f_1 and card(B) = card(C) via a function f_2 then card(A) = card(C) via the function $f_2 \circ f_1$.

Example 0.70: Let A and B be sets. If there is a one-to-one function from A to B but no bijection between A and B, then we will write

$$\text{card}(A) < \text{card}(B).$$

This notation is read "A has smaller cardinality than B."
We use the notation

$$\text{card}(A) \leq \text{card}(B)$$

to mean that either card(A) < card(B) or card(A) = card(B).

Example 0.71: An extremely simple example of this last concept is given by $A = \{1, 2, 3\}$ and $B = \{a, b, c, d, e\}$. Then, the function

$$
\begin{aligned}
f : A &\rightarrow B \\
1 &\mapsto a \\
2 &\mapsto b \\
3 &\mapsto c
\end{aligned}
$$

is a one-to-one function from A to B. But there is no one-to-one function from B to A. We write

$$\text{card}(A) < \text{card}(B).$$

We shall see more profound applications, involving infinite sets, in our later discussions. □

Notice that $\text{card}(A) \leq \text{card}(B)$ and $\text{card}(B) \leq \text{card}(C)$ imply that $\text{card}(A) \leq \text{card}(C)$. Moreover, if $A \subset B$, then the inclusion map $i(a) = a$ is a one-to-one function of A into B; therefore $\text{card}(A) \leq \text{card}(B)$.

The next theorem gives a useful method for comparing the cardinality of two sets.

Theorem 0.72: (Schroeder-Bernstein) *Let A, B, be sets. If there is a one-to-one function $f : A \to B$ and a one-to-one function $g : B \to A$, then A and B have the same cardinality.*

Proof: It is convenient to assume that A and B are disjoint; we may do so by replacing A by $\{(a, 0) : a \in A\}$ and B by $\{(b, 1) : b \in B\}$. Let D be the image of f and C be the image of g. Let us define a *chain* to be a sequence of elements of either A or B—that is, a function $\phi : \mathbb{N} \to (A \cup B)$—such that

- $\phi(1) \in B \setminus D$;
- If for some j we have $\phi(j) \in B$, then $\phi(j + 1) = g(\phi(j))$;
- If for some j we have $\phi(j) \in A$, then $\phi(j + 1) = f(\phi(j))$.

We see that a chain is a sequence of elements of $A \cup B$ such that the first element is in $B \setminus D$, the second in A, the third in B, and so on. Obviously, each element of $B \setminus D$ occurs as the first element of at least one chain.

Define $S = \{a \in A : a$ is some term of some chain$\}$. It is helpful to note that

$$S = \{x : x \text{ can be written in the form}$$
$$g(f(g(\cdots g(y)\ldots))) \text{ for some } y \in B \setminus D\}. \tag{0.72.1}$$

We set

$$k(x) = \begin{cases} f(x) & \text{if } x \in A \setminus S \\ g^{-1}(x) & \text{if } x \in S \end{cases}$$

Note that the second half of this definition makes sense because $S \subseteq C$. Then, $k : A \to B$. We shall show that in fact k is a bijection.

First, notice that f and g^{-1} are one-to-one. This is not quite enough to show that k is one-to-one, but we now reason as follows: If $f(x_1) = g^{-1}(x_2)$ for some $x_1 \in A \setminus S$ and some $x_2 \in S$, then $x_2 = g(f(x_1))$. But, by (0.72.1), the fact that $x_2 \in S$ now implies that $x_1 \in S$. That is a contradiction. Hence, k is one-to-one.

It remains to show that k is onto. Fix $b \in B$. We seek an $x \in A$ such that $k(x) = b$.

Case A: If $g(b) \in S$, then $k(g(b)) \equiv g^{-1}(g(b)) = b$; hence, the x that we seek is $g(b)$.

Case B: If $g(b) \notin S$, then we claim that there is an $x \in A$ such that $f(x) = b$. Assume this claim for the moment.

Now the x that we found in the last paragraph must lie in $A \setminus S$. For if not then x would be in some chain. Then, $f(x)$ and $g(f(x)) = g(b)$ would also lie in that chain. Hence, $g(b) \in S$, and that is a contradiction. But, $x \in A \setminus S$ tells us that $k(x) = f(x) = b$. That completes the proof that k is onto. Hence, k is a bijection.

To prove the claim in Case B, notice that if there is no x with $f(x) = b$, then $b \in B \setminus D$. Thus, some chain would begin at b. So, $g(b)$ would be a term of that chain. Hence, $g(b) \in S$ and that is a contradiction.

The proof of the Schroeder–Bernstein theorem is complete.   □

Remark 0.73: Let us reiterate some of the earlier ideas in light of the Schroeder–Bernstein theorem. If A and B are sets and if there is a one-to-one function $f: A \to B$, then we know that $\mathrm{card}(A) \leq \mathrm{card}(B)$. If there is no one-to-one function $g: B \to A$, then we may write $\mathrm{card}(A) < \mathrm{card}(B)$. But if instead there is a one-to-one function $g: B \to A$, then $\mathrm{card}(B) \leq \mathrm{card}(A)$ and the Schroeder–Bernstein theorem guarantees therefore that $\mathrm{card}(A) = \mathrm{card}(B)$.   □

Now, it is time to look at some specific examples.

Remark 0.74: Let E be the set of all even integers and O the set of all odd integers. Then,

$$\mathrm{card}(E) = \mathrm{card}(O).$$

Indeed, the function

$$f(j) = j + 1$$

is a bijection from E to O.   □

Example 0.75: Let E be the set of even integers. Then,

$$\mathrm{card}(E) = \mathrm{card}(\mathbb{Z}).$$

The function

$$g(j) = j/2$$

is a bijection from E to $\mathbb{Z}$. □

This last example is a bit surprising, for it shows that the set $\mathbb{Z}$ can be put in one-to-one correspondence with a proper subset E of itself. In other words, we are saying that the integers $\mathbb{Z}$ "have the same number of elements" as a proper subset of $\mathbb{Z}$. Such a phenomenon cannot occur with finite sets.

Example 0.76: We have

$$\text{card}(\mathbb{Z}) = \text{card}(\mathbb{N}).$$

We define the function f from $\mathbb{Z}$ to $\mathbb{N}$ as follows:

- $f(j) = -(2j + 1)$ if j is negative
- $f(j) = 2j + 2$ if j is positive or zero

The values that f takes on the negative numbers are $1, 3, 5, \ldots$, on the positive numbers are $4, 6, 8, \ldots$, and $f(0) = 2$. Thus, f is one-to-one and onto. □

Definition 0.77: If a set A has the same cardinality as $\mathbb{N}$ then we say that A is *countable*.

By putting together the preceding examples, we see that the set of even integers, the set of odd integers, and the set of all integers are examples of countable sets.

Example 0.78: The set of all ordered pairs of positive integers

$$S = \{(j, k) : j, k \in \mathbb{N}\}$$

is countable.

To see this, we will use the Schroeder–Bernstein theorem. The function

$$f(j) = (j, 1)$$

is a one-to-one function from $\mathbb{N}$ to S. Also, the function $g(j, k) = 2^j \cdot 3^k$ is a one-to-one function from S to $\mathbb{N}$. By the Schroeder–Bernstein theorem, S and $\mathbb{N}$ have the same cardinality; hence, S is countable. □

Remark 0.79: You may check for yourself that the function $F(j, k) = 2^{j-1} \cdot (2k - 1)$ is an explicit bijection from S to $\mathbb{N}$.

Since there is a bijection of the set of *all* integers with the set $\mathbb{N}$, it follows from the last example that the set of all pairs of integers (positive *and* negative) is countable.

Notice that the word "countable" is a good descriptive word: if S is a countable set then we can think of S as having a first element (the one corresponding to $1 \in \mathbb{N}$), a second element (the one corresponding to $2 \in \mathbb{N}$), and so forth. Thus, we write $S = \{s(1), s(2), \ldots\} = \{s_1, s_2, \ldots\}$.

Definition 0.80: A nonempty set S is called *finite* if there is a bijection of S with a set of the form $\{1, 2, \ldots, n\}$ for some positive integer n. If no such bijection exists, then the set is called *infinite*.

An important property of the natural numbers $\mathbb{N}$ is that any subset $S \subset \mathbb{N}$ has a least element. This is known as the Well Ordering Principle, and is studied in a course on logic. In this text, we take the properties of the natural numbers as given. We use some of these properties in the next proposition.

Proposition 0.81: *If S is a countable set and R is a subset of S then either R is empty or R is finite or R is countable.*

Proof: Assume that R is not empty. Write $S = \{s_1, s_2, \ldots\}$. Let j_1 be the least positive integer such that $s_{j_1} \in R$. Let j_2 be the least integer following j_1 such that $s_{j_2} \in R$. Continue in this fashion. If the process terminates at the nth step, then R is finite and has n elements.

If the process does not terminate, then we obtain an enumeration of the elements of R:

$$1 \leftrightarrow s_{j_1}$$
$$2 \leftrightarrow s_{j_2}$$
$$\ldots$$

etc.

All elements of R are enumerated in this fashion since $j_\ell \geq \ell$. Therefore, R is countable.   □

A set is called *denumerable* if it is either empty, finite, or countable. Notice that the word "denumerable" is not the same as "countable." In fact, "countable" is just one instance of denumerable.

The set $\mathbb{Q}$ of all rational numbers consists of all expressions

$$\frac{a}{b},$$

where a and b are integers and $b \neq 0$. Thus, $\mathbb{Q}$ can be identified with the set of all ordered pairs (a, b) of integers with $b \neq 0$. After discarding duplicates, such as $\frac{2}{4} = \frac{1}{2}$, and using the examples mentioned earlier, we find that the set $\mathbb{Q}$ is countable.

Theorem 0.82: *Let S_1, S_2 be countable set s. Set $S = S_1 \cup S_2$. Then S is countable.*

Proof: Let us write

$$S_1 = \{s_1^1, s_2^1, \ldots\}$$
$$S_2 = \{s_1^2, s_2^2, \ldots\}.$$

If $S_1 \cap S_2 = \emptyset$, then the function

$$s_j^k \mapsto (j, k)$$

is a bijection of S with a subset of $\{(j, k) : j, k \in \mathbb{N}\}$. We proved earlier (Example A2.55) that the set of ordered pairs of elements of $\mathbb{N}$ is countable. By Proposition A2.58, S is countable as well.

If there exist elements that are common to S_1, S_2 then discard any duplicates. The same argument (use the preceding proposition) shows that S is countable. □

Theorem 0.83: *If S and T are each countable sets, then so is*

$$S \times T \equiv \{(s, t) : s \in S, t \in T\}.$$

Proof: Since S is countable there is a bijection f from S to $\mathbb{N}$. Likewise, there is a bijection g from T to $\mathbb{N}$. Therefore, the function

$$(f \times g)(s, t) = (f(s), g(t))$$

is a bijection of $S \times T$ with $\mathbb{N} \times \mathbb{N}$, the set of order pairs of positive integers. But we saw in Example A2.55 that the latter is a countable set. Hence, so is $S \times T$. □

Remark 0.84: We used the theorem as a vehicle for defining the concept of *set-theoretic product*: If A and B are sets, then

$$A \times B \equiv \{(a, b) : a \in A, b \in B\}.$$

More generally, if $A_1, A_2, ..., A_k$ are sets then

$$A_1 \times A_2 \times \cdots \times A_k = \{(a_1, a_2, ..., a_k) : a_j \subset A_j \text{ for all } j = 1, ..., k\}.$$

□

Corollary 0.85: *If $S_1, S_2, ..., S_k$ are each countable sets, then so is the set*

$$S_1 \times S_2 \times \cdots \times S_k = \{(s_1, ..., s_k) : s_1 \in S_1, ..., s_k \in S_k\}$$

consisting of all ordered k–tuples $(s_1, s_2, ..., s_k)$ with $s_j \in S_j$.

Proof: We may think of $S_1 \times S_2 \times S_3$ as $(S_1 \times S_2) \times S_3$. Since $S_1 \times S_2$ is countable (by the theorem) and S_3 is countable, then so is $(S_1 \times S_2) \times S_3 = S_1 \times S_2 \times S_3$ countable. Continuing in this fashion, we can see that any finite product of countable sets is also a countable set. □

We are accustomed to the union $A \cup B$ of two sets or, more generally, the union $A_1 \cup A_2 \cup \cdots \cup A_k$ of finitely many sets. But sometimes we wish to consider the union of infinitely many sets. Let $S_1, S_2, ...$ be countably many sets. We say that x is an element of

$$\bigcup_{j=1}^{\infty} S_j$$

if x is an element of at least one of the S_j.

Corollary 0.86: *The countable union of countable sets is countable.*

Proof: Let $A_1, A_2, ...$ each be countable sets. If the elements of A_j are enumerated as $\{a_k^j\}$ and if the sets A_j are pairwise disjoint then the correspondence

$$a_k^j \leftrightarrow (j, k)$$

is one-to-one between the union of the sets A_j and the countable set $\mathbb{N} \times \mathbb{N}$. This proves the result when the sets A_j have no common element. If some of the A_j have elements in common, then we discard duplicates in the union and use Proposition A2.58. □

Proposition 0.87: *The collection $\mathcal{P}$ of all polynomials with integer coefficients is countable.*

Proof: Let $\mathcal{P}_k$ be the set of polynomials of degree k with integer coefficients. A polynomial p of degree k has the form

$$p(x) = p_0 + p_1 x + p_2 x^2 + \cdots + p_k x^k.$$

The identification

$$p(x) \leftrightarrow (p_0, p_1, \ldots, p_k)$$

identifies the elements of $\mathcal{P}_k$ with the $(k + 1)$-tuples of integers. By Corollary 0.85, it follows that $\mathcal{P}_k$ is countable. But then Corollary 0.86 implies that

$$\mathcal{P} = \bigcup_{j=0}^{\infty} \mathcal{P}_j$$

is countable. □

Georg Cantor's remarkable discovery is that *not all infinite sets are countable.* We next give an example of this phenomenon.

In what follows, a *sequence* on a set S is a function from $\mathbb{N}$ to S. We usually write such a sequence as $s(1), s(2), s(3), \ldots$ or as $s_1, s_2, s_3, \ldots$.

Example 0.88: There exists an infinite set which is not countable (we call such a set *uncountable*). Our example will be the set S of all sequences on the set $\{0, 1\}$. In other words, S is the set of all infinite sequences of 0s and 1s. To see that S is uncountable, assume the contrary. Then, there is a first sequence

$$\mathcal{S}^1 = \{s_j^1\}_{j=1}^{\infty},$$

a second sequence

$$\mathcal{S}^2 = \{s_j^2\}_{j=1}^{\infty},$$

and so forth. This will be a complete enumeration of all the members of S. But, now consider the sequence $\mathcal{T} = \{t_j\}_{j=1}^{\infty}$, which we construct as follows:

- If $s_1^1 = 0$ then set $t_1 = 1$; if $s_1^1 = 1$ then set $t_1 = 0$;
- If $s_2^2 = 0$ then set $t_2 = 1$; if $s_2^2 = 1$ then set $t_2 = 0$;
- If $s_3^3 = 0$ then set $t_3 = 1$; if $s_3^3 = 1$ then set $t_3 = 0$;

$$\cdots$$

- If $s_j^j = 0$ then set $t_j = 1$; if $s_j^j = 1$ then set $t_j = 0$;

etc.

Now, the sequence $\mathcal{T}$ differs from the first sequence $\mathcal{S}^1$ in the first element: $t_1 \neq s_1^1$.

The sequence $\mathcal{T}$ differs from the second sequence $\mathcal{S}^2$ in the second element: $t_2 \neq s_2^2$.

And so on: the sequence $\mathcal{T}$ differs from the jth sequence $\mathcal{S}^j$ in the jth element: $t_j \neq s_j^j$. So the sequence $\mathcal{T}$ is not in the set S. But $\mathcal{T}$ is *supposed* to be in the set S because it is a sequence of 0s and 1s and all of these have been hypothesized to be enumerated.

This contradicts our assumption, so S must be uncountable.   □

Example 0.89: Consider the set of all decimal representations of numbers—both terminating and nonterminating. Here, a terminating decimal is one of the form

$$27.43926$$

while a nonterminating decimal is one of the form

$$3.14159265\ldots.$$

In the case of the nonterminating decimal, no repetition is implied; the decimal simply continues without cease.   □

Now, the set of all those decimals containing only the digits 0 and 1 can be identified in a natural way with the set of sequences containing only 0 and 1 (just put commas between the digits). And, we just saw that the set of such sequences is uncountable.

Since the set of all decimal numbers is an even bigger set, it must be uncountable also.

As you may know, the set of all decimals identifies with the set of all real numbers. We find then that the set $\mathbb{R}$ of all real numbers is uncountable. (Contrast this with the situation for the rationals.) In Section 1.1, we shall learn about how the real number system is constructed using just elementary set theory.   □

It is an important result of set theory (due to Cantor) that, given any set S, the set of all subsets of S (called the *power set* of S) has strictly greater cardinality than the set S itself. As a simple example, let $S = \{a, b, c\}$. Then, the set of all subsets of S is

$$\{\varnothing, \{a\}, \{b\}, \{c\}, \{a, b\}, \{a, c\}, \{b, c\}, \{a, b, c\}\}.$$

The set of all subsets has eight elements, while the original set has just three.

Even more significant is the fact that if S is an infinite set, then the set of all its subsets has greater cardinality than S itself. This is a famous theorem of Cantor. Thus, there are infinite sets of arbitrarily large cardinality.

In some of the examples in this section, we constructed a bijection between a given set (such as $\mathbb{Z}$) and a proper subset of that set (such as E, the even integers). It follows from the definitions that this is possible only when the sets involved are infinite.

Exercises

1. Let q be a rational number. Construct a sequence $\{x_j\}$ of irrational numbers such that $x_j \to q$. This means that, for each $\varepsilon > 0$, there is a positive integer K such that if $j > K$, then $|x_j - q| < \varepsilon$.

2. Let S be a set of real numbers with the property that, whenever $x, y \in S$ and $x < t < y$, then $t \in S$. Can you give a simple description of the set S?

3. Let $a_1 < a_2 < \cdots$ be real numbers. Prove that either there is a real number α such that $a_j \to \alpha$ (refer to Exercise 6.4 for this notation) *or else* the sequence $\{a_j\}$ increases without bound.

4. Prove that subtraction is well defined in the integers.

5. Give a careful discussion of the failure of the operation of division in the integers.

6. Prove that addition and subtraction are well defined in the rational number system $\mathbb{Q}$.

7. Determine whether $\sqrt{2} + \sqrt{3}$ is rational or irrational.

8. Prove that every nonzero complex number $z \in \mathbb{C}$ has two distinct square roots in $\mathbb{C}$.

9. The complex number $1 = 1 + 0i$ has three cube roots. Use any means to find them, and sketch them on an Argand diagram (refer to Exercise 6.21 for terminology).

10. If $z = x + iy \in \mathbb{C}$ is any nonzero complex number, then let

$$r^2 = |z|^2 = x^2 + y^2.$$

The number r is the distance of z to the origin in the Argand plane (Exercise 6.21). It is also the modulus of z. Set $\xi = z/r$. Show that $|\xi| = 1$. Now apply Exercise 6.25 to conclude that

$$z = r\, e^{i\theta},$$

some $0 \le \theta < 2\pi$. This is called the *polar form* of the complex number z.

11. Prove that the complex numbers cannot be made into an ordered field (as discussed in Section 6.7).

12. Prove that addition and multiplication are commutative in the complex number system.

13. Let p be a polynomial and assume that $\alpha \in \mathbb{C}$ is a root of p. Prove that $(z - \alpha)$ evenly divides $p(z)$ with no remainder.

Note

1 Some textbooks use the word "codomain" instead of "range." We shall use only the word "range."

1

Real and Complex Numbers

1.1 The Real Numbers

This is a book about analysis in the real number system. Such a study must be founded on a careful consideration of *what the real numbers are* and *how they are constructed*. In this section, we give a careful treatment of the real number system. In the next, we consider the complex numbers.

We know from real numbers calculus that, for many purposes, the rational numbers are inadequate. It is important to work in a number system that is closed with respect to the operations we shall perform. This includes the limiting operations. While the rationals are closed under the usual arithmetic operations (addition, subtraction, multiplication, and division), they are *not* closed under the limits mathematical operation of taking *limits*. For instance, the sequence of rational numbers 3, 3.1, 3.14, 3.141, ... consists of terms that seem to be getting closer and closer together, *seem* to tend to some limit, and yet there is no rational number that will serve as a limit (of course, it turns out that the limit is π—an "irrational" number).

We will now deal with the real number system, a system that contains all limits of sequences of rational numbers (as well as all limits of sequences of real numbers!). In fact, our plan will be as follows: in this section, we shall discuss all the requisite properties of the reals. The actual construction of the reals is rather subtle, and we shall put that in an Appendix to Section 1.1.

Definition 1.1: Let A be an ordered set and X a subset of A. The set X is called *bounded above* if there is an element $b \in A$ such that $x \le b$ for all $x \in X$. We call the element b an *upper bound* for the set X.

Example 1.2: Let $A = \mathbb{Q}$ (the rational numbers) with the usual ordering. The set $X = \{x \in \mathbb{Q} : 2 < x < 4\}$ is bounded above. For example, 15 is an upper bound for X. So are the numbers 12 and 4. It is interesting to observe that no element of this particular X can actually be an upper bound for X. The number 4 is a good candidate, but 4 is not an element of X. In fact, if $b \in X$, then $(b + 4)/2 \in X$ and $b < (b + 4)/2$, so b could not be an upper bound for X. □

DOI: 10.1201/9781003222682-2

It turns out that the most convenient way to formulate the notion that the real numbers have "no holes" (that is, that all sequences which seem to be converging actually have something to converge to) is in terms of upper bounds.

Definition 1.3: Let A be an ordered set and X a subset of A. An element $b \in A$ is called a *least upper bound* (or *supremum*) for X if b is an upper bound for X and $b \leq b'$ for every upper bound b' for X. We denote the supremum of X by sup X. The supremum is also sometimes called the *least upper bound* and denoted by lub X.

By its very definition, if a least upper bound exists, then it is unique. Notice that we *could have* phrased the definition as "The point b is the least upper bound for X if, whenever $c < b$, then c cannot be an upper bound for X."

Example 1.4: In the last example, we considered the set X of rational numbers strictly between 2 and 4. We observed there that 4 is the least upper bound for X. Note that this least upper bound is not an element of the set X.

The set $Y = \{y \in \mathbb{Z} : -9 \leq y \leq 7\}$ has least upper bound 7. In this case, the least upper bound *is* an element of the set Y.   □

Notice that we may define a lower bound for a subset of an ordered set in a fashion similar to that for an upper bound:

Definition 1.5: A point $\ell \in A$ is a lower bound for $X \subseteq A$ if $\ell \leq x$ for all $x \in X$. A *greatest lower bound* (or *infimum*) for X is then defined to be a lower bound c such that $c \geq c'$ for every lower bound c' for X. We denote the infimum of X by infimum X. The infimum is also sometimes called the *greatest lower bound* and denoted by glb X.

As with the least upper bound, we may note that the definition of greatest lower bound could be phrased in this way: "the point c is the greatest lower bound for X if, whenever $e > c$, then e cannot be a lower bound for X."

Example 1.6: The set $X = \{x \in \mathbb{Q} : 2 < x < 4\}$ in the last two examples has lower bounds $-20, 0, 1, 2$, for instance. The greatest lower bound is 2, which is *not* an element of the set.

The set $Y = \{y \in \mathbb{Z} : -9 \leq y \leq 7\}$ in the last example has lower bounds— among others—given by $-53, -22, -10$, and -9. The number -9 is the greatest lower bound. It *is* an element of Y.   □

The purpose that the real numbers will serve for us is as follows: they will contain the rationals, they will still be an ordered field (a field is a set with arithmetic operations $+$ and $\cdot$ —see the Appendix at the end of Section 1.1), and *every subset which has an upper bound will have a least upper bound*. (See (KRA1) for a thorough treatment of the concept of ordered field.) We formulate this result as a theorem.

Theorem 1.7: *There exists an ordered field* $\mathbb{R}$, *which* **(i)** *contains* $\mathbb{Q}$ *and* **(ii)** *has the property that any nonempty subset of* $\mathbb{R}$—*which has an upper bound—has a least upper bound (in the number system* $\mathbb{R}$).

The last property described in this theorem is called the *least upper bound property* of the real numbers. As mentioned previously, this theorem will be proved in the Appendix to Section 1.1. Of course, the least upper bound property, in and of itself, is something of a technicality. However, we shall see that a great many interesting and powerful properties of the real numbers can be derived from it.

Now, we begin to realize why it is so important to *construct* the number systems that we will use. We are endowing $\mathbb{R}$ with a great many properties. Why do we have any right to suppose that there exists a set with all these properties? We must produce one! We do so in the Appendix to Section 1.1.

Let us begin to explore the richness of the real numbers. The next theorem states a property, which is not shared by the rationals. It is fundamental in its importance.

Theorem 1.8: *Let x be a positive real number. Then, there is a positive real number y such that* $y^2 = y \cdot y = x$.

Proof: We will use throughout this proof the fact that if $0 < a < b$, then $a^2 < b^2$.

Let

$$S = \{s \in \mathbb{R} : s > 0 \text{ and } s^2 < x\}.$$

Then, S is not empty since $x/2 \in S$ if $x < 2$ and $1 \in S$ otherwise. Also, S is bounded above since $x + 1$ is an upper bound for S. By Theorem 1.6, the set S has a least upper bound. Call it y. Obviously, $0 < \min\{x/2, 1\} \leq y$; hence, y is positive. We claim that $y^2 = x$. To see this, we eliminate the other two possibilities.

If $y^2 < x$, then set $\epsilon = (x - y^2)/[4(x + 1)]$. Then, $\epsilon > 0$ and

$$
\begin{aligned}
(y + \epsilon)^2 &= y^2 + 2 \cdot y \cdot \epsilon + \epsilon^2 \\
&= y^2 + 2 \cdot y \cdot \frac{x - y^2}{4(x+1)} + \frac{x - y^2}{4(x+1)} \cdot \frac{x - y^2}{4(x+1)} \\
&< y^2 + 2 \cdot y \cdot \frac{x - y^2}{4y} + \frac{x - y^2}{4(x+1)} \cdot \frac{x - y^2}{4(x+1)} \\
&< y^2 + \frac{x - y^2}{2} + \frac{x - y^2}{4} \cdot \frac{x}{4x} \\
&< y^2 + (x - y^2) \\
&= x.
\end{aligned}
$$

Thus, $y + \epsilon \in S$, and y cannot be an upper bound for S. This contradiction tells us that $y^2 \not< x$.

Similarly, if it were the case that $y^2 > x$, then we set $\epsilon = (y^2 - x)/[4(x + 1)]$. A calculation like the one we just did (see Exercise 5) then shows that $(y - \epsilon)^2 \geq x$. Hence, $y - \epsilon$ is also an upper bound for S, and y is therefore not the *least* upper bound. This contradiction shows that $y^2 \not> x$.

The only remaining possibility is that $y^2 = x$.   □

Remark 1.9: The theorem tells us in particular that $\sqrt{2}, \sqrt{5}, \sqrt{8}, \sqrt{11}$, etc. all exist in the real number system. And, each of these numbers is irrational (see Theorem A1.23 where it is shown that $\sqrt{2}$ is irrational). In fact, the only square roots of integers that are *not* irrational are the square roots of the perfect squares 1, 4, 9, 16, 25, and so on.   □

A similar proof shows that if n is a positive integer and x a positive real number, then there is a positive real number y such that $y^n = x$. Exercise 14 asks you to provide the details.

We next use the least upper bound property of the real numbers to establish two important qualitative properties of the real numbers.

Theorem 1.10: *The set $\mathbb{R}$ of real numbers satisfies the Archimedean Property:*

> Let a and b be positive real numbers. Then, there is a natural number n
> such that $na > b$.

The set $\mathbb{Q}$ of rational numbers satisfies the following Density Property:

> Let $c < d$ be real numbers. Then, there is a rational number q with $c < q < d$.

Proof: Suppose the Archimedean Property to be false. Then, $S = \{na : n \in \mathbb{N}\}$ has b as an upper bound. Therefore, S has a finite supremum β. Since $a > 0$, it follows that $\beta - a < \beta$. So, $\beta - a$ is not an upper bound for S, and there must be a natural number n' such that $n' \cdot a > \beta - a$. But then, $(n' + 1)a > \beta$, and β cannot be the supremum for S. This contradiction proves the first assertion.

For the second property, let $\lambda = d - c > 0$. By the Archimedean Property, choose a positive integer N such that $N \cdot \lambda > 1$. Again, the Archimedean Property gives a natural number P such that $P > N \cdot c$ and another Q such that $Q > -N \cdot c$. Thus, we see that Nc falls between the integers $-Q$ and P; therefore, there must be an integer M between $-Q$ and P such that

$$M - 1 \leq Nc < M.$$

Thus, $c < M/N$. Also,

$$M \leq Nc + 1 \text{ hence } \frac{M}{N} \leq c + \frac{1}{N} < c + \lambda = d.$$

So M/N is a rational number lying between c and d. □

Remark 1.11: The density property stated earlier says that between any two real numbers is a rational number. Even more can be said. In fact, **(i)** between every two irrational numbers is a rational number and **(ii)** between every two rational numbers is an irrational number. □

In Appendix II at the end of the book, we establish that the set of all decimal representations of numbers is uncountable. It follows that the set of all real numbers is uncountable. In fact, the same proof shows that the set of all real numbers in the interval $(0, 1)$, or in any nonempty open interval (c, d), is uncountable.

The set $\mathbb{R}$ of real numbers is uncountable (see Section A2.7 in Appendix II), yet the set $\mathbb{Q}$ of rational numbers is countable. It follows that the set $\mathbb{R} \backslash \mathbb{Q}$ of *irrational* numbers is uncountable. In particular, it is nonempty. Thus, we may see with very little effort that there exist a great many real numbers that cannot be expressed as a quotient of integers. However, it can be quite difficult to see whether any particular real number (such as π or e or $\sqrt[5]{2}$) is irrational.

We conclude by recalling the "absolute value" notation: absolute value

Definition 1.12: Let x be a real number. We define

$$|x| = \begin{cases} x & \text{if } x > 0 \\ 0 & \text{if } x = 0 \\ -x & \text{if } x < 0 \end{cases}$$

It is left as an exercise for you to verify the important *triangle inequality*:

$$|x + y| \leq |x| + |y|.$$

(**Hint:** It is convenient to verify that the square of the left-hand side is less than or equal to the square of the right-hand side. See Exercise 7.)

Appendix: Construction of the Real Numbers

There are several techniques for constructing the real number system $\mathbb{R}$ from the rational numbers system $\mathbb{Q}$ (see Appendix I for a discussion of the origin of the rational numbers). We use the method of Dedekind (Julius W. R. Dedekind, 1831–1916) cuts because it uses a minimum of new ideas and is fairly brief.

The number system that we shall be constructing is an instance of a *field* (the complex numbers, in the next section, also form a field). The definition is as follows:

Definition 1.13: A set S is called a *field* if it is equipped with a binary field operation (usually called addition and denoted "+") and a second binary operation (called multiplication and denoted "·") such that the following axioms are satisfied: (Here, A stands for "addition," M stands for "multiplication," and D stands for "distributive law.")

A1. S is closed under addition: if $x, y \in S$, then $x + y \in S$.

A2. Addition is commutative: if $x, y \in S$, then $x + y = y + x$.

A3. Addition is associative: if $x, y, z \in S$, then $x + (y + z) = (x + y) + z$.

A4. There exists an element, called 0, in S which is an additive identity: if $x \in S$, then $0 + x = x$.

A5. Each element of S has an additive inverse: if $x \in S$, then there is an element $-x \in S$ such that $x + (-x) = 0$.

M1. S is closed under multiplication: if $x, y \in S$, then $x \cdot y \in S$.

M2. Multiplication is commutative: if $x, y \in S$, then $x \cdot y = y \cdot x$.

M3. Multiplication is associative: if $x, y, z \in S$, then $x \cdot (y \cdot z) = (x \cdot y) \cdot z$.

M4. There exists an element, called 1, which is a multiplicative identity: if $x \in S$, then $x \cdot 1 = x$.

M5. Each nonzero element of S has a multiplicative inverse: if $0 \neq x \in S$, then there is an element $x^{-1} \in S$ such that $x \cdot (x^{-1}) = 1$. The element x^{-1} is sometimes denoted $1/x$.

D1. Multiplication distributes over addition: if $x, y, z \in S$, then

$$x \cdot (y + z) = x \cdot y + x \cdot z.$$

Definition 1.14: A *cut* is a subset C of $\mathbb{Q}$ with the following properties:

- $C \neq \emptyset$
- If $s \in C$ and $t < s$, then $t \in C$
- If $s \in C$, then there is a $u \in C$ such that $u > s$
- There is a rational number x such that $c < x$ for all $c \in C$.

You should think of a cut C as the set of all rational numbers to the left of some point in the real line. Since we have not constructed the real line yet, we cannot define a cut in that simple way; we have to make the construction more indirect. But, if you consider the four properties of a cut, they describe a set that looks like a "rational halfline."

Notice that if C is a cut and $s \notin C$, then any rational $t > s$ is also not in C. Also, if $r \in C$ and $s \notin C$, then it must be that $s > r$.

Definition 1.15: If C and $\mathcal{D}$ are cuts, then we say that $C < \mathcal{D}$ provided that C is a subset of $\mathcal{D}$ but $C \neq \mathcal{D}$.

Check for yourself that "<" is an ordering on the set of all cuts.

Now, we introduce operations of addition and multiplication, which will turn the set of all cuts into a field.

Definition 1.16: If C and $\mathcal{D}$ are cuts, then we define

$$C + \mathcal{D} = \{c + d : c \in C, d \in \mathcal{D}\}.$$

We define the cut $\hat{0}$ to be the set of all negative rationals.

The cut $\hat{0}$ will play the role of the additive identity. We are now required to check that field axioms **A1–A5** hold.

For **A1**, we need to see that $C + \mathcal{D}$ is a cut. Obviously, $C + \mathcal{D}$ is not empty. If s is an element of $C + \mathcal{D}$ and t is a rational number less than s, write $s = c + d$, where $c \in C$ and $d \in \mathcal{D}$. Then, $t - c < s - c = d \in \mathcal{D}$; so, $t - c \in \mathcal{D}$; and $c \in C$. Hence, $t = c + (t - c) \in C + \mathcal{D}$. A similar argument shows that there is an $r > s$ such that $r \in C + \mathcal{D}$. Finally, if x is a rational upper bound for C and y is a rational upper bound for $\mathcal{D}$, then $x + y$ is a rational upper bound for $C + \mathcal{D}$. We conclude that $C + \mathcal{D}$ is a cut.

Since addition of rational numbers is commutative, it follows immediately that addition of cuts is commutative. Associativity follows in a similar fashion.

Now, we show that if C is a cut, then $C + \hat{0} = C$. For, if $c \in C$ and $z \in \hat{0}$, then $c + z < c + 0 = c$; hence, $C + \hat{0} \subseteq C$. Also, if $c' \in C$, then choose a $d' \in C$ such that $c' < d'$. Then, $c' - d' < 0$; so, $c' - d' \in \hat{0}$. And, $c' = d' + (c' - d')$. Hence, $C \subseteq C + \hat{0}$. We conclude that $C + \hat{0} = C$.

Finally, for Axiom **A5**, we let C be a cut and set $-C$ to be equal to $\{d \in \mathbb{Q} : c + d < 0 \text{ for all } c \in C\}$. If x is a rational upper bound for C and $c \in C$, then $-x \in -C$; so, $-C$ is not empty. By its very definition, $C + (-C) \subseteq \hat{0}$. Further, if $z \in \hat{0}$ and $c \in C$, we set $c' = z - c$. Then, $c' \in -C$ and $z = c + c'$. Hence, $\hat{0} \subseteq C + (-C)$. We conclude that $C + (-C) = \hat{0}$.

Having verified the axioms for addition, we turn now to multiplication.

Definition 1.17: If C and $\mathcal{D}$ are cuts, then we define the product $C \cdot \mathcal{D}$ as follows:

- If $C, \mathcal{D} > \hat{0}$, then $C \cdot \mathcal{D} = \{q \in \mathbb{Q} : q < c \cdot d$ for some $c \in C, d \in \mathcal{D}$ with $c > 0, d > 0\}$

- If $C > \hat{0}$, $\mathcal{D} < \hat{0}$, then $C \cdot \mathcal{D} = -(C \cdot (-\mathcal{D}))$
- If $C < \hat{0}$, $\mathcal{D} > \hat{0}$, then $C \cdot \mathcal{D} = -((-C) \cdot \mathcal{D})$
- If $C, \mathcal{D} < \hat{0}$, then $C \cdot \mathcal{D} = (-C) \cdot (-\mathcal{D})$
- If either $C = \hat{0}$ or $\mathcal{D} = \hat{0}$, then $C \cdot \mathcal{D} = \hat{0}$.

Notice that, for convenience, we have defined multiplication of negative numbers just as we did in high school. The reason is that the definition that we use for the product of two positive numbers cannot work when one of the two factors is negative (exercise).

It is now a routine exercise to verify that the set of all cuts, with this definition of multiplication, satisfies field axioms **M1–M5**. The proofs follow those for **A1–A5** rather closely.

For the distributive property, one first checks the case when all the cuts are positive, reducing it to the distributive property for the rationals. Then, one handles negative cuts on a case-by-case basis.

We now know that the collection of all cuts forms an ordered field. Denote this field by the symbol $\mathbb{R}$. We next verify the crucial property of $\mathbb{R}$ that sets it apart from $\mathbb{Q}$.

Theorem 1.18: *The ordered field $\mathbb{R}$ satisfies the least upper bound property.*

Proof: Let S be a subset of $\mathbb{R}$, which is bounded above. Define

$$S^* = \bigcup_{C \in S} C.$$

Then, S^* is clearly nonempty, and it is therefore a cut since it is a union of cuts. It is also clearly an upper bound for S since it contains each element of S. It remains to check that S^* is the least upper bound for S.

In fact, if $\mathcal{T} < S^*$, then $\mathcal{T} \subsetneq S^*$ and there is a rational number q in $S^* \setminus \mathcal{T}$. But, by the definition of S^*, it must be that $q \in C$ for some $C \in S$. So, $C > \mathcal{T}$, and $\mathcal{T}$ cannot be an upper bound for S. Therefore, S^* is the least upper bound for S, as desired.   □

We have shown that $\mathbb{R}$ is an ordered field that satisfies the least upper bound property. It remains to show that $\mathbb{R}$ contains (a copy of) $\mathbb{Q}$ in a natural way. In fact, if $q \in \mathbb{Q}$, we associate to it the element $\varphi(q) = C_q \equiv \{x \in \mathbb{Q} : x < q\}$. Then, C_q is obviously a cut. It is also routine to check that

$$\varphi(q + q') = \varphi(q) + \varphi(q') \text{ and } \varphi(q \cdot q') = \varphi(q) \cdot \varphi(q').$$

Therefore, we see that φ represents $\mathbb{Q}$ as a subfield of $\mathbb{R}$.

Exercises

1. Give an example of a set of real numbers that contains its least upper bound but not its greatest lower bound. Give an example of a set that contains its greatest lower bound but not its least upper bound.

2. Give an example of a set of real numbers that does *not* have a least upper bound. Give an example of a set of real numbers that does *not* have a greatest lower bound.

3. A set A in the reals with least upper bound equal to its greatest lower bound. What does that tell you about the set?

4. What is the least upper bound of the set $(-\infty, 0)$? What is the greatest lower bound of the set $(0, \infty)$?

5. Let A be a set of real numbers that is bounded above and set $\alpha = \sup A$. Let $B = \{-a : a \in A\}$. Prove that $\inf B = -\alpha$. Prove the same result with the roles of infimum and supremum reversed.

6. What is the least upper bound of the set

$$S = \{x : x^2 < 2\}?$$

Explain why this question has a sensible answer in the real number system but not in the rational number system.

7. Prove that the least upper bound and greatest lower bound for a set of real numbers is each unique.

8. Consider the unit circle C (the circle with center the origin in the plane and radius 1). Let

$$S = \{\alpha : 2\alpha < (\text{the circumference of } C)\}.$$

Show that S is bounded above. Let p be the least upper bound of S. Say explicitly what the number p is. This exercise works in the real number system, but not in the rational number system. Why?

9. Prove the triangle inequality.

10. Let $\varnothing$ be the empty set—the set with no elements. Prove that $\sup \varnothing = -\infty$ and $\inf \varnothing = +\infty$.

11. Prove that addition of the real numbers (as constructed in the Appendix) is commutative. Now, prove that it is associative.

12. Complete the calculation in the proof of Theorem 1.7.

13. Describe a countable set of nonrational real numbers between 0 and 1.

*14. Let f be a function with domain the reals and range the reals. Assume that f has a local minimum at each point x in its domain.

(This means that, for each $x \in \mathbb{R}$, there is an $\epsilon = \epsilon_x > 0$ such that, whenever $|x - t| < \epsilon$ then $f(x) \leq f(t)$.) *Do not assume that f is differentiable, or continuous, or anything nice like that.* Prove that the image of f is countable. (**Hint:** When I solved this problem as a student my solution was ten pages long; however, there is a one-line solution due to Michael Spivak.)

*15. Let λ be a positive irrational real number. If n is a positive integer, chose by the Archimedean Property an integer k such that $k\lambda \leq n < (k + 1)\lambda$. Let $\varphi(n) = n - k\lambda$. Prove that the set of all *varphi* (n), $n > 0$, is dense in the interval $[0, \lambda]$. (**Hint:** Examine the proof of the density of the rationals in the reals.)

*16. Let n be a natural number and x a positive real number. Prove that there is a positive real number y such that $y^n = x$. Is y unique?

17. Suppose that A and B are sets of real numbers with the same upper bound and same lower bound. What does that tell you about the sets?

1.2 The Complex Numbers

When we first learn about the complex number system, the most troublesome point is the very beginning: "let's pretend that the number -1 has a square root. Call it i." What gives us the right to "pretend" in this fashion? The answer is that we have no such right.[1] If -1 has a square root, then we should be able to construct a number system in which that is the case. That is what we shall do in this section.

Definition 1.19: The system of *complex numbers*, denoted by the symbol $\mathbb{C}$, consists of all ordered pairs (a, b) of real numbers. We add two complex numbers (a, b) and $(\tilde{a}, \tilde{b})$ by the formula

$$(a, b) + (\tilde{a}, \tilde{b}) = (a + \tilde{a}, b + \tilde{b}).$$

We multiply two complex numbers by the formula

$$(a, b) \cdot (\tilde{a}, \tilde{b}) = (a \cdot \tilde{a} - b \cdot \tilde{b}, a \cdot \tilde{b} + \tilde{a} \cdot b).$$

Remark 1.20: If you are puzzled by this definition of multiplication, do not worry. In a few moments, you will see that it gives rise to the notion of multiplication of complex numbers that you are accustomed to. Perhaps more importantly, a naive rule for multiplication like $(a, b) \cdot (\tilde{a}, \tilde{b}) = (a\tilde{a}, b\tilde{b})$ gives rise to nonsense like $(1, 0) \cdot (0, 1) = (0, 0)$. It is really necessary for us to use the initially counterintuitive definition of multiplication that is presented here. □

Example 1.21: Let $z = (3, -2)$ and $w = (4, 7)$ be two complex numbers. Then,

$$z + w = (3, -2) + (4, 7) = (3 + 4, -2 + 7) = (7, 5).$$

Also,

$$z \cdot w = (3, -2) \cdot (4, 7) = (3 \cdot 4 - (-2) \cdot 7, 3 \cdot 7 + 4 \cdot (-2)) = (26, 13).$$ □

As usual, we ought to check that addition and multiplication are commutative, associative, that multiplication distributes over addition, and so forth. We shall leave these tasks to the exercises. Instead, we develop some of the crucial, and more interesting, properties of our new number system.

Theorem 1.22: The following properties hold for the number system $\mathbb{C}$.

    a. *The number $1 \equiv (1, 0)$ is the multiplicative identity: $1 \cdot z = z \cdot 1 = z$ for any $z \in \mathbb{C}$.*

    b. *The number $0 \equiv (0, 0)$ is the additive identity: $0 + z = z + 0 = z$ for any $z \in \mathbb{C}$.*

    c. *Each complex number $z = (x, y)$ has an additive inverse $-z = (-x, -y)$: it holds that $z + (-z) = (-z) + z = 0$.*

    d. *The number $i \equiv (0, 1)$ satisfies $i \cdot i = -1$; in other words, i is a square root of -1.*

Proof: These are direct calculations, but it is important for us to work out these facts.

First, let $z = (x, y)$ be any complex number. Then,

$$1 \cdot z = (1, 0) \cdot (x, y) = (1 \cdot x - 0 \cdot y, 1 \cdot y + x \cdot 0) = (x, y) = z.$$

This proves the first assertion.

For the second, we have

$$0 + z = (0, 0) + (x, y) = (0 + x, 0 + y) = (x, y) = z.$$

With z as given, set $-z = (-x, -y)$. Then,

$$z + (-z) = (x, y) + (-x, -y) = (x + (-x), y + (-y)) = (0, 0) = 0.$$

Finally, we calculate

$$i \cdot i = (0, 1) \cdot (0, 1) = (0 \cdot 0 - 1 \cdot 1, 0 \cdot 1 + 0 \cdot 1) = (-1, 0) = -1.$$

Thus, as asserted, i is a square root of -1.   □

Proposition 1.23: *If $z \in \mathbb{C}$, $z \neq 0$, then there is a complex number w such that $z \cdot w = 1$.*

Proof: You might be thinking, "Well, of course, $w = 1/z$. But this is nonsense. The expression $1/z$ is *not* written in the form of a complex number!"
    Write $z = (x, y)$ and set

$$w = \left(\frac{x}{x^2 + y^2}, \frac{-y}{x^2 + y^2} \right).$$

Since $z \neq 0$, $x^2 + y^2 \neq 0$, so this definition makes sense. Then, it is straightforward to verify that $z \cdot w = 1$:

$$
\begin{aligned}
z \cdot w &= (x, y) \cdot \left(\frac{x}{x^2 + y^2}, \frac{-y}{x^2 + y^2} \right) \\
&= \left(x \cdot \frac{x}{x^2 + y^2} - y \cdot \frac{-y}{x^2 + y^2}, x \cdot \frac{-y}{x^2 + y^2} + \frac{x}{x^2 + y^2} \cdot y \right) \\
&= \left(\frac{x^2 + y^2}{x^2 + y^2}, \frac{-yx + xy}{x^2 + y^2} \right) \\
&= (1, 0) \\
&= 1.
\end{aligned}
$$
                                                                                   □

Example 1.24: Consider the complex number $z = 2 + 3i$. According to the proposition,

$$w = \left(\frac{2}{2^2 + 3^2}, \frac{-3}{2^2 + 3^2} \right) = \left(\frac{2}{13}, \frac{-3}{13} \right)$$

will be the multiplicative inverse of z. And, indeed,

$$z \cdot w = (2, 3) \cdot \left(\frac{2}{13}, \frac{-3}{13} \right) = \left(\frac{4}{13} + \frac{9}{13}, \frac{-6}{13} + \frac{6}{13} \right) = \left(\frac{13}{13}, 0 \right) = (1, 0) = 1. \quad □$$

Of course, we interpret the quotient z/w of complex numbers to mean $z \cdot (1/w)$. This will be a new complex number.

Thus, every nonzero complex number has a multiplicative inverse. The other field axioms for $\mathbb{C}$ are easy to check. We conclude that the number system $\mathbb{C}$ forms a field. You will prove in the exercises that it is not possible to order this field. If α is a real number, then we associate α with the complex number $(\alpha, 0)$. Thus, we have the natural "embedding"

$$\mathbb{R} \ni \alpha \mapsto (\alpha, 0) \in \mathbb{C}.$$

In this way, we can think of the real numbers as a *subset* of the complex numbers. In fact, the real fleld $\mathbb{R}$ is a *subfield* of the complex field $\mathbb{C}$. This means that if $\alpha, \beta \in \mathbb{R}$ and $(\alpha, 0)$, $(\beta, 0)$ are the corresponding elements in $\mathbb{C}$, then $\alpha + \beta$ corresponds to $(\alpha + \beta, 0)$ and $\alpha \cdot \beta$ corresponds to $(\alpha \cdot \beta, 0)$. These assertions are explored more thoroughly in the exercises.

With the remarks in the preceding paragraph, we can sometimes ignore the distinction between the real numbers and the complex numbers. For example, we can write

$$5 \cdot i$$

and understand that it means $(5, 0) \cdot (0, 1) = (0, 5)$. Likewise, the expression

$$5 \cdot 1$$

can be interpreted as $5 \cdot 1 = 5$ or as $(5, 0) \cdot (1, 0) = (5, 0)$ without any danger of ambiguity.

Theorem 1.25: *Every complex number can be written in the form $a + b \cdot i$, where aand bare real numbers. In fact, if $z = (x, y) \in \mathbb{C}$, then*

$$z = x + y \cdot i.$$

Proof: With the identification of real numbers as a subfield of the complex numbers, we have that

$$x + y \cdot i = (x, 0) + (y, 0) \cdot (0, 1) = (x, 0) + (0, y) = (x, y) = z$$

as claimed. □

Now that we have constructed the complex number field, we will adhere to the usual custom of writing complex numbers as $z = a + b \cdot i$ or, more simply, $a + bi$. We call a the *real part* of z, denoted by Re z, and b the *imaginary part* of z, denoted Im z. We have

$$(a + bi) + (\tilde{a} + \tilde{b}i) = (a + \tilde{a}) + (b + \tilde{b})i$$

and

$$(a + bi) \cdot (\tilde{a} + \tilde{b}i) = (a \cdot \tilde{a} - b \cdot \tilde{b}) + (a \cdot \tilde{b} + \tilde{a} \cdot b)i.$$

Example 1.26: Let $z = 3 - 7i$ and $w = 4 + 6i$. Then,

$$\begin{aligned}
z + w &= (3 - 7i) + (4 + 6i) = 7 - i, \\
z \cdot w &= (3 - 7i) \cdot (4 + 6i) \\
&= (3 \cdot 4 - (-7) \cdot 6) + (3 \cdot 6 + (-7) \cdot 4)i \\
&= 54 - 10i.
\end{aligned}$$
□

If $z = a + bi$ is a complex number, then we define its *complex conjugate* to be the number $\bar{z} = a - bi$. We record some elementary facts about the complex conjugate.

Proposition 1.27: *If z, w are complex numbers, then*

1. $\overline{z + w} = \bar{z} + \bar{w}$;
2. $\overline{z \cdot w} = \bar{z} \cdot \bar{w}$;
3. $z + \bar{z} = 2 \cdot \text{Re } z$;
4. $z - \bar{z} = 2 \cdot i \cdot \text{Im } z$;
5. $z \cdot \bar{z} \geq 0$, *with equality holding if and only if $z = 0$.*

Proof: Write $z = a + bi$, $w = c + di$. Then,

$$\begin{aligned}
\overline{z + w} &= \overline{(a + c) + (b + d)i} \\
&= (a + c) - (b + d)i \\
&= (a - bi) + (c - di) \\
&= \bar{z} + \bar{w}.
\end{aligned}$$

This proves **(1)**. Assertions **(2)**, **(3)**, and **(4)** are proved similarly. For **(5)**, notice that

$$z \cdot \bar{z} = (a + bi) \cdot (a - bi) = a^2 + b^2 \geq 0.$$

Clearly, equality holds if and only if $a = b = 0$.
□

Example 1.28: Let $z = -2 + 4i$ and $w = 5 - 3i$. Then,

$$\bar{z} = -2 - 4i.$$

Also,

$$\overline{z \cdot w} = \overline{(-2 + 4i)(5 - 3i)} = \overline{2 + 26i} = 2 - 26i$$

while

$$\bar{z} \cdot \bar{w} = (-2 - 4i) \cdot (5 + 3i) = (-10 + 12) + (-6 - 20)i = 2 - 26i. \qquad \square$$

The expression $|z|$ is defined to be the nonnegative square root of $z \cdot \bar{z}$:

$$|z| = +\sqrt{z \cdot \bar{z}} = \sqrt{x^2 + y^2}$$

when $z = x + iy$. It is called the *modulus* of z and plays the same role for the complex field that absolute value plays for the real field. It is the distance of z to the origin. The modulus has the following properties:

Proposition 1.29: *If $z, w \in \mathbb{C}$ then*

1. $|z| = |\bar{z}|$;
2. $|z \cdot w| = |z| \cdot |w|$;
3. $|\operatorname{Re} z| \le |z|$, $|\operatorname{Im} z| \le |z|$;
4. $|z + w| \le |z| + |w|$;

Proof: Write $z = a + bi$, $w = c + di$. Then, **(1)**, **(2)**, **(3)** are immediate. For **(4)**, we calculate that

$$
\begin{aligned}
|z + w|^2 &= (z + w) \cdot (\overline{z + w}) \\
&= z \cdot \bar{z} + z \cdot \bar{w} + w \cdot \bar{z} + w \cdot \bar{w} \\
&= |z|^2 + 2\operatorname{Re}(z \cdot \bar{w}) + |w|^2 \\
&\le |z|^2 + 2|z \cdot \bar{w}| + |w|^2 \\
&= |z|^2 + 2|z| \cdot |w| + |w|^2 \\
&= (|z| + |w|)^2.
\end{aligned}
$$

Taking square roots proves **(4)**. $\qquad \square$

Example 1.30: Let $z = 3 + 4i$ and $w = -5 + 2i$. Then,

$$|z| = \sqrt{3^2 + 4^2} = \sqrt{25} = 5 \text{ and } |w| = \sqrt{(-5)^2 + 2^2} = \sqrt{29}.$$

Also,

$$|z \cdot w| = |(3 + 4i)(-5 + 2i)| = |-23 - 14i| = \sqrt{23^2 + 14^2} = \sqrt{725} = 5\sqrt{29}$$

while

$$|z| \cdot |w| = 5 \cdot \sqrt{29}.$$
□

Observe that, if z is real, then $z = a + 0i$ and the modulus of z equals the absolute value of a. Likewise, if $z = 0 + bi$ is pure imaginary, then the modulus of z equals the absolute value of b. In particular, the fourth part of the proposition reduces, in the real case, to the triangle inequality

$$|a + b| \leq |a| + |b|.$$

If z is any nonzero complex number, then let $r = |z|$. Now define $\xi = z/r$. We see that ξ is a complex number of modulus 1. Thus, ξ lies on the unit circle, so it subtends an angle θ with the positive x-axis. Then, $\xi = \cos \theta + i \sin \theta$. It is shown in Section 9.3 that

$$e^{i\theta} = \xi = \cos \theta + i \sin \theta.$$

(**Hint:** You may verify this formula for yourself by writing out the power series for the exponential and writing out the power series for cosine and sine.) We often call

$$z = re^{i\theta}$$

the *polar form* of z.

Example 1.31: Let us find all cube roots of the complex number $z = -1 + i$. Using the notation of the preceding paragraph, we see that $r = \sqrt{(-1)^2 + 1^2} = \sqrt{2}$. Thus, $\xi = z/r = -1/\sqrt{2} + (1/\sqrt{2})i$. Examining Figure 1.1, we see that $\theta = 3\pi/4$. We have learned then that

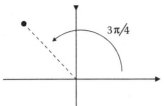

FIGURE 1.1
The polar form of $-1 + i$.

$$z = -1 + i = \sqrt{2}\,e^{i3\pi/4}.$$

For the first cube root w_1 of z, we write $w_1 = se^{i\psi}$, and we solve for s and ψ. We know that

$$(w_1)^3 = z$$

so

$$(se^{i\psi})^3 = \sqrt{2}\,e^{i3\pi/4}$$

or

$$s^3 e^{i3\psi} = \sqrt{2}\,e^{i3\pi/4}.$$

It is natural then to conclude that

$$s^3 = \sqrt{2}$$

and

$$3\psi = 3\pi/4.$$

We conclude that $s = 2^{1/6}$ and $\psi = \pi/4$. We have found that

$$w_1 = 2^{1/6}e^{i\pi/4}$$

is a cube root of z. But, this is not the only cube root! There are three cube roots in total.

We next notice that z can also be written

$$z = \sqrt{2}\,e^{i((3\pi/4)+2\pi)}.$$

(Observe that there is some ambiguity built into the polar form of a complex number, just as there is ambiguity in the polar coordinates that you learned about in calculus. The reason is that the cosine and sine functions are 2π-periodic.)

Now, we repeat the calculation given earlier with this new form for the complex number z. We know that

$$(w_2)^3 = z$$

so,

$$(se^{i\psi})^3 = \sqrt{2}\,e^{i((3\pi/4)+2\pi)}$$

or

$$s^3 e^{i3\psi} = \sqrt{2}\,e^{i11\pi/4}.$$

It is natural then to conclude that

$$s^3 = \sqrt{2}$$

and

$$3\psi = 11\pi/4.$$

We conclude that $s = 2^{1/6}$ and $\psi = 11\pi/12$. We have found that

$$w_2 = 2^{1/6}e^{i11\pi/12}$$

is a cube root of z.

Let us do the calculation one more time with z now written as

$$z = \sqrt{2}\,e^{i((3\pi/4)+4\pi)}$$

(again we exploit the periodicity of sine and cosine). We know that

$$(w_3)^3 = z$$

so,

$$(se^{i\psi})^3 = \sqrt{2}\,e^{i((3\pi/4)+4\pi)}$$

or,

$$s^3 e^{i3\psi} = \sqrt{2}\, e^{i19\pi/4}.$$

It is natural then to conclude that

$$s^3 = \sqrt{2}$$

and

$$3\psi = 19\pi/4.$$

We conclude that $s = 2^{1/6}$ and $\psi = 19\pi/12$. We have found that

$$w_3 = 2^{1/6} e^{i19\pi/12}$$

is a cube root of z.

There is no sense to repeat these calculations any further. It is true that $z = \sqrt{2}\, e^{i(3\pi/4 + 6\pi)}$. But, performing our calculations for this form of z would simply cause us to rediscover w_1. We have found three cube roots of z, and that is the end of the calculation. □

We conclude this discussion by recording the most important basic fact about the complex numbers. Carl Friedrich Gauss gave five proofs of this theorem (the Fundamental Theorem of Algebra) in his doctoral dissertation.

Theorem 1.32: *Let $p(z)$ be any polynomial of degree at least 1. Then p has a root $\alpha \in \mathbb{C}$ such that $p(\alpha) = 0$.*

Using a little algebra, one can in fact show that a polynomial of degree k has k roots (counting multiplicity).

Exercises

1. Show that, if z is a nonzero complex number, then its multiplicative inverse is given by

$$w = \frac{\bar{z}}{|z|^2}.$$

2. Refer to Exercise 1. If $z, w \in \mathbb{C}$, then prove that $\overline{z/w} = \bar{z}/\bar{w}$.

3. Find all cube roots of the complex number $1 + i$.

4. Taking the commutative, associative, and distributive laws of addition and multiplication for the real number system for granted, establish these laws for the complex numbers.

5. Find a complex number z so that $e^z = i$.

6. Say something about the uniqueness of the complex number z in Exercise 5.

7. Consider the function $\phi : \mathbb{R} \to \mathbb{C}$ given by $\phi(x) = x + i \cdot 0$. Prove that ϕ respects addition and multiplication in the sense that $\phi(x + x') = \phi(x) + \phi(x')$ and $\phi(x \cdot x') = \phi(x) \cdot \phi(x')$.

8. Prove that the field of complex numbers cannot be made into an *ordered* field. (**Hint:** Since $i \neq 0$, then either $i > 0$ or $i < 0$. Both lead to a contradiction.)

9. Prove that the complex roots of a polynomial with real coefficients occur in complex conjugate pairs.

10. Calculate the square roots of i.

11. Prove that the set of all complex numbers is uncountable.

12. Prove that any nonzero complex number z has kth roots $r_1, r_2, \ldots, r_k$. That is, prove that there are k of them.

13. In the complex plane, draw a picture of

$$S = \{z \in \mathbb{C} : |z - 1| + |z + 1| = 2\}.$$

14. Refer to Exercise 9. Show that the kth roots of z all lie on a circle centered at the origin, and that they are equally spaced.

15. Find all the cube roots of $1 + i$.

16. Find all the square roots of $-1 - i$.

17. Prove that the set of all complex numbers with rational real part is uncountable.

18. Prove that the set of all complex numbers with both real and imaginary parts rational is countable.

19. Prove that the set $\{z \in \mathbb{C} : |z| = 1\}$ is uncountable.

*20. In the complex plane, draw a picture of

$$T = \{z \in \mathbb{C} : |z + \bar{z}| - |z - \bar{z}| = 2\}.$$

*21. Use the Fundamental Theorem of Algebra to prove that any polynomial of degree k has k (not necessarily distinct) roots. ([**Hint:** Use the Euclidean algorithm.)

22. Let $\alpha \neq 0$ be a complex number. Prove that α has *exactly* three cube roots—not more and not less.

Note

1 The complex numbers were initially developed so that we would have a number system in which all polynomial equations are solvable. One of the reasons, historically, that mathematicians had trouble accepting the complex numbers is that they did not believe that they really existed—they were just made up. This is, in part, how they came to be called "imaginary" and "complex." Mathematicians had similar trouble accepting negative numbers; for a time, negative numbers were called "forbidden."

2

Sequences

2.1 Convergence of Sequences

A *sequence* of real numbers is a function $\varphi: \mathbb{N} \to \mathbb{R}$. We often write the sequence as $\varphi(1)$, $\varphi(2)$, ... or, more simply, as φ_1, φ_2, ... A sequence of complex numbers is defined similarly, with $\mathbb{R}$ replaced by $\mathbb{C}$.

Example 2.1: The function $\varphi(j) = 1/j$ is a sequence of real numbers. We will often write such a sequence as $\varphi_j = 1/j$ or as $\{1, 1/2, 1/3, \ldots\}$ or as $\{1/j\}_{j=1}^{\infty}$. The function $\psi(j) = \cos j + i \sin j$ is a sequence of complex numbers.

Do not be misled into thinking that a sequence must form a pattern, or be given by a formula. Obviously the ones which are given by formulas are easy to write down, but they are not typical. For example, the coefficients in the decimal expansion of π, $\{3, 1, 4, 1, 5, 9, 2, 6, 5, \ldots\}$, fit our definition of sequence—but they are not given by any obvious pattern. □

The most important question about a sequence is whether it converges. We define this notion as follows.

Definition 2.2: A sequence $\{a_j\}$ of real (resp. complex) numbers is said to *converge* to a real (resp. complex) number α if, for each $\epsilon > 0$, there is an integer $N > 0$ such that, if $j > N$, then $|a_j - \alpha| < \epsilon$. We call α the *limit* of the sequence $\{a_j\}$. We write $\lim_{j \to \infty} a_j = \alpha$. We also sometimes write $a_j \to \alpha$.

If a sequence $\{a_j\}$ does not converge then we frequently say that it *diverges*.

Example 2.3: Let $a_j = 1/j$, $j = 1, 2, \ldots$. Then the sequence converges to 0. For let $\epsilon > 0$. Choose N to be the next integer after $1/\epsilon$ (we use here the Archimedean principle). If $j > N$ then

$$|a_j - 0| = |a_j| = \frac{1}{j} < \frac{1}{N} < \epsilon,$$

proving the claim.

DOI: 10.1201/9781003222682-3

Let $b_j = (-1)^j$, $j = 1, 2, \ldots$. Then the sequence *does not converge*. To prove this assertion, suppose to the contrary that it does. Suppose that the sequence converges to a number α. Let $\epsilon = 1/2$. By definition of convergence, there is an integer $N > 0$ such that, if $j > N$, then $|b_j - \alpha| < \epsilon = 1/2$. For such j we have

$$|b_j - b_{j+1}| = |(b_j - \alpha) + (\alpha - b_{j+1})| \leq |b_j - \alpha| + |\alpha - b_{j+1}|$$

(by the triangle inequality—see the end of Section 1.1). But this last is

$$< \epsilon + \epsilon = 1.$$

On the other hand,

$$|b_j - b_{j+1}| = |(-1)^j - (-1)^{j+1}| = 2.$$

The last two lines yield that $2 < 1$, a clear contradiction. So the sequence $\{b_j\}$ has no limit. $\qquad\square$

We begin with a few intuitively appealing properties of convergent sequences which will be needed later. First, a definition.

Definition 2.4: A sequence a_j is said to be *bounded* if there is a number $M > 0$ such that $|a_j| \leq M$ for every j.

Now we have

Proposition 2.5: Let $\{a_j\}$ *be a convergent sequence*. Then *we have*:

- *The limit of the sequence is unique.*
- *The sequence is bounded.*

Proof: Suppose that the sequence has two limits α and $\tilde{\alpha}$. Let $\epsilon > 0$. Then there is an integer $N > 0$ such that for $j > N$ we have the inequalitye $|a_j - \alpha| < \epsilon/2$. Likewise, there is an integer $\tilde{N} > 0$ such that for $j > \tilde{N}$ we have $|a_j - \tilde{\alpha}| < \epsilon/2$.
Let $N_0 = \max\{N, \tilde{N}\}$. Then, for $j > N_0$, we have

$$|\alpha - \tilde{\alpha}| = |(\alpha - a_j) + (a_j - \tilde{\alpha})| \leq |\alpha - a_j| + |a_j - \tilde{\alpha}| < \epsilon/2 + \epsilon/2 = \epsilon.$$

Since this inequality holds for any $\epsilon > 0$ we have that $\alpha = \tilde{\alpha}$.

Next, with α the limit of the sequence and $\epsilon = 1$, we choose an integere $N > 0$ such that $j > N$ implies that $|a_j - \alpha| < \epsilon = 1$. For such j we have that

$$|a_j| = |(a_j - \alpha) + \alpha| \le |a_j - \alpha| + |\alpha| < 1 + |\alpha| \equiv P.$$

Let $Q = \max\{|a_1|, |a_2|, ..., |a_N|\}$. If j is any natural number then either $1 \le j \le N$ (in which case $|aj| \le Q$) or else $j > N$ (in which case $|aj| \le P$).

Set $M = \max\{P, Q\}$. Then $|aj| \le M$ for all j, as desired. So the sequence is bounded. $\qquad\square$

The next proposition records some elementary properties of limits of sequences.

Proposition 2.6: *Let $\{aj\}$ be a sequence of real or complex numbers with limit α and $\{b_j\}$ be a sequence of real or complex numbers with limit β. Then we have:*

1. *If c is a constant then the sequence $\{c{\cdot}a_j\}$ converges to $c{\cdot}\alpha$;*
2. *The sequence $\{a_j + b_j\}$ converges to $\alpha + \beta$;*
3. *The sequence $a_j \cdot b_j$ converges to $\alpha{\cdot}\beta$;*
4. *If $b_j \ne 0$ for all j and $\beta \ne 0$ then the sequence a_j/b_j converges to α/β.*

Proof: For part **(1)**, we may assume that $c \ne 0$ (for when $c = 0$ there is nothing to prove). Let $\epsilon > 0$. Choose an integer $N > 0$ such that for $j > N$ it holds that

$$|aj - \alpha| < \frac{\epsilon}{|c|}.$$

For such j we have that

$$|c \cdot a_j - c \cdot \alpha| = |c| \cdot |a_j - \alpha| < |c| \cdot \frac{\epsilon}{|c|} = \epsilon.$$

This proves the first assertion.

The proof of part **(2)** is similar, and we leave it as an exercise.

For part **(3)**, notice that the sequence $\{aj\}$ is bounded (by the second part of Proposition 2.5): say that $|a_j| \le M$ every j. Let $\epsilon > 0$. Choose an integer $N > 0$ so that $|a_j - \alpha| < \epsilon/(2M + 2|\beta|)$ when $j > N$. Also choose an integer $\tilde{N} > 0$ such that $|b_j - \beta| < \epsilon/(2M + 2|\beta|)$ when $j > \tilde{N}$. Then for $j > \max\{N, \tilde{N}\}$, we have that

$$|a_j b_j - \alpha\beta| = |a_j(b_j - \beta) + \beta(a_j - \alpha)|$$
$$\leq |a_j(b_j - \beta)| + |\beta(a_j - \alpha)|$$
$$< M \cdot \frac{\epsilon}{2M + 2\,|\beta|} + |\beta| \cdot \frac{\epsilon}{2M + 2\,|\beta|}$$
$$\leq \frac{\epsilon}{2} + \frac{\epsilon}{2}$$
$$= \epsilon.$$

So the sequence $\{aj \cdot bj\}$ converges to $\alpha\beta$.

Part (4) is proved in a similar fashion and we leave the details as an exercise.   □

Remark 2.7: You were probably puzzled by the choice of N and $\tilde{N}$ in the proof of part (3) of Proposition 2.6—where did the number $\epsilon/(2M + 2|\beta|)$ come from? The answer of course becomes obvious when we read on further in the proof. So the lesson here is that a proof is constructed backward: you look to the end of the proof to see what you need to specify earlier on. Skill in these matters can come only with practice.   □

Example 2.8: Let $a_j = \sin(1/j)/(1/j)$ and $b_j = j^2/(2j^2+j)$. Then $a_j \to 1$ and $b_j \to 1/2$ as $j \to \infty$. Let us say a few words about why this is true. You learned in your calculus class that

$$\lim_{x \to 0} \frac{\sin x}{x} = 1.$$

Letting $x = 1/j$ and $j \to \infty$ then yields that

$$\lim_{j \to \infty} \frac{\sin(1/j)}{1/j} = 1.$$

For the second limit, write

$$b_j = \frac{j^2}{2j^2 + j} = \frac{j^2/j^2}{(2j^2 + 1)/j^2} = \frac{1}{2 + 1/j^2}.$$

Now it is evident that

$$\lim_{j \to \infty} b_j = \frac{1}{2}.$$

From the above we may conclude, using Proposition 2.6, that

$$\lim_{j \to \infty} 5a_j = 5,$$

$$\lim_{j \to \infty} (a_j + b_j) = 1 + \frac{1}{2} = \frac{3}{2},$$

$$\lim_{j \to \infty} a_j \cdot b_j = 1 \cdot \frac{1}{2} = \frac{1}{2}.$$

and

$$\lim_{j \to \infty} \frac{a_j}{b_j} = \frac{1}{1/2} = 2.$$

$\square$

When discussing the convergence of a sequence, we often find it inconvenient to deal with the definition of convergence as given. For this definition makes reference to the number to which the sequence is supposed to converge, and we often do not know this number in advance. Would it not be useful to be able to decide whether a series converges *without knowing to what limit it converges?*

Definition 2.9: Let $\{a_j\}$ be a sequence of real (resp. complex) numbers. We say that the sequence satisfies the *Cauchy criterion* (A. L. Cauchy, 1789–1857)—more briefly, that the sequence is *Cauchy*—if, for each $\epsilon > 0$, there is an integer $N > 0$ such that if $j, k > N$ then $|a_j - a_k| < \epsilon$.

As you study this definition, you will see that it mandates that the elements of the sequence *get close together and stay close together.*

Example 2.10: Let $a_j = 1/j$. Of course we know intuitively that this sequence converges to 0. But let us, just for practice, verify that the sequence is Cauchy.

Let $\epsilon > 0$. By the Archimedean principle, choose a positive integer $N > 1/\epsilon$. Then, for $j > k > N$, we have

$$|a_j - a_k| = |1/j - 1/k| = \frac{|j - k|}{jk} < \frac{j}{jk} = \frac{1}{k} < \frac{1}{N} < \epsilon.$$

This shows that the sequence $\{a_j\}$ satisfies the Cauchy criterion. $\square$

Notice that the concept of a sequence being Cauchy simply makes precise the notion of the elements of the sequence (i) *getting* closer together and (ii) *staying* close together.

Lemma 2.11: *Every Cauchy sequence is bounded.*

Proof: Let $\epsilon = 1 > 0$. There is an integer $N > 0$ such that $|a_j - a_k| < \epsilon = 1$ whenever $j, k > N$. Thus, if $j \geq N + 1$, we have

$$\begin{aligned}
|a_j| &\leq |a_{N+1} + (a_j - a_{N+1})| \\
&\leq |a_{N+1}| + |a_j - a_{N+1}| \\
&\leq |a_{N+1}| + 1 \equiv K.
\end{aligned}$$

Let $L = \max\{|a_1|, |a_2|, \ldots, |a_N|\}$. If j is any natural number, then either $1 \leq j \leq N$, in which case $|a_j| \leq L$, or else $j > N$, in which case $|a_j| \leq K$.
Set $M = \max\{K, L\}$. Then, for any j, $|a_j| \leq M$ as required.   □

In what follows we shall use an interesting and not entirely obvious version of the triangle inequality. You know the triangle inequality as

$$|a + b| \leq |a| + |b|.$$

But let us instead write

$$|a| = |(a + b) - b| = |(a + b) + (-b)| \leq |a + b| + |-b| = |a + b| + |b|.$$

From this we conclude that

$$|a + b| \geq |a| - |b|.$$

A similar argument allows us to analyze $|a - b| < c$. This means that

$$-c < a - b < c$$

or

$$b - c < a$$

and

$$a < b + c.$$

Hence

$$b - c < a < b + c.$$

Theorem 2.12: Let {aj} be a sequence of real numbers. The sequence is Cauchy if and only if it converges to some limit α.

Proof: First assume that the sequence converges to a limit α. Let $\epsilon > 0$. Choose, by definition of convergence, an integer $N > 0$ such that if $j > N$ then $|a_j - \alpha| < \epsilon/2$. If $j, k > N$ then

$$|a_j - a_k| \leq |a_j - \alpha| + |\alpha - a_k| < \frac{\epsilon}{2} + \frac{\epsilon}{2} = \epsilon.$$

So the sequence is Cauchy.

Conversely, suppose that the sequence is Cauchy. Define

$$S = \{x \in \mathbb{R}:$$
$$x < a_j \text{ for all but finitely many } j\}.$$

[**Hint:** You might find it helpful to think of this set as

$$S = \{x \in \mathbb{R}: \text{there is a positive integer } k \text{ such that } x < a_j \text{ for all } j \geq k\}.]$$

By the lemma, the sequence {a_j} is bounded by some positive number M. If x is a real number less than $-M$, then $x \in S$, so S is nonempty. Also S is bounded above by M. Let $\alpha = \sup S$. Then α is a well-defined real number, and we claim that α is the limit of the sequence {a_j}.

To see this, let $\epsilon > 0$. Choose an integer $N > 0$ such that $|a_j - a_k| < \epsilon/2$ whenever $j, k > N$. Notice that this last inequality implies that

$$|a_j - a_{N+1}| < \epsilon/2 \text{ when } j \geq N + 1 \qquad (2.12.1)$$

hence (by the discussion preceding the statement of the theorem)

$$a_j > a_{N+1} - \epsilon/2 \text{ when } j \geq N + 1.$$

Thus $a_{N+1} - \epsilon/2 \in S$ and it follows that

$$\alpha \geq a_{N+1} - \epsilon/2. \qquad (2.12.2)$$

Line (2.12.1) also shows that

$$a_j < a_{N+1} + \epsilon/2 \text{ when } j \geq N + 1.$$

Thus $a_{N+1} + \epsilon/2 \notin S$ and

$$\alpha \le a_{N+1} + \epsilon/2. \tag{2.12.3}$$

Combining lines (2.12.2) and (2.12.3) gives

$$|\alpha - a_{N+1}| \le \epsilon/2. \tag{2.12.4}$$

But then line (2.12.4) yields, for $j > N$, that

$$|\alpha - a_j| \le |\alpha - a_{N+1}| + |a_{N+1} - a_j| < \epsilon/2 + \epsilon/2 = \epsilon.$$

This proves that the sequence $\{a_j\}$ converges to α, as claimed. □

Corollary 2.13: *Let $\{a_j\}$ be a sequence of complex numbers. The sequence is Cauchy if and only if it is convergent.*

Proof: Write $a_j = a_j + ib_j$, with a_j, b_j real. Then $\{a_j\}$ is Cauchy if and only if $\{a_j\}$ and $\{b_j\}$ are Cauchy. Also $\{a_j\}$ is convergent to a complex limit α if and only if $\{a_j\}$ converges to Re α and $\{b_j\}$ converges to Im α. These observations, together with the theorem, prove the corollary. □

Definition 2.14: Let $\{a_j\}$ be a sequence of real numbers. The sequence is said to be *increasing* if $a_1 \le a_2 \le \dots$. It is decreasing if $a_1 \ge a_2 \ge \dots$.

A sequence is said to be *monotone* if it is either increasing or decreasing.

Example 2.15: Let $a_j = j/(j + 1)$. We see that

$$a_j < a_j + 1$$

just because this is the same as

$$\frac{j}{j+1} < \frac{j+1}{j+2}$$

or

$$j(j + 2) < (j + 1)^2$$

or

$$j^2 + 2j < j^2 + 2j + 1$$

and that is definitely true. Hence the sequence $\{a_j\}$ is increasing. In fact it increases to 1.

On the other hand, let $b_j = (j + 1)/j$. We see that

$$b_j > b_{j+1}$$

just because this is the same as

$$\frac{j+1}{j} > \frac{j+2}{j+1}$$

or

$$(j+1)^2 > j(j+2)$$

or

$$j^2 + 2j + 1 > j^2 + 2j$$

and that is definitely true. Hence the sequence $\{b_j\}$ is decreasing. Indeed it decreases to 1. □

Proposition 2.16: *If $\{a_j\}$ is an increasing sequence which is bounded above—$a_j \leq M < \infty$ for all j—then $\{a_j\}$ is convergent. If $\{b_j\}$ is a decreasing sequence which is bounded below—$b_j \geq K > -\infty$ for all j—then $\{b_j\}$ is convergent.*

Proof: Let $\epsilon > 0$. Let $\alpha = \sup a_j < \infty$. By definition of supremum, there is an integer N so that $|a_N - \alpha| < \epsilon$. Then, if $\ell \geq N + 1$, we have $a_N \leq a\ell \leq \alpha$ hence $|a\ell - \alpha| < \epsilon$. Thus the sequence converges to α.

The proof for decreasing sequences is similar and we omit it. □

Example 2.17: Let $a_1 = \sqrt{2}$ and set $a_{j+1} = \sqrt{2 + a_j}$ for $j \geq 1$. You can verify that $\{a_j\}$ is increasing and bounded above (by 4 for example). What is its limit (which is guaranteed to exist by the proposition)? □

A proof very similar to that of the proposition gives the following useful fact:

Corollary 2.18: *Let S be a nonempty set of real numbers which is bounded above and below. Let β be its supremum and α its infimum. If $\epsilon > 0$ then there are $s, t \in S$ such that $|s - \beta| < \epsilon$ and $|t - \alpha| < \epsilon$.*

Proof: This is a restatement of the proof of the proposition. □

Example 2.19: Let S be the set $(0, 2) \subseteq \mathbb{R}$. Then the supremum of S is 2.

And, if $\epsilon > 0$ is small, then the point $\alpha_\epsilon = 2 - \epsilon/2$ lies in the set. Note that $|\alpha_\epsilon - 2| < \epsilon$.

Likewise, the infimum of S is 0. And, if $\epsilon > 0$ is small, then the point $\beta_\epsilon = 0 + \epsilon/2$ lies in the set. Note that $|\beta_\epsilon - 0| < \epsilon$.   □

We conclude the section by recording one of the most useful results for calculating the limit of a sequence:

Proposition 2.20: (The Pinching Principle) *Let $\{a_j\}$, $\{b_j\}$, and $\{c_j\}$ be sequences of real numbers satisfying*

$$a_j \leq b_j \leq c_j$$

for every j sufficiently large. If

$$\lim_{j \to \infty} a_j = \lim_{j \to \infty} c_j = \alpha$$

for some real number α, then

$$\lim_{j \to \infty} b_j = \alpha.$$

Proof: This proof is requested of you in the exercises.   □

Example 2.21: Define

$$a_j = \frac{\sin j \, \cos 2j}{j^2}.$$

Then

$$0 \leq |a_j| \leq \frac{1}{j^2}.$$

It is clear that

$$\lim_{j \to \infty} 0 = 0$$

and

$$\lim_{j \to \infty} \frac{1}{j^2} = 0.$$

Therefore

$$\lim_{j \to \infty} |a_j| = 0$$

so that

$$\lim_{j \to \infty} a_j = 0.$$ □

****

Augustin-Louis Cauchy

Baron Augustin-Louis Cauchy (1789–1857) was a French mathematician, engineer, and physicist who made pioneering contributions to several branches of mathematics, including mathematical analysis and continuum mechanics. He almost singlehandedly founded complex analysis and the study of permutation groups in abstract algebra.

Hans Freudenthal stated: "More concepts and theorems have been named for Cauchy than for any other mathematician (in elasticity alone there are sixteen concepts and theorems named for Cauchy)." Cauchy was a prolific writer; he wrote approximately eight hundred research articles and five complete textbooks on a variety of topics in the fields of mathematics and mathematical physics.

Cauchy had two brothers: Alexandre Laurent Cauchy, who became a president of a division of the court of appeal in 1847 and a judge of the court of cassation in 1849, and Eugene Franc͵ois Cauchy, a publicist who also wrote several mathematical works.

Cauchy married Aloise de Bure in 1818. She was a close relative of the publisher who published most of Cauchy's works. Cauchy's father was a high official in the Parisian Police of the Ancien Regime, but lost this position due to the French Revolution, which broke out one month before Augustin-Louis was born. The Cauchy family survived the revolution and the following Reign of Terror by escaping to Arcueil. After the execution of Robespierre, it was safe for the family to return to Paris. When Napoleon Bonaparte came to power, Louis-Franc͵ois Cauchy was further promoted, and became Secretary-General of the Senate, working directly under Laplace (who is now better known for his work on mathematical physics). The famous mathematician Lagrange was also a friend of the Cauchy family.

On Lagrange's advice, Augustin-Louis was enrolled in the école Centrale du Panthéon, the best secondary school of Paris at that time, in the fall of 1802. Cauchy did very well in this new school. In spite of these successes, Augustin-Louis chose an engineering career, and prepared himself for the entrance examination to the école Polytechnique.

In 1805, he placed second of 293 applicants on this exam and was admitted. The school functioned under military discipline, which caused the young and pious Cauchy some problems in adapting. Nevertheless, he finished the

Polytechnique in 1807, at the age of 18, and went on to the École des Ponts et Chaussées (School for Bridges and Roads).

After finishing school in 1810, Cauchy accepted a job as a junior engineer in Cherbourg, where Napoleon intended to build a naval base. Although he had an extremely busy managerial job, he still found time to prepare three mathematical manuscripts, which he submitted to the Premi´ere Classe (First Class) of the Institut de France.

In September 1812, now 23 years old, Cauchy returned to Paris after becoming ill from overwork. In Paris, he would have a much better chance to find a mathematics related position. The next three years Augustin-Louis was mainly on unpaid sick leave, and spent his time quite fruitfully, working on mathematics (on the related topics of symmetric functions, the symmetric group and the theory of higher-order algebraic equations).

In 1815 Napoleon was defeated at Waterloo, and the newly installed Bourbon king Louis XVIII took the restoration in hand. The Acadmie des Sciences was reestablished in March 1816; Lazare Carnot and Gaspard Monge were removed from this Academy for political reasons, and the king appointed Cauchy to take the place of one of them.

In November 1815, Louis Poinsot, who was an associate professor at the cole Polytechnique, asked to be exempted from his teaching duties for health reasons. Cauchy was by then a rising mathematical star, who certainly merited a professorship. One of his great successes at that time was the proof of Fermat's polygonal number theorem. However, the fact that Cauchy was known to be very loyal to the Bourbons doubtless also helped him in becoming the successor of Poinsot. In 1816, this Bonapartist, non-religious school was reorganized, and several liberal professors were fired; the reactionary Cauchy was promoted to full professor.

When Cauchy was 28 years old, he was still living with his parents. His father found it high time for his son to marry; he found him a suitable bride, Aloise de Bure, five years his junior. In 1819 the couple's first daughter, Marie Franqise Alicia, was born, and in 1823 the second and last daughter, Marie Mathilde.

The conservative political climate that lasted until 1830 suited Cauchy perfectly. He received cross-appointments at the Collége de France, and the Faculté des Sciences de Paris.

In 1830 the July revolution marked a turning point in Cauchy's life, and a break in his mathematical productivity. Cauchy, shaken by the fall of the government, and moved by a deep hatred of the liberals who were taking power, left Paris to go abroad, leaving his family behind. In 1831 Cauchy went to the Italian city of Turin, and after some time there, he accepted an offer from the King of Sardinia (who ruled Turin and the surrounding Piedmont region) for a chair of theoretical physics, which was created especially for him.

In August 1833 Cauchy left Turin for Prague, to become the science tutor of the thirteen-year-old Duke of Bordeaux Henri d'Artois (1820–1883), the exiled Crown Prince and grandson of Charles X. As a professor of the Ecolé

Polytechnique, Cauchy had been a notoriously bad lecturer, assuming levels of understanding that only a few of his best students could reach, and cramming his allotted time with too much material.

During his civil engineering days, Cauchy once had been briefly in charge of repairing a few of the Parisian sewers, and he made the mistake of mentioning this to his pupil; with great malice, the young Duke went about saying Mister Cauchy started his career in the sewers of Paris. The only good that came out of this episode was Cauchy's promotion to baron, a title by which Cauchy set great store.

Cauchy returned to Paris and his position at the Academy of Sciences late in 1838.[8] He could not regain his teaching positions, because he still refused to swear an oath of allegiance.

IIn November 1839 Cauchy was elected to the Bureau des Longitudes, and discovered immediately that the matter of the oath was not so easily dispensed with. He was not a formal member of the Bureau, did not receive payment, could not participate in meetings, and could not submit papers. Still Cauchy refused to take any oaths.

After losing control of the public education system, the Catholic Church sought to establish its own branch of education and found in Cauchy a staunch and illustrious ally. When a chair of mathematics became vacant at the Collége de France in 1843, Cauchy applied for it, but received just three of 45 votes.

Not unexpectedly, the idea came up in bureaucratic circles that it would be useful to again require a loyalty oath from all state functionaries, including university professors. This time a cabinet minister was able to convince the Emperor to exempt Cauchy from the oath. Cauchy remained a professor at the University until his death at the age of 67. He received the Last Rites and died of a bronchial condition at 4 a.m. on 23 May 1857.

His name is one of the 72 names inscribed on the Eiffel Tower.

****

Exercises

1. Suppose a sequence $\{a_j\}$ has the property that, for every natural number N, there is a j_N such that $a_{j_N} = a_{j_N+1} = \cdots = a_{j_N+N}$. In other words, the sequence has arbitrarily long repetitive strings. Does it follow that the sequence converges?

2. Let α be an irrational real number and let a_j be a sequence of rational numbers converging to α. Suppose that each a_j is a fraction expressed in lowest terms: $a_j = \alpha_j/\beta_j$. Prove that the β_j are unbounded.

3. Let $\{a_j\}$ be a sequence of rational numbers all of which have denominator a power of 2. What are the possible limits of such a sequence?

4. Redo Exercise **3** with the additional hypothesis that all of the denominators are less than or equal to 2^{10}.

5. Use the integral of $1/(1 + t^2)$, together with Riemann sums (ideas which you know from calculus, and which we shall treat rigorously later in the book), to develop a scheme for calculating the digits of π.

6. Prove Corollary 2.18.

7. Prove Proposition 2.20.

8. Prove parts **(2)** and **(4)** of Proposition 2.6.

9. Give an example of a decreasing sequence that converges to π.

10. Prove the following result, which we have used without comment in the text: Let S be a set of real numbers which is bounded above and let $t = \sup S$. For any $\epsilon > 0$ there is an element $s \in S$ such that $t - \epsilon < s \leq t$. **Remark:** Notice that this result makes good intuitive sense: the elements of S should become arbitrarily close to the supremum t, otherwise there would be enough room to decrease the value of t and make the supremum even smaller.) Formulate and prove a similar result for the infimum.

11. Let $\{a_j\}$ be a sequence of real or complex numbers. Suppose that every subsequence has itself a subsequence which converges to a given number α. Prove that the full sequence converges to α.

*12. Let $\{a_j\}$ be a sequence of complex numbers. Suppose that, for every pair of integers $N > M > 0$, it holds that $|a_M - a_{M+1}| + |a_{M+1} - a_{M+2}| + \cdots + |a_{N-1} - a_N| \leq 1$. Prove that $\{a_j\}$ converges.

13. Let $a_1, a_2 > 0$ and for $j \geq 3$ define $a_j = a_{j-1} + a_{j-2}$. Show that this sequence cannot converge to a finite limit.

14. Give an example of a sequence $\{a_j\}$ which diverges but so that $\{a_j^2\}$ and $\{a_j^4\}$ converge.

15. Suppose that $\{a_j\}$ is a sequence of real numbers such that $\{a_j^2\}$ converges and $\{a_j^3\}$ converges. Does it follow that $\{a_j\}$ converges?

2.2 Subsequences

Let $\{a_j\}$ be a given sequence. If

$$0 < j_1 < j_2 < \cdots$$

are positive integers then the function

$$k \mapsto a_{jk}$$

is called a *subsequence* of the given sequence. We usually write the sub-sequence as

$$\{a_{j_k}\}_{k=1}^{\infty} \text{ or } \{a_{j_k}\}.$$

Example 2.22: Consider the sequence

$$\{2^j\} = \{2, 4, 8, \ldots\}.$$

Then the sequence

$$\{2^{2k}\} = \{4, 16, 64, \ldots\} \tag{2.22.1}$$

is a subsequence. Notice that the subsequence contains a subcollection of elements of the original sequence *in the same order*. In this example, $j_k = 2k$. Another subsequence is

$$\{2^{(2k)}\} = \{4, 16, 256, \ldots\}. \tag{2.22.2}$$

In this instance, it holds that $j_k = 2^k$. Notice that this new subsequence is in fact a subsequence of the first subsequence (2.22.1). That is, it is a sub-subsequence of the original sequence $\{2^j\}$. □

Proposition 2.23: *If $\{a_j\}$ is a convergent sequence with limit α, then every subsequence converges to the limit α.*

Conversely, if a sequence $\{b_j\}$ has the property that each of its subsequences is convergent, then $\{b_j\}$ itself is convergent.

Proof: Assume $\{a_j\}$ is convergent to a limit α, and let $\{a_{j_k}\}$ be a subsequence. Let $\epsilon > 0$ and choose $N > 0$ such that $|a_j - \alpha| < \epsilon$ whenever $j > N$. Now if $k > N$, then $j_k > N$ hence $|a_{j_k} - \alpha| < \epsilon$. Therefore, by definition, the subsequence $\{a_{j_k}\}$ also converges to α.

The converse is trivial, simply because the sequence is a subsequence of itself. □

Now we present one of the most fundamental theorems of basic real ana-lysis (due to B. Bolzano, 1781–1848, and K. Weierstrass, 1815–1897).

Theorem 2.24: (Bolzano–Weierstrass) *Let $\{a_j\}$ be a bounded sequence in $\mathbb{R}$. Then there is a subsequence which converges.*

Proof: Suppose that $|a_j| \leq M$ for every j. We may assume that $M > 0$. It is convenient to formulate our hypothesis as $a_j \in [-M, M]$ for every j.

One of the two intervals $[-M, 0]$ and $[0, M]$ must contain infinitely many elements of the sequence. Assume that $[0, M]$ does. Choose a_{j_1} to be one of the infinitely many sequence elements in $[0, M]$.

Next, one of the intervals $[0, M/2]$ and $[M/2, M]$ must contain infinitely many elements of the sequence. Suppose that it is $[0, M/2]$. Choose an element a_{j_2}, with $j_2 > j_1$, from $[0, M/2]$. Continue in this fashion, halving the interval, choosing a half with infinitely many sequence elements, and selecting the next subsequential element from that half.

Let us analyze the resulting subsequence. Notice that $|a_{j_1} - a_{j_2}| \leq M$ since both elements belong to the interval $[0, M]$. Likewise, $|a_{j_2} - a_{j_3}| \leq M/2$ since both elements belong to $[0, M/2]$. In general, $|a_{j_k} - a_{j_{k+1}}| \leq 2^{-k+1} \cdot M$ for each $k \in \mathbb{N}$.

Now let $\epsilon > 0$. Choose an integer $N > 0$ such that $2^{-N} < \epsilon/(4M)$. Then, for any $m > l > N$ we have

$$
\begin{aligned}
|a_{j_l} - a_{j_m}| &= |(a_{j_l} - a_{j_{l+1}}) + (a_{j_{l+1}} - a_{j_{l+2}}) + \ldots + (a_{j_{m-1}} - a_{j_m})| \\
&\leq |a_{j_l} - a_{j_{l+1}}| + |a_{j_{l+1}} - a_{j_{l+2}}| + \ldots + |a_{j_{m-1}} - a_{j_m}| \\
&\leq 2^{-l+1} \cdot M + 2^{-l} \cdot M + \cdots + 2^{-m+2} \cdot M \\
&= (2^{-l+1} + 2^{-l} + \cdots + 2^{-m+2}) \cdot M \\
&= ((2^{-l+2} - 2^{-l+1}) + (2^{-l+1} - 2^{-l}) + \cdots + (2^{-m+3} - 2^{-m+2})) \cdot M \\
&= (2^{-l+2} - 2^{-m+2}) \cdot M \\
&< 2^{-l+2} \cdot M \\
&< 4 \cdot \frac{\epsilon}{4M} \cdot M \\
&= \epsilon.
\end{aligned}
$$

We see that the subsequence $\{a_{j_k}\}$ is Cauchy, so it converges. $\square$

Remark 2.25: The Bolzano–Weierstrass theorem is a generalization of our result from the last section about increasing sequences which are bounded above (resp. decreasing sequences which are bounded below). For such a sequence is surely bounded above and below (why?). So it has a convergent subsequence. And thus it follows easily that the entire sequence converges. Details are left as an exercise.

It is a fact—which you can verify for yourself—that *any* real sequence has a monotone subsequence. This observation implies Bolzano–Weierstrass. $\square$

Example 2.26: In this text we have not yet given a rigorous definition of the function sinx (see Section 9.3). However, just for the moment, use the definition you learned in calculus class and consider the sequence $\{\sin j\}_{j=1}^{\infty}$. Notice that the sequence is bounded in absolute value by 1. The Bolzano–Weierstrass theorem guarantees that there is a convergent subsequence, even though it would be very difficult to say precisely what that convergent subsequence is. $\square$

Corollary 2.27: *Let $\{\alpha_j\}$ be a bounded sequence of complex numbers. Then there is a convergent subsequence.*

Proof: Write $\alpha_j = a_j + ib_j$, with $a_j, b_j \in \mathbb{R}$. The fact that $\{\alpha_j\}$ is bounded implies that $\{a_j\}$ is bounded. By the Bolzano–Weierstrass theorem, there is a convergent subsequence $\{a_{jk}\}$.

Now the sequence $\{b_{jk}\}$ is bounded. So it has a convergent subsequence $\{b_{jk_l}\}$. Then the sequence $\{\alpha_{jk_l}\}$ is convergent, and is a subsequence of the original sequence $\{\alpha_j\}$. □

In earlier parts of this chapter we have discussed sequences that converge to a finite number. Such a sequence is, by Proposition 2.5, bounded. However, in some mathematical contexts, it is useful to speak of a sequence "diverging[1] to infinity." We now will treat briefly the idea of "divergence to infinity."

Definition 2.28: We say that a sequence $\{a_j\}$ of real numbers *diverges to $+\infty$* if, for every $M > 0$, there is an integer $N > 0$ such that $a_j > M$ whenever $j > N$. We write $a_j \to +\infty$.

We say that $\{a_j\}$ *diverges to $-\infty$* if, for every $K > 0$, there is an integer $N > 0$ such that $a_j < -K$ whenever $j > N$. We write $a_j \to -\infty$.

Remark 2.29: Notice that the statement $a_j \to +\infty$ means that we can make a_j become arbitrarily large and positive and *stay* large and positive just by making j large enough.

Likewise, the statement $a_j \to -\infty$ means that we can force a_j to be arbitrarily large and negative, and *stay* large and negative, just by making j large enough. □

Example 2.30: The sequence $\{j^2\}$ diverges to $+\infty$. The sequence $\{-2j + 18\}$ diverges to $-\infty$. The sequence $\{j + (-1)^j \cdot j\}$ has no infinite limit and no finite limit. However, the subsequence $\{0, 0, 0, \ldots\}$ converges to 0 and the subsequence $\{4, 8, 12 \ldots\}$ diverges to $+\infty$. □

With the new language provided by Definition 2.28, we may generalize Proposition 2.16:

Proposition 2.31: *Let $\{a_j\}$ be an increasing sequence of real numbers. Then the sequence has a limit—either a finite number or $+\infty$.*

Let $\{b_j\}$ be a decreasing sequence of real numbers. Then the sequence has a limit—either a finite number or $-\infty$. □

In the same spirit as the last definition, we also have the following:

Definition 2.32: If S is a set of real numbers which is *not* bounded above, we say that its supremum (or least upper bound) is $+\infty$.

If T is a set of real numbers which is not bounded below, then we say that its infimum (or greatest lower bound) is $-\infty$.   □

Exercises

1. Use the Bolzano–Weierstrass theorem to show that every decreasing sequence that is bounded below converges.

2. Give an example of a sequence of rational numbers with the property that, for any real number α, or for $\alpha = +\infty$ or $\alpha = -\infty$, there is a subsequence approaching α.

3. Prove that if $\{a_j\}$ has a subsequence diverging to $\pm\infty$ then $\{a_j\}$ cannot converge.

4. Let $x_1 = 2$. For $j \geq 1$, set

$$x_{j+1} = x_j - \frac{x_j^2 - 2}{2x_j}.$$

   Show that the sequence $\{x_j\}$ is decreasing and bounded below. What is its limit?

5. The sequence

$$a_j = (1 + 1/2 + 1/3 + \cdots + 1/j) - \log j$$

   is a famous example. It is known to converge, but nobody knows whether the limit is rational or irrational. Draw a picture which shows that the sequence converges.

6. Provide the details of the proof of Proposition 2.31.

*7. Provide the details of the assertion that the sequence $\{\cos j\}$ is dense in the interval $[-1, 1]$.

*8. Let n be a positive integer. Consider $n, n + 1, \ldots$ modulo π. This means that you subtract from each number the greatest multiple of π that does not exceed it. Prove that this collection of numbers is dense in $[0, \pi]$. That is, the numbers get arbitrarly close to any element of this interval.

*9. Let $S = \{0, 1, 1/2, 1/3, 1/4, \ldots\}$. Give an example of a sequence $\{a_j\}$ with the property that, for each $s \in S$, there is a subsequence converging to s, but no subsequence converges to any limit not in S.

*10. Give another proof of the Bolzano–Weierstrass theorem as follows. If $\{a_j\}$ is a bounded sequence let $b_j = \inf\{a_j, a_j^{+1}, \ldots\}$. Then each b_j is finite, $b_1 \leq b_2 \leq \ldots$, and $\{b_j\}$ is bounded above. Now use Proposition 2.16.

*11. Prove that the sequence

$$a_N = \sum_{m=1}^{N} \frac{\sin m}{m}$$

converges.

*12. Prove that the sequence

$$a_N = \sum_{m=1}^{N} \frac{\sin^2 m}{m}$$

diverges.

13. Let $\{a_j\}$ be a sequence of real numbers with the property that every subsequence has itself a subsequence that converges. Prove that the original sequence $\{a_j\}$ converges.

14. Suppose that $\{a_j\}$ is a sequence of real numbers with the property $\{\sin a_j\}$ converges. Does it follow that $\{a_j\}$ converges?

2.3 Lim sup and Lim inf

Convergent sequences are useful objects, but the unfortunate truth is that most sequences do not converge. Nevertheless, we would like to have a language for discussing the asymptotic behavior of *any* real sequence $\{a_j\}$ as $j \rightarrow \infty$. That is the purpose of the concepts of "limit superior" (or "upper limit") and "limit inferior" (or "lower limit").

Definition 2.33: Let $\{a_j\}$ be a sequence of real numbers. For each j let

$$A_j = \inf\{a_j, a_{j+1}, a_{j+2}, \ldots\}.$$

Then $\{A_j\}$ is an increasing sequence (since, as j becomes large, we are taking the infimum of a smaller set of numbers), so it has a limit (either a finite limit or $\pm\infty$). We define the *limit infimum* of $\{a_j\}$ to be

$$\lim\inf a_j = \lim_{j \to \infty} A_j.$$

It is common to refer to this number as the lim inf of the sequence. Likewise, let

$$B_j = \sup\{a_j, a_{j+1}, a_{j+2}, \cdots\ \}.$$

Then $\{B_j\}$ is a decreasing sequence (since, as j becomes large, we are taking the supremum of a smaller set of numbers), so it has a limit (either a finite limit of $\pm\infty$). We define the *limit supremum* of $\{a_j\}$ to be

$$\lim \sup\ a_j = \lim_{j\to\infty} B_j.$$

It is common to refer to this number as the limsup of the sequence.   □

Notice that the lim sup or lim inf of a sequence can be $\pm\infty$.

Remark 2.34: What is the intuitive content of this definition? For each j, A_j picks out the greatest lower bound of the sequence in the jth position or later. So the sequence $\{A_j\}$ should tend to the *smallest* possible limit of any subsequence of $\{a_j\}$.

Likewise, for each j, B_j picks out the least upper bound of the sequence in the jth position or later. So the sequence $\{B_j\}$ should tend to the *greatest* possible limit of any subsequence of $\{a_j\}$. We shall make these remarks more precise in Proposition 2.36 below.

Notice that it is implicit in the definition that *every* real sequence has a limit supremum and a limit infimum.

Example 2.35: Consider the sequence $\{(-1)^j\}$. Of course this sequence does not converge. Let us calculate its lim sup and lim inf.

Referring to the definition, we have that $A_j = -1$ for every j. So

$$\lim \inf(-1)^j = \lim(-1) = -1.$$

Similarly, $B_j = +1$ for every j. Therefore

$$\lim \sup(-1)^j = \lim(+1) = +1.$$

As we predicted in the remark, the lim inf is the least subsequential limit, and the lim sup is the greatest subsequential limit.   □

Now let us prove the characterizing property of lim sup and lim inf to which we have been alluding.

Proposition 2.36: *Let $\{a_j\}$ be a sequence of real numbers. Let $\beta = \lim \sup_{j\to\infty} a_j$ and $\alpha = \lim \inf_{j\to\infty} a_j$. If $\{a_{j_\ell}\}$ is any subsequence of the given sequence then*

$$\alpha \leq \lim_{\ell \to \infty} \inf a_{j\ell} \leq \lim_{\ell \to \infty} \sup a_{j\ell} \leq \beta.$$

Moreover, there is a subsequence $\{a_{jk}\}$ such that

$$\lim_{k \to \infty} a_{jk} = \alpha$$

and another sequence $\{a_{jm}\}$ such that

$$\lim_{k \to \infty} a_{jm} = \beta.$$

Proof: For simplicity in this proof we assume that the lim sup and lim inf are finite. The case of infinite lim sups and lim infs is treated in the exercises.

We begin by considering the lim inf. There is a $j_1 \geq 1$ such that $|A_1 - a_{j1}| < 2^{-1}$. We choose j_1 to be as small as possible. Next, we choose j_2, necessarily greater than j_1, such that j_2 is as small as possible and $|a_{j2} - A_2| < 2^{-2}$. Continuing in this fashion, we select $j_k > j_{k-1}$ such that $|a_{jk} - A_k| < 2^{-k}$, etc.

Recall that $A_k \to \alpha = \lim \inf_{j \to \infty} a_j$. Now fix $\epsilon > 0$. If N is an integer so large that $k > N$ implies that $|A_k - \alpha| < \epsilon/2$ and also that $2^{-N} < \epsilon/2$ then, for such k, we have

$$\begin{aligned}
|a_{jk} - \alpha| &\leq |a_{jk} - A_k| + |A_k - \alpha| \\
&< 2^{-k} + \frac{\epsilon}{2} \\
&< \frac{\epsilon}{2} + \frac{\epsilon}{2} \\
&= \epsilon.
\end{aligned}$$

Thus the subsequence $\{a_{jk}\}$ converges to α, the lim inf of the given sequence. A similar construction gives a (different) subsequence $\{a_{nk}\}$ converging to β, the lim sup of the given sequence.

Now let $\{a_{j\ell}\}$ be *any* subsequence of the sequence $\{a_j\}$. Let β^* be the lim sup of this subsequence. Then, by the first part of the proof, there is a subsequence $\{a_{j\ell_m}\}$ such that

$$\lim_{m \to \infty} a_{j\ell_m} = \beta^*$$

But $a_{j\ell_m} \leq B_{j\ell_m}$ by the very definition of the Bs. Thus

$$\beta^* = \lim_{m \to \infty} a_{j\ell_m} \leq \lim_{m \to \infty} B_{j\ell_m} = \beta$$

or

$$\lim_{\ell \to \infty} \sup a_{j\ell} \leq \beta,$$

as claimed. A similar argument shows that

$$\liminf_{l \to \infty} a_{jl} \geq \alpha.$$

This completes the proof of the proposition.   □

Corollary 2.37: *If $\{a_j\}$ is a sequence and $\{a_{jk}\}$ is a convergent subsequence then*

$$\liminf_{l \to \infty} a_j \leq \lim_{k \to \infty} a_{jk} \leq \liminf_{j \to \infty} a_j.$$

Example 2.38: Consider the sequence

$$a_j = \frac{(-1)^j j^2}{j^2 + j}.$$

It is helpful to rewrite this sequence as

$$a_j = (-1)^j \cdot \left(1 - \frac{j}{j^2 + j}\right).$$

Then, looking at the terms of even index, it is easy to see that the lim sup of this sequence is +1. And, looking at the terms of odd index, it is easy to see that the lim inf of this sequence is −1.

Every convergent subsequence of $\{a_j\}$ will have limit lying between −1 and +1.   □

We close this section with a fact that is analogous to one for the supremum and infimum. Its proof is analogous to arguments we have seen before.

Proposition 2.39: *Let $\{a_j\}$ be a sequence and set lim sup $a_j = \beta$ and lim inf $a_j = \alpha$. Assume that α, β are finite real numbers. Let $\epsilon > 0$. Then there are arbitrarily large j such that $a_j > \beta - \epsilon$. Also there are arbitrarily large k such that $a_k < \alpha + \epsilon$.*

Example 2.40: Consider the sequence $\{a_j\}$ in the last example. Let $\epsilon > 0$. Choose j even so that $j > (1 - \epsilon)/\epsilon$. Then

$$\frac{j^2}{j^2 + j} > 1 - \epsilon.$$

Now again choose $\epsilon > 0$. Choose j odd so that $j > (1 - \epsilon)/\epsilon$. Then

$$-\frac{j^2}{j^2 + j} < -1 + \epsilon.$$

□

Exercises

1. Consider $\{a_j\}$ both as a sequence and as a set. How are the lim sup and the sup related? How are the lim inf and the inf related? Give examples.

2. Let $\{a_j\}$ be a sequence of positive numbers. How are the lim sup and lim inf of $\{a_j\}$ related to the lim sup and lim inf of $\{1/a_j\}$?

3. How are the lim sup and lim inf of $\{a_j\}$ related to the lim sup and lim inf of $\{-a_j\}$?

4. Let $\{a_j\}$ be a real sequence. Prove that if

$$\lim \inf a_j = \lim \sup a_j$$

then the sequence $\{a_j\}$ converges. Prove the converse as well.

*5. What is the lim sup of the sequence $\{\sin j\}$?

6. What is the lim inf of the sequence in Exercise 5?

7. Let $a < b$ be real numbers. Give an example of a real sequence whose lim sup is b and whose lim inf is a.

8. Explain why we can make no sense of the concepts of lim sup and lim inf for complex sequences.

9. Let $\{a_j\}$, $\{b_j\}$ be sequences of real numbers. Prove the inequality lim sup$(a_j + b_j) \leq$ lim sub $a_j +$ lim sub b_j. How are the lim infs related? How is the quantity (lim sub a_j) · (lim sub b_j) related to lim sup$(a_j \cdot b_j)$? How are the lim infs related?

10. Give an example of a sequence whose lim sup and lim inf differ by 1.

11. Prove Corollary 2.37.

12. Prove Proposition 2.39.

13. Prove a version of Proposition 2.36 when the indicated lim sup and/or lim inf are $\pm\infty$.

14. Prove a version of Proposition 2.39. when the indicated limsup and/or liminf are $\pm\infty$.

*15. Find the lim sup and lim inf of the sequences

$$\{|\sin j|^{\sin j}\} \text{ and } \{|\cos j|^{\cos j}\}.$$

16. If the liminf of a sequence is precisely 1 less than the limsup of that sequence, then what does this tell you about the sequence?

2.4 Some Special Sequences

We often obtain information about a new sequence by comparison with a sequence that we already know. Thus it is well to have a catalogue of fundamental sequences which provide a basis for comparison.

Example 2.41: Fix a real number a. The sequence $\{a^j\}$ is called a *power sequence* If $-1 < a < 1$ then the sequence converges to 0. If $a = 1$ then the sequence is a constant sequence and converges to 1. If $a > 1$ then the sequence diverges to $+\infty$. Finally, if $a \leq -1$ then the sequence diverges.   □

Recall that, in Section 1.1, we discussed the existence of nth roots of positive real numbers. If $\alpha > 0$, $m \in \mathbb{Z}$, and $n \in \mathbb{N}$ then we may define

$$\alpha^{m/n} = (\alpha^m)^{1/n}.$$

Thus we may talk about rational powers of a positive number. Next, if $\beta \in \mathbb{R}$ then we may define

$$\alpha^\beta = \sup\{\alpha^q : q \in \mathbb{Q}, q < \beta\}.$$

Thus we can define *any real power* of a positive real number. The exercises ask you to verify several basic properties of these exponentials.

Lemma 2.42: *If $\alpha > 1$ is a real number and $\beta > 0$ then $\alpha^\beta > 1$.*

Proof: Let q be a positive rational number which is less than β. Suppose that $q = m/n$, with m, n integers. It is obvious that $\alpha^m > 1$ and hence that $(\alpha^m)^{1/n} > 1$. Since α^β majorizes this last quantity, we are done.   □

Example 2.43: Fix a real number α and consider the sequence $\{j^\alpha\}$. If $\alpha > 0$ then it is easy to see that $j^\alpha \to +\infty$: to verify this assertion fix $M > 0$ and take the number N to be the first integer after $M^{1/\alpha}$.

If $\alpha = 0$ then j^α is a constant sequence, identically equal to 1.

If $\alpha < 0$ then $j^\alpha = 1/j^{-\alpha}$. The denominator of this last expression tends to $+\infty$ hence the sequence j^α tends to 0.   □

Example 2.44: The sequence $\{j^{1/j}\}$ converges to 1. In fact, consider the expressions $\alpha_j = j^{1/j} - 1 > 0$. We have that

$$j = (\alpha_j + 1)^j \geq \frac{j(j-1)}{2}(\alpha_j)^2,$$

(the latter being just one term from the binomial expansion). Thus

$$0 < \alpha_j \leq \sqrt{2/(j-1)}$$

as long as $j \geq 2$. It follows that $\alpha_j \to 0$ or $j^{1/j} \to 1$. □

Example 2.45: Let α be a positive real number. Then the sequence $\alpha^{1/j}$ converges to 1. To see this, first note that the case $\alpha = 1$ is trivial, and the case $\alpha > 1$ implies the case $\alpha < 1$ (by taking reciprocals). So we concentrate on $\alpha > 1$. But then we have

$$1 < \alpha^{1/j} < j^{1/j}$$

when $j > \alpha$. Since $j^{1/j}$ tends to 1, Proposition 2.20 applies and the proof is complete. □

Example 2.46: Let $\lambda > 1$ and let α be real. Then the sequence

$$\left\{ \frac{j^\alpha}{\lambda^j} \right\}_{j=1}^\infty$$

converges to 0.

To see this, fix an integer $k > \alpha$ and consider $j > 2k$. [Notice that k is fixed once and for all but j will be allowed to tend to $+\infty$ at the appropriate moment.] Writing $\lambda = 1 + \mu$, $\mu > 0$, we have that

$$\lambda^j = (1 + \mu)^j > \frac{j(j-1)(j-2)\cdots(j-k+1)}{k(k-1)(k-2)\cdots 2\cdot 1} \cdot 1^{j-k} \cdot \mu^k.$$

Of course this comes from picking out the kth term of the binomial expansion for $(1 + \mu)^j$. Notice that, since $j > 2k$, then each of the expressions $j, (j-1), \ldots (j-k+1)$ in the numerator on the right exceeds $j/2$. Thus

$$\lambda^j > \frac{j^k}{2^k \cdot k!} \cdot \mu^k$$

and

$$0 < \frac{j^\alpha}{\lambda^j} < j^\alpha \cdot \frac{2^k \cdot k!}{j^k \cdot \mu^k} = \frac{j^{\alpha-k} \cdot 2^k \cdot k!}{\mu^k}.$$

Since $\alpha - k < 0$, the right side tends to 0 as $j \to \infty$.   □

Example 2.47: The sequence

$$\left\{ \left(1 + \frac{1}{j} \right)^j \right\}$$

converges. In fact it is increasing and bounded above. Use the Binomial Expansion to prove this assertion. The limit of the sequence is the number that we shall later call e (in honor of Leonhard Euler, 1707–1783, who first studied it in detail). We shall study this sequence later in the book.   □

Example 2.48: The sequence

$$\left(1 - \frac{1}{j} \right)^j$$

converges to $1/e$, where the definition of e is given in the last example. More generally, the sequence

$$\left(1 + \frac{x}{j} \right)^j$$

converges to e^x (here e^x is defined as in the discussion following Example 2.41 above).   □

Exercises
1. Let α be a positive real number and let $p/q = m/n$ be two different representations of the same rational number r. Prove that

$$(\alpha^m)^{1/n} = (\alpha^p)^{1/q}.$$

Also prove that

$$(\alpha^{1/n})^m = (\alpha^m)^{1/n}.$$

If β is another positive real and γ is any real then prove that

$$(\alpha \cdot \beta)^\gamma = \alpha^\gamma \cdot \beta^\gamma.$$

2. Discuss the convergence of the sequence $\{(1/j)^{1/j}\}_{j=1}^\infty$.
3. Discuss convergence or divergence of the sequence $\{[\log j]^{1/j}\}$.
4. Discuss convergence or divergence of the sequence $\{[2^j]^{1/2^j}\}$.
5. Discuss the convergence of the sequence $\{(j^j)/(2j)!\}_{j=2}^\infty$.
6. Prove that the exponential, as defined in this section, satisfies

$$(a^b)^c = a^{bc} \quad \text{and} \quad a^b a^c = a^{b+c}.$$

*7. Refer to Exercise 5 in Section 2.2. Consider the sequence given by

$$a_j = \left[1 + \frac{1}{2} + \frac{1}{3} + \cdots + \frac{1}{j}\right] - \log j.$$

Then $\{a_j\}$ converges to a limit γ. This number was first studied by Euler. It arises in many different contexts in analysis and number theory. As a challenge problem, show that

$$|a_j - \gamma| \le \frac{C}{j}$$

for some universal constant $C > 0$. It is not known whether γ is rational or irrational.

*8. Give a recursive definition of the Fibonacci sequence. Find a generating function for the Fibonacci sequence and use it to derive an explicit formula for the nth term of the sequence.

9. A sequence is defined by the rule $a_0 = 2$, $a_1 = 1$, and $a_j = 3a_{j-1} - a_{j-2}$. Find a formula for a_j.

10. A sequence is defined by the rule $a_0 = 4$, $a_1 = -1$, and $a_j = -a_{j-1} + 2a_{j-2}$. Find a formula for a_j.

*11. Consider the sequence

$$a_j = \left(1 + \frac{1}{1^2}\right) \cdot \left(1 + \frac{1}{2^2}\right) \cdot \left(1 + \frac{1}{3^2}\right) \cdots \left(1 + \frac{1}{j^2}\right).$$

Discuss convergence and divergence.

*12. Prove that

$$\left(1 + \frac{x}{j}\right)^{j}$$

converges to e^x for any real number x.

*13. Give an example of a sequence of rational numbers that converges to π.

Note

1 Some books say "converging to infinity," but this terminology can be confusing.

3

Series of Numbers

3.1 Convergence of Series

In this section we will use standard summation notation:

$$\sum_{j=m}^{n} a_j \equiv a_m + a_{m+1} + \cdots + a_n.$$

A series is an infinite sum. One of the most effective ways to handle an infinite process in mathematics is with a limit. This consideration leads to the following definition:

Definition 3.1: The formal expression

$$\sum_{j=1}^{\infty} a_j,$$

where the a_js are real or complex numbers, is called a *series*. For $N = 1, 2, 3, \ldots,$ the expression

$$S_N = \sum_{j=1}^{N} a_j = a_1 + a_2 + \ldots a_N$$

is called the Nth *partial sum* of the series. In case

$$\lim_{N \to \infty} S_N$$

exists and is finite we say that the series *converges*. The limit of the partial sums is called the *sum* of the series. If the series does not converge, then we say that the series *diverges*.

DOI: 10.1201/9781003222682-4

Notice that the question of convergence of a series, which should be thought of as an *addition process*, reduces to a question about the *sequence* of partial sums.

Example 3.2: Consider the series

$$\sum_{j=1}^{\infty} 2^{-j}.$$

The Nth partial sum for this series is

$$S_N = 2^{-1} + 2^{-2} + \cdots + 2^{-N}.$$

In order to determine whether the sequence $\{S_N\}$ has a limit, we rewrite S_N as

$$S_N = (2^{-0} - 2^{-1}) + (2^{-1} - 2^{-2}) + \cdots$$
$$(2^{-N+1} - 2^{-N}).$$

The expression on the right of the last equation telescopes (i.e., successive pairs of terms cancel) and we find that

$$S_N = 2^{-0} - 2^{-N}.$$

Thus

$$\lim_{N \to \infty} S_N = 2^{-0} = 1.$$

We conclude that the series converges.   □

Example 3.3: Let us examine the series

$$\sum_{j=1}^{\infty} \frac{1}{j}$$

for convergence or divergence. (This series is commonly called the *harmonic series* because it describes the harmonics in music.) Now

$$S_1 = 1 = \frac{2}{2}$$

$$S_2 = 1 + \frac{1}{2} = \frac{3}{2}$$

$$S_4 = 1 + \frac{1}{2} + \left(\frac{1}{3} + \frac{1}{4}\right)$$

$$\geq 1 + \frac{1}{2} + \left(\frac{1}{4} + \frac{1}{4}\right) \geq 1 + \frac{1}{2} + \frac{1}{2} = \frac{4}{2}$$

$$S_8 = 1 + \frac{1}{2} + \left(\frac{1}{3} + \frac{1}{4}\right) + \left(\frac{1}{5} + \frac{1}{6} + \frac{1}{7} + \frac{1}{8}\right)$$

$$\geq 1 + \frac{1}{2} + \left(\frac{1}{4} + \frac{1}{4}\right) + \left(\frac{1}{8} + \frac{1}{8} + \frac{1}{8} + \frac{1}{8}\right)$$

$$= \frac{5}{2}.$$

In general this argument shows that

$$S_{2^k} \geq \frac{k + 2}{2}.$$

The sequence of S_Ns is increasing since the series contains only positive terms. The fact that the partial sums $S_1, S_2, S_4, S_8, \ldots$ increases without bound shows that the entire sequence of partial sums must increase without bound. We conclude that the series diverges. □

Just as with sequences, we have a Cauchy criterion for series:

Proposition 3.4: *The series $\sum_{j=1}^{\infty} a_j$ converges if and only if, for every $\epsilon > 0$, there is an integer $N \geq 1$ such that, if $n \geq m > N$, then*

$$\left| \sum_{j=m}^{n} a_j \right| < \epsilon. \tag{3.4.1}$$

The condition (3.4.1) is called the Cauchy criterion for series.

Proof: Suppose that the Cauchy criterion holds. Pick $\epsilon > 0$ and choose N so large that (3.4.1) holds. If $n \geq m > N$, then

$$|S_n - S_m| = \left| \sum_{j=m+1}^{n} a_j \right| < \epsilon$$

by hypothesis. Thus the sequence $\{S_N\}$ is Cauchy in the sense discussed for sequences in Section 2.1. We conclude that the sequence $\{S_N\}$ converges; by definition, therefore, the series converges.

Conversely, if the series converges, then, by definition, the sequence $\{S_N\}$ of partial sums converges. In particular, the sequence $\{S_N\}$ must be Cauchy.

Thus, for any $\epsilon > 0$, there is a number $N > 0$ such that if $n \geq m > N$, then

$$|S_n - S_m| < \epsilon.$$

This just says that

$$\left| \sum_{j=m+1}^{n} a_j \right| < \epsilon,$$

and this last inequality is the Cauchy criterion for series.   □

Example 3.5: Let us use the Cauchy criterion to verify that the series

$$\sum_{j=1}^{\infty} \frac{1}{j \cdot (j+1)}$$

converges.

Notice that, if $n \geq m > 1$, then

$$\left| \sum_{j=m}^{n} \frac{1}{j \cdot (j+1)} \right| = \left(\frac{1}{m} - \frac{1}{m+1} \right) + \left(\frac{1}{m+1} - \frac{1}{m+2} \right) + \ldots + \left(\frac{1}{n} - \frac{1}{n+1} \right).$$

The sum on the right plainly telescopes and we have

$$\left| \sum_{j=m}^{n} \frac{1}{j \cdot (j+1)} \right| = \frac{1}{m} - \frac{1}{n+1}.$$

Let $\epsilon > 0$. Let us choose N to be the next integer after $1/\epsilon$. Then, for $n \geq m > N$, we may conclude that

$$\left| \sum_{j=m}^{n} \frac{1}{j \cdot (j+1)} \right| = \frac{1}{m} - \frac{1}{n+1} < \frac{1}{m} < \frac{1}{N} < \epsilon.$$

This is the desired conclusion.   □

The next result gives a necessary condition for a series to converge. It is a useful device for detecting divergent series, although it can never tell us that a series converges.

Proposition 3.6: (The Zero Test) *If the series*

$$\sum_{j=1}^{\infty} a_j$$

converges, then the terms a_j tend to zero as $j \to \infty$.

Proof: Since we are assuming that the series converges, then it must satisfy the Cauchy criterion. Let $\epsilon > 0$, then >0. Then there is an integer $N \geq 1$ such that, if $n \geq m > N$, then

$$\left| \sum_{j=m}^{n} a_j \right| < \epsilon. \tag{3.6.1}$$

We take $n = m$ and $m > N$. Then, (3.6.1) becomes

$$|a_m| < \epsilon.$$

But this is precisely the conclusion that we desire. □

Example 3.7: The series $\sum_{j=1}^{\infty} (-1)^j$ must diverge, *even though its terms appear to be cancelling each other out.* The reason is that the summands do not tend to zero; hence the preceding proposition applies.

Write out several partial sums of this series to see more explicitly that the partial sums are $-1, +1, -1, +1,\dots$ and hence that the series diverges. □

We conclude this section with a necessary and sufficient condition for convergence of a series of nonnegative terms. As with some of our other results on series, it amounts to little more than a restatement of a result on sequences.

Proposition 3.8: *A series*

$$\sum_{j=1}^{\infty} a_j$$

with all $a_j \geq 0$ is convergent if and only if the sequence of partial sums is bounded.

Proof: Notice that, because the summands are nonnegative, we have

$$S_1 = a_1 \le a_1 + a_2 = S_2,$$

$$S_2 = a_1 + a_2 \le a_1 + a_2 + a_3 = S_3,$$

and in general

$$S_N \le S_N + a_{N+1} = S_{N+1}.$$

Thus the sequence $\{S_N\}$ of partial sums forms a increasing sequence. We know that such a sequence is convergent to a finite limit if and only if it is bounded above (see Section 2.1). This completes the proof.   □

Example 3.9: The series $\sum_{j=1}^{\infty} 1$ is divergent since the summands are nonnegative and the sequence of partial sums $\{S_N\} = \{N\}$ is unbounded.

Referring back to Example 3.3, we see that the series $\sum_{j=1}^{\infty} \frac{1}{j}$ diverges because its partial sums are unbounded.

We see from the first example that the series $\sum_{j=1}^{\infty} 2^{-j}$ converges because its partial sums are all bounded above by 1.   □

It is frequently convenient to begin a series with summation at $j = 0$ or some other term instead of $j = 1$. All of our convergence results still apply to such a series because of the Cauchy criterion. In other words, the convergence or divergence of a series will depend only on the behavior of its "tail."

Exercises

1. Discuss convergence or divergence for each of the following series:

(a) $\displaystyle\sum_{j=1}^{\infty} \frac{(2^j)^2}{j!}$ (b) $\displaystyle\sum_{j=1}^{\infty} \frac{(2j)!}{(3j)!}$

(c) $\displaystyle\sum_{j=1}^{\infty} \frac{j!}{j^j}$ (d) $\displaystyle\sum_{j=1}^{\infty} \frac{(-1)^j}{3j^2 - 5j + 6}$

(e) $\displaystyle\sum_{j=1}^{\infty} \frac{2j - 1}{3j^2 - 2}$ (f) $\displaystyle\sum_{j=1}^{\infty} \frac{2j - 1}{3j^3 - 2}$

(g) $\displaystyle\sum_{j=1}^{\infty} \frac{\log(j + 1)}{[1 + \log j]^j}$ (h) $\displaystyle\sum_{j=12}^{\infty} \frac{1}{j \, \log^3 j}$

(i) $\displaystyle\sum_{j=2}^{\infty} \frac{\log(2)}{\log j}$ (j) $\displaystyle\sum_{j=2}^{\infty} \frac{1}{j \log^{1.1} j}$

2. If $b_j > 0$ for every j and if $\sum_{j=1}^{\infty} b_j$ converges then prove that $\sum_{j=1}^{\infty} (b_j)^2$ converges. Prove that the assertion is false if the positivity hypothesis is omitted. How about third powers?

3. If $b_j > 0$ for every j and if $\sum_{j=1}^{\infty} b_j$ converges then prove that $\sum_{j=1}^{\infty} \frac{1}{1+b_j}$ diverges.

4. If $b_j > 0$ and $\sum_j b_j$ converges, then what can you say about $\sum_j \sin b_j$?

5. If $b_j > 0$ and $\sum_j b_j$ converges, then what can you say about $\sum_j e^{b_j}$?

6. Let $\sum_{j=1}^{\infty} a_j$ be a divergent series of positive terms. Prove that there exist numbers b_j, $0 < b_j < a_j$, such that $\sum_{j=1}^{\infty} b_j$ diverges.

   Similarly, let $\sum_{j=1}^{\infty} c_j$ be a convergent series of positive terms. Prove that there exist numbers d_j, $0 < c_j < d_j$, such that $\sum_{j=1}^{\infty} d_j$ converges.

   Thus we see that there is no "smallest" divergent series and no "largest" convergent series.

7. TRUE or FALSE: If $a_j > c > 0$ and $\sum 1/a_j$ converges, then $\sum a_j$ converges.

8. If $b_j > 0$ and $\sum_j b_j$ converges then what can you say about $\sum_j b_j/(1 + b_j)$?

9. If $a_j > 0$, $b_j > 0$, $\sum_j a_j^2$ converges, and $\sum_j b_j^2$ converges, then what can you say about $\sum_j a_j b_j$?

10. If $b_j > 0$ and $\sum_j b_j$ diverges, then what can you say about $\sum_j 2^{-j} b_j$?

11. If $b_j > 0$ and $\sum_j b_j$ converges, then what can you say about $\sum_j b_j/j^2$?

12. If $a_j > 0$ and $\sum_j a_j^2$ converges, then what can you say about $\sum_j a_j^4$? How about $\sum_j a_j^3$?

13. Let α and β be positive real numbers. Discuss convergence and divergence for the series

$$\sum_{j=2}^{\infty} \frac{1}{j^{\alpha} \, |\log j|^{\beta}}.$$

*14. Let k be a positive integer. Discuss convergence or divergence for the series

$$\sum_{j=1}^{\infty} \frac{j^k}{2^j}.$$

15. If $b_j > 0$ and $\sum_j (1/(1 + b_j)$ converges, then what can you say about $\sum_j b_j$?

3.2 Elementary Convergence Tests

As previously noted, a series may converge because its terms are non-negative and diminish in size fairly rapidly (thus causing its partial sums to grow slowly) or it may converge because of cancellation among the terms. The tests which measure the first type of convergence are the most obvious and these are the "elementary" ones that we discuss in the present section.

Proposition 3.10: (The Comparison Test) *Suppose that $\sum_{j=1}^{\infty} a_j$ is a convergent series of nonnegative terms. If $\{b_j\}$ are real or complex numbers and if $|b_j| \le a_j$ for every j then the series $\sum_{j=1}^{\infty} b_j$ converges.*

Proof: Because the first series converges, its satisfies the Cauchy criterion for series. Hence, given $\epsilon > 0$, there is an N so large that if $n \ge m > N$ then

$$\left| \sum_{j=m}^{n} a_j \right| < \epsilon.$$

But then

$$\left| \sum_{j=m}^{n} b_j \right| \le \sum_{j=m}^{n} |b_j| \le \sum_{j=m}^{n} a_j < \epsilon.$$

It follows that the series $\sum b_j$ satisfies the Cauchy criterion for series. Therefore it converges.   □

Corollary 3.11: *If $\sum_{j=1}^{\infty} a_j$ is as in the proposition and if $0 \le b_j \le a_j$ for every j then the series $\sum_{j=1}^{\infty} b_j$ converges.*

Proof: Obvious. Simply notice that $|b_j| = b_j$.   □

Example 3.12: The series $\sum_{j=1}^{\infty} 2^{-j} \sin j$ is seen to converge by comparing it with the series $\sum_{j=1}^{\infty} 2^{-j}$.   □

Theorem 3.13: (The Cauchy Condensation Test) *Assume that $a_1 \ge a_2 \ge \cdots \ge a_j \ge \cdots 0$. The series*

$$\sum_{j=1}^{\infty} a_j$$

converges if and only if the series

$$\sum_{k=1}^{\infty} 2^k \cdot a_{2^k}$$

converges.

Proof: First assume that the series $\sum_{j=1}^{\infty} a_j$ converges. Notice that, for each $k \geq 1$,

$$2^{k-1} \cdot a_{2^k} = \underbrace{a_{2^k} + a_{2^k} + \ldots + a_{2^k}}_{2^{k-1} \text{ times}}$$

$$\leq a_{2^{k-1}+1} + a_{2^{k-1}+2} + \ldots + a_{2^k}.$$

$$= \sum_{m=2^{k-1}+1}^{2^k} a_m$$

Therefore

$$\sum_{k=1}^{N} 2^{k-1} \cdot a_{2^k} \leq \sum_{k=1}^{N} \sum_{m=2^{k-1}+1}^{2^k} a_m = \sum_{m=2}^{2^N} a_m.$$

Since the partial sums on the right are bounded (because the series of a_js converges), so are the partial sums on the left. It follows that the series

$$\sum_{k=1}^{\infty} 2^k \cdot a_{2^k}$$

converges.

For the converse, assume that the series

$$\sum_{k=1}^{\infty} 2^k \cdot a_{2^k} \tag{3.13.1}$$

converges. Observe that, for $k \geq 1$,

$$\sum_{m=2^{k-1}+1}^{2^k} a_j = a_{2^{k-1}+1} + a_{2^{k-1}+2} + \cdots + a_{2^k}$$

$$\leq \underbrace{a_{2^{k-1}} + a_{2^{k-1}} + \cdots + a_{2^{k-1}}}_{2^{k-1} \text{ times}}$$

$$= 2^{k-1} \cdot a_{2^{k-1}}.$$

It follows that

$$\sum_{m=2}^{2^N} a_j = \sum_{k=1}^{N} \sum_{m=2^{k-1}+1}^{2^k} a_m$$

$$\le \sum_{k=1}^{N} 2^{k-1} \cdot a_{2^{k-1}}.$$

By the hypothesis that the series (3.13.1) converges, the partial sums on the right must be bounded. But then the partial sums on the left are bounded as well. Since the summands a_j are nonnegative, the series on the left converges.   □

Example 3.14: We apply the Cauchy condensation test to the harmonic series

$$\sum_{j=1}^{\infty} \frac{1}{j}.$$

It leads us to examine the series

$$\sum_{k=1}^{\infty} 2^k \cdot \frac{1}{2^k} = \sum_{k=1}^{\infty} 1.$$

Since the latter series diverges, the harmonic series diverges as well.   □

Proposition 3.15: (Geometric Series) *Let α be a complex number. The series*

$$\sum_{j=0}^{\infty} \alpha^j$$

is called a geometric series. It converges if and only if $|\alpha| < 1$. In this circumstance, the sum of the series (that is, the limit of the partial sums) is $1/(1 - \alpha)$.

Proof: Let S_N denote the Nth partial sum of the geometric series. Then

$$\alpha \cdot S_N = \alpha(1 + \alpha + \alpha^2 + \ldots \alpha^N)$$
$$= \alpha + \alpha^2 + \ldots \alpha^{N+1}.$$

It follows that $\alpha \cdot S_N$ and S_N are nearly the same: in fact

$$\alpha \cdot S_N + 1 - \alpha^{N+1} = S_N.$$

Solving this equation for the quantity S_N yields

$$S_N = \frac{1 - \alpha^{N+1}}{1 - \alpha}$$

when $\alpha \neq 1$.

If $|\alpha| < 1$ then $\alpha^{N+1} \to 0$, hence the sequence of partial sums tends to the limit $1/(1 - \alpha)$. If $|\alpha| > 1$ then α^{N+1} diverges, hence the sequence of partial sums diverges. This completes the proof for $|\alpha| \neq 1$. But the divergence in case $|\alpha| = 1$ follows because the summands will not tend to zero. □

Example 3.16: The series

$$\sum_{j=0}^{\infty} 3^{-j}$$

is a geometric series. Writing it as

$$\sum_{j=0}^{\infty} \left(\frac{1}{3}\right)^j,$$

we see that the sum is

$$\frac{1}{1 - 1/3} = \frac{3}{2}.$$

The series

$$\sum_{j=2}^{\infty} \left(\frac{3}{4}\right)^j$$

is not quite a geometric series because the summation process does not begin at $j = 0$. But this situation is easily repaired. We write the series as

$$\left(\frac{3}{4}\right)^2 \sum_{j=0}^{\infty} \left(\frac{3}{4}\right)^j$$

and then we see that the sum is

$$\frac{9}{16} \cdot \frac{1}{1 - 3/4} = \frac{9}{16} \cdot 4 = \frac{9}{4}.$$ □

Corollary 3.17: *Let r be a real number. The series*

$$\sum_{j=1}^{\infty} \frac{1}{j^r}$$

converges if r exceeds 1 and diverges otherwise.

Proof: When $r > 1$ we can apply the Cauchy Condensation Test. This leads us to examine the series

$$\sum_{k=1}^{\infty} 2^k \cdot 2^{-kr} = \sum_{k=1}^{\infty} (2^{1-r})^k.$$

This last is a geometric series, with the role of α played by the quantity $\alpha = 2^{1-r}$. When $r > 1$ then $|\alpha| < 1$ so the series converges. Otherwise it diverges.   □

Example 3.18: The series

$$\sum_{j=1}^{\infty} \frac{1}{j^{3/2}}$$

converges because $3/2 > 1$.
    The series

$$\sum_{j=1}^{\infty} \frac{1}{j^{2/3}}$$

diverges because $2/3 < 1$.   □

Theorem 3.19: (The Root Test) *Consider the series*

$$\sum_{j=1}^{\infty} a_j.$$

If

$$\limsup_{j \to \infty} |a_j|^{1/j} < 1$$

then the series converges.

Proof: Refer again to the discussion of the concept of limit superior in Chapter 2. By our hypothesis, there is a number $0 < \beta < 1$ and an integer $N > 1$ such that, for all $j > N$, it holds that

$$|a_j|^{1/j} < \beta.$$

In other words,

$$|a_j| < \beta^j.$$

Since $0 < \beta < 1$ the sum of the terms on the right constitutes a convergent geometric series. By the Comparison Test, the sum of the terms on the left converges. □

Theorem 3.20: (The Ratio Test) *Consider a series*

$$\sum_{j=1}^{\infty} a_j.$$

If

$$\limsup_{j \to \infty} \left| \frac{a_{j+1}}{a_j} \right| < 1$$

then the series converges.

Proof: It is possible to supply a proof similar to that of the Root Test. We leave such a proof for the exercises, and instead supply an argument which relates the two tests in an interesting fashion.

Let

$$\lambda = \limsup_{j \to \infty} \left| \frac{a_{j+1}}{a_j} \right| < 1.$$

Select a real number μ such that $\lambda < \mu < 1$. By the definition of lim sup, there is an N so large that if $j > N$ then

$$\left| \frac{a_{j+1}}{a_j} \right| < \mu.$$

This may be rewritten as

$$|a_{j+1}| < \mu \cdot |a_j|, \quad j \geq N.$$

Thus (much as in the proof of the Root Test) we have for $k \geq 0$ that

$$\left|a_{N+k}\right| \le \mu \cdot \left|a_{N+k-1}\right| \le \mu \cdot \mu \cdot \left|a_{N+k-2}\right| \le \cdots \le \mu^k \cdot \left|a_N\right|.$$

It is convenient to denote $N + k$ by n, $n \ge N$. Thus the last inequality reads

$$\left|a_n\right| < \mu^{n-N} \cdot \left|a_N\right|$$

or

$$\left|a_n\right|^{1/n} < \mu^{(n-N)/n} \cdot \left|a_N\right|^{1/n}.$$

Remembering that N has been fixed once and for all, we pass to the lim sup as $n \to \infty$. The result is

$$\limsup_{n \to \infty} \left|a_n\right|^{1/n} \le \mu.$$

Since $\mu < 1$, we find that our series satisfies the hypotheses of the Root Test. Hence it converges.   □

Remark 3.21: The proof of the Ratio Test shows that *if* a series passes the Ratio Test then it passes the Root Test (the converse is not true, as you will learn in Exercise **2**). Put another way, the Root Test is a better test than the Ratio Test because it will give information whenever the Ratio Test does and also in some circumstances when the Ratio Test does not.

Why do we therefore learn the Ratio Test? The answer is that there are circumstances when the Ratio Test is easier to apply than the Root Test.

Example 3.22: The series

$$\sum_{j=1}^{\infty} \frac{2^j}{j!}$$

is easily studied using the Ratio Test (recall that $j! \equiv j \cdot (j - 1) \cdot \ldots 2 \cdot 1$). Indeed $a_j = 2^j/j!$ and

$$\left|\frac{a_{j+1}}{a_j}\right| = \frac{2^{j+1}/(j + 1)!}{2^j/j!}.$$

We can perform the division to see that

$$\left|\frac{a_{j+1}}{a_j}\right| = \frac{2}{j + 1}.$$

The lim sup of the last expression is 0. By the Ratio Test, the series converges.

Notice that in this example, while the Root Test applies in principle, it would be difficult to use in practice.   □

Example 3.23: We apply the Root Test to the series

$$\sum_{j=1}^{\infty} \frac{j^2}{2^j}.$$

Observe that

$$a_j = \frac{j^2}{2^j}$$

hence that

$$|a_j|^{1/j} = \frac{(j^{1/j})^2}{2}.$$

As $j \to \infty$, we see that

$$\lim_{j\to\infty} \sup |a_j|^{1/j} = \frac{1}{2}.$$

By the Root Test, the series converges.   □

It is natural to ask whether the Ratio and Root Tests can detect divergence. Neither test is necessary and sufficient: there are series which elude the analysis of both tests. However, the arguments that we used to establish Theorems 3.19 and 3.20 can also be used to establish the following (the proofs are left as exercises):

Theorem 3.24: (The Root Test for Divergence) *Consider the series*

$$\sum_{j=1}^{\infty} a_j$$

of nonzero terms. If

$$\lim_{j\to\infty} \inf |a_j|^{1/j} > 1$$

then the series diverges.

Theorem 3.25: (The Ratio Test for Divergence) *Consider the series*

$$\sum_{j=1}^{\infty} a_j.$$

If

$$\liminf_{j \to \infty} \left| \frac{a_{j+1}}{a_j} \right| > 1,$$

then the series diverges.

In both the Root Test and the Ratio Test, if the lim sup or lim inf is equal to 1, then no conclusion is possible. The exercises give examples of series, some of which converge and some of which do not, in which these tests give lim sup or lim inf equal to 1.

Example 3.26: Consider the series

$$\sum_{j=1}^{\infty} \frac{j!}{2^j}.$$

We apply the Ratio Test:

$$\frac{a_{j+1}}{a_j} = \frac{(j+1)!/2^{j+1}}{j!/2^j} = \frac{j+1}{2}.$$

This expression is > 2 for $j > 3$. Therefore, by the Ratio Test for Divergence, the series diverges. □

Example 3.27: Consider the series

$$\sum_{j=1}^{\infty} \frac{j^j}{4^j}.$$

We apply the Root Test:

$$\sqrt[j]{a_j} = \sqrt[j]{j^j/4^j} = j/4.$$

This expression is >2 for $j > 8$. Therefore the lim sup is >1 and the series diverges. □

We conclude this section by saying a word about the integral test.

Proposition 3.28: (The Integral Test) *Let f be a continuous function on $[0, \infty)$ that is monotonically decreasing. The series*

$$\sum_{j=1}^{\infty} f(j)$$

converges if and only if the integral

$$\int_{1}^{\infty} f(x)\,dx$$

converges.

We have not treated the integral yet in this book, so we shall not prove the result here.

Example 3.29: Consider the harmonic series

$$\sum_{j=1}^{\infty} \frac{1}{j}.$$

The terms of this series satisfy the hypothesis of the integral test. Also

$$\int_{1}^{\infty} \frac{1}{x}\,dx = \lim_{N \to +\infty} \log x \Big|_{1}^{N} = \lim_{N \to +\infty} [\log N - \log 1] = +\infty$$

Therefore the series diverges. □

Exercises

1. Let p be a polynomial with no constant term. If $b_j > 0$ for every j and if $\sum_{j=1}^{\infty} b_j$ converges then prove that the series $\sum_{j=1}^{\infty} p(b_j)$ converges.

2. Examine the series

$$\frac{1}{3} + \frac{1}{5} + \frac{1}{3^2} + \frac{1}{5^2} + \frac{1}{3^3} + \frac{1}{5^3} + \frac{1}{3^4} + \frac{1}{5^4} + \cdots$$

Prove that the Root Test shows that the series converges while the Ratio Test gives no information.

3. Check that both the Root Test and the Ratio Test give no information for the series $\sum_{j=1}^{\infty} \frac{1}{j}$, $\sum_{j=1}^{\infty} \frac{1}{j^2}$. However, one of these series is divergent and the other is convergent.

4. Let a_j be a sequence of real numbers. Define

$$m_j = \frac{a_1 + a_2 + \ldots a_j}{j}.$$

Prove that if $\lim_{j\to\infty} a_j = \ell$ then $\lim_{j\to\infty} m_j = \ell$. Give an example to show that the converse is not true.

5. Imitate the proof of the Root Test to give a direct proof of the Ratio Test.

6. Let $\sum_j a_j$ and $\sum_j b_j$ be series of positive terms. Prove that, if there is a constant $C > 0$ such that

$$\frac{1}{C} \le \frac{a_j}{b_j} \le C$$

for all j large, then either both series diverge or both series converge.

7. Prove that if a series of positive terms passes the Ratio Test, then it also passes the Root Test.

8. TRUE or FALSE: If the a_j are positive and $\sum a_j$ converges then $\sum a_j/j$ converges.

9. TRUE or FALSE: If a_j and b_j are positive and $\sum_j a_j$ and $\sum b_j$ both converge, then $\sum_j a_j b_j$ converges.

10. Prove Theorem 3.24.

11. Prove Theorem 3.25.

12. Derive the Raabe-Duhamel Test: Let $a_j > 0$. Set

$$b_j = j\left(\frac{a_j}{a_{j+1}} - 1\right).$$

Now let

$$L = \lim_{j\to\infty} b_j.$$

If $L > 1$ then the series $\sum_j a_j$ converges. If $L < 1$ then the series $\sum_j a_j$ diverges. If $L = 1$ then the test is inconclusive.

$$b_j = \log j\left(j\left(\frac{a_j}{a_{j+1}} - 1\right) - 1\right).$$

13. Derive Bertrand's Test: Let $a_j > 0$. Set

Now let

$$L = \lim_{j \to \infty} b_j.$$

If $L > 1$ then the series $\sum_j a_j$ converges. If $L < 1$ then the series $\sum_j a_j$ diverges. If $L = 1$ then the test is inconclusive.

14. Derive Gauss's Test: Let $a_j > 0$. Write

$$\frac{a_j}{a_{j+1}} = 1 + \frac{\alpha}{j} + \mathcal{E},$$

where $\mathcal{E}$ is an error of size $1/j^\beta$ and $\beta > 1$. Then $\sum_j a_j$ converges if $\alpha > 1$ and diverges if $\alpha \leq 1$.

3.3 Advanced Convergence Tests

In this section we consider convergence tests for series which depend on cancellation among the terms of the series. One of the most profound of these depends on a technique called *summation by parts*. You may wonder whether this process is at all related to the "integration by parts" procedure that you learned in calculus—it has a similar form. Indeed it will turn out (and we shall see the details of this assertion as the book develops) that summing a series and performing an integration are two aspects of the same limiting process. The summation by parts method is merely our first glimpse of this relationship.

Proposition 3.30: (Summation by Parts) Let $\{a_j\}_{j=0}^\infty$ and $\{b_j\}_{j=0}^\infty$ be two sequences of real or complex numbers. For $N = 0, 1, 2, \ldots$ set

$$A_N = \sum_{j=0}^{N} a_j$$

(we adopt the convention that $A_{-1} = 0$). Then, for any $0 \leq m \leq n < \infty$, it holds that

$$\sum_{j=m}^{n} a_j \cdot b_j = [A_n \cdot b_n - A_{m-1} \cdot b_m]$$
$$+ \sum_{j=m}^{n-1} A_j \cdot (b_j - b_{j+1}).$$

Proof: We write

$$\Sigma_{j=m}^{n} a_j \cdot b_j = \Sigma_{j=m}^{n} (A_j - A_{j-1}) \cdot b_j$$
$$= \Sigma_{j=m}^{n} A_j \cdot b_j - \Sigma_{j=m}^{n} A_{j-1} \cdot b_j$$
$$= \Sigma_{j=m}^{n} A_j \cdot b_j - \Sigma_{j=m-1}^{n-1} A_j \cdot b_{j+1}$$
$$= \Sigma_{j=m}^{n-1} A_j \cdot (b_j - b_{j+1}) + A_n \cdot b_n - A_{m-1} \cdot b_m.$$

This is what we wished to prove.   □

Now we apply summation by parts to prove a convergence test due to Niels Henrik Abel (1802–1829).

Theorem 3.31: (Abel's Convergence Test) *Consider the series*

$$\sum_{j=0}^{\infty} a_j \cdot b_j.$$

Suppose that

1. *The partial sums $A_N = \Sigma_{j=0}^{N} a_j$ form a bounded sequence;*
2. $b_0 \geq b_1 \geq b_2 \geq \dots;$
3. $\lim_{j \to \infty} b_j = 0.$

Then the original series

$$\sum_{j=0}^{\infty} a_j \cdot b_j$$

converges.

Proof: Suppose that the partial sums A_N are bounded in absolute value by a number K. Pick $\epsilon > 0$ and choose an integer N so large that $b_N < \epsilon/(2K)$. For $N < m \leq n < \infty$ we use the partial summation formula to write

$$|\Sigma_{j=m}^{n} a_j \cdot b_j| = |A_n \cdot b_n - A_{m-1} \cdot b_m + \Sigma_{j=m}^{n-1} A_j \cdot (b_j - b_{j+1})|$$
$$\leq K \cdot |b_n| + K \cdot |b_m| + K \cdot \Sigma_{j=m}^{n-1} |b_j - b_{j+1}|.$$

Now we take advantage of the facts that $b_j \geq 0$ for all j and that $b_j \geq b_{j+1}$ for all j to estimate the last expression by

$$K \cdot \left[b_n + b_m + \sum_{j=m}^{n-1} (b_j - b_{j+1}) \right].$$

[Notice that the expressions $b_j - b_{j+1}$, b_m, and b_n are all nonnegative.] Now the sum collapses and the last line is estimated by

$$K \cdot [b_n + b_m - b_n + b_m] = 2 \cdot K \cdot b_m.$$

By our choice of N the right side is smaller than ϵ. Thus our series satisfies the Cauchy criterion and therefore converges. □

Example 3.32: (The Alternating Series Test) As a first application of Abel's convergence test, we examine alternating series. Consider a series of the form

$$\sum_{j=1}^{\infty} (-1)^j \cdot b_j, \tag{3.32.1}$$

with $b_1 \geq b_2 \geq b_3 \geq \cdots \geq 0$ and $b_j \to 0$ as $j \to \infty$. We set $a_j = (-1)^j$ and apply Abel's test. We see immediately that all partial sums A_N are either -1 or 0. In particular, this sequence of partial sums is bounded. And the b_js are decreasing and tending to zero. By Abel's convergence test, the alternating series (3.32.1) converges. □

Proposition 3.33: *Let $b_1 \geq b_2 \geq \ldots$ and assume that $b_j \to 0$. Consider the alternating series $\sum_{j=1}^{\infty} (-1)^j b_j$ as in the last example. It is convergent: let S be its sum. Then the partial sums S_N satisfy $|S - S_N| \leq b_{N+1}$.*

Proof: Observe that

$$|S - S_N| = |b_{N+1} - b_{N+2} + b_{N+3} - +\ldots|.$$

But

$$b_{N+2} - b_{N+3} + -\ldots \leq b_{N+2} + (-b_{N+3} + b_{N+3})$$
$$+ (-b_{N+5} + b_{N+5}) +\ldots$$
$$= b_{N+2}$$

and

$$b_{N+2} - b_{N+3} + -\ldots \geq (b_{N+2} - b_{N+2}) + (b_{N+4} - b_{N+4}) + \ldots$$
$$= 0 .$$

It follows that

$$|S - S_N| \leq |b_{N+1}|$$

as claimed.   □

Example 3.34: Consider the series

$$\sum_{j=1}^{\infty} (-1)^j \frac{1}{j} .$$

Then the partial sum $S_{100} = -.688172$ is within 0.01 (in fact within $1/101$) of the full sum S and the partial sum $S_{10000} = -.6930501$ is within 0.0001 (in fact within $1/10001$) of the sum S.   □

Example 3.35: Next we examine a series which is important in the study of Fourier analysis. Consider the series

$$\sum_{j=1}^{\infty} \frac{\sin j}{j} . \tag{3.35.1}$$

We already know that the series $\sum \frac{1}{j}$ diverges. However, the expression $\sin j$ changes sign in a rather sporadic fashion. We might hope that the series (3.35.1) converges because of cancellation of the summands. We take $a_j = \sin j$ and $b_j = 1/j$. Abel's test will apply if we can verify that the partial sums A_N of the a_js are bounded. To see this we use a trick:

Observe that

$$\cos(j + 1/2) = \cos j \cdot \cos 1/2 - \sin j \cdot \sin 1/2$$

and

$$\cos(j - 1/2) = \cos j \cdot \cos 1/2 + \sin j \cdot \sin 1/2 .$$

Subtracting these equations and solving for $\sin j$ yields that

$$\sin j = \frac{\cos(j - 1/2) - \cos(j + 1/2)}{2 \cdot \sin 1/2} .$$

We conclude that

$$A_N = \sum_{j=1}^{N} a_j = \sum_{j=1}^{N} \frac{\cos(j - 1/2) - \cos(j + 1/2)}{2 \cdot \sin 1/2}.$$

Of course this sum collapses and we see that

$$A_N = \frac{-\cos(N + 1/2) + \cos 1/2}{2 \cdot \sin 1/2}.$$

Thus

$$|A_N| \le \frac{2}{2 \cdot \sin 1/2} = \frac{1}{\sin 1/2},$$

independent of N.

Thus the hypotheses of Abel's test are verified and the series

$$\sum_{j=1}^{\infty} \frac{\sin j}{j}$$

converges. □

Remark 3.36: It is interesting to notice that both the series

$$\sum_{j=1}^{\infty} \frac{|\sin j|}{j} \quad \text{and} \quad \sum_{j=1}^{\infty} \frac{\sin^2 j}{j}$$

diverge. The proofs of these assertions are left as exercises for you. □

We turn next to the topic of absolute and conditional convergence.

Definition 3.37: A series of real or complex numbers

$$\sum_{j=1}^{\infty} a_j$$

is said to be absolutely convergent if

$$\sum_{j=1}^{\infty} |a_j|$$

converges.

We have:

Proposition 3.38: *If the series $\sum_{j=1}^{\infty} a_j$ is absolutely convergent, then it is convergent.*

Proof: This is an immediate corollary of the Comparison Test.   □

Definition 3.39: A series $\sum_{j=1}^{\infty} a_j$ is said to be *conditionally convergent* if $\sum_{j=1}^{\infty} a_j$ converges, but it does not converge absolutely.

We see that absolutely convergent series are convergent but the next example shows that the converse is not true.

Example 3.40: The series

$$\sum_{j=1}^{\infty} \frac{(-1)^j}{j}$$

converges by the Alternating Series Test. However, it is not absolutely convergent because the harmonic series

$$\sum_{j=1}^{\infty} \frac{1}{j}$$

diverges.   □

There is a remarkable robustness result for absolutely convergent series that fails dramatically for conditionally convergent series. This result is enunciated in the next theorem. We first need a definition.

Definition 3.41: Let $\sum_{j=1}^{\infty} a_j$ be a given series. Let $\{p_j\}_{j=1}^{\infty}$ be a sequence in which every positive integer occurs once and only once (but not necessarily in the usual order). We call $\{p_j\}$ a *permutation* of the natural numbers.

Then the series

$$\sum_{j=1}^{\infty} a_{p_j}$$

is said to be a *rearrangement* of the given series.

Theorem 3.42: (Riemann, Weierstrass) *If the series $\sum_{j=1}^{\infty} a_j$ of real numbers is absolutely convergent and if the sum of the series is ℓ , then every rearrangement of the series converges also to ℓ.*

If the real series $\sum_{j=1}^{\infty} b_j$ is conditionally convergent and if β is any real number or $\pm\infty$ then there is a rearrangement of the series that converges to β.

Proof: We prove the first assertion here and explore the second in the exercises.

Let us choose a rearrangement of the given series and denote it by $\sum_{j=1}^{\infty} a_{p_j}$, where p_j is a permutation of the positive integers. Pick $\epsilon > 0$. By the hypothesis that the original series converges absolutely we may choose an integer $N > 0$ such that $N < m \leq n < \infty$ implies that

$$\sum_{j=m}^{n} |a_j| < \epsilon. \tag{3.42.1}$$

[The presence of the absolute values in the left side of this inequality will prove crucial in a moment.] Choose a positive integer M such that $M \geq N$ and the integers $1, \ldots, N$ are all contained in the list $p_1, p_2, \ldots, p_M$. If $K > M$ then the partial sum $\sum_{j=1}^{K} a_j$ will trivially contain the summands $a_1, a_2, \ldots a_N$. Also the partial sum $\sum_{j=1}^{K} a_{p_j}$ will contain the summands $a_1, a_2, \ldots a_N$. It follows that

$$\sum_{j=1}^{K} a_j - \sum_{j=1}^{K} a_{p_j}$$

will contain only summands *after* the Nth one in the original series. By inequality (3.42.1) we may conclude that

$$\left| \sum_{j=1}^{K} a_j - \sum_{j=1}^{K} a_{p_j} \right| \leq \sum_{j=N+1}^{\infty} |a_j| \leq \epsilon.$$

We conclude that the rearranged series converges; and it converges to the same sum as the original series. □

Exercises

1. If $1/2 > b_j > 0$ for every j and if $\sum_{j=1}^{\infty} b_j$ converges then prove that $\sum_{j=1}^{\infty} \frac{b_j}{1 - b_j}$ converges.

2. Follow these steps to give another proof of the Alternating Series Test: a) Prove that the odd partial sums form an increasing sequence; b) Prove that the even partial sums form a decreasing sequence; c) Prove that every even partial sum majorizes all subsequent odd partial sums; d) Use a pinching principle.

3. What can you say about the convergence or divergence of

$$\sum_{j=1}^{\infty} \frac{(2j + 3)^{1/2} - (2j)^{1/2}}{j^{3/4}}?$$

4. For which exponents k and ℓ does the series

$$\sum_{j=2}^{\infty} \frac{1}{j^k \mid \log j \mid^{\ell}}$$

converge?

5. Let p be a polynomial with integer coefficients and degree at least 1. Let $b_1 \geq b_2 \geq \cdots \geq 0$ and assume that $b_j \to 0$. Prove that if $(-1)^{p(j)}$ is not always positive and not always negative then in fact it will alternate in sign so that $\sum_{j=1}^{\infty} (-1)^{p(j)} \cdot b_j$ will converge.

6. Explain in words how summation by parts is analogous to integration by parts.

7. If $\gamma_j > 0$ and $\sum_{j=1}^{\infty} \gamma_j$ converges then prove that

$$\sum_{j=1}^{\infty} (\gamma_j)^{1/2} \cdot \frac{1}{j^{\alpha}}$$

converges for any $\alpha > 1/2$. Give an example to show that the assertion is false if $\alpha = 1/2$.

*8. Assume that $\sum_{j=1}^{\infty} b_j$ is a convergent series of positive real numbers. Let $s_j = \sum_{\ell=1}^{j} b_{\ell}$. Discuss convergence or divergence for the series $\sum_{j=1}^{\infty} s_j \cdot b_j$. Discuss convergence or divergence for the series $\sum_{j=1}^{\infty} \frac{b_j}{1 + s_j}$.

*9. If $b_j > 0$ for every j and if $\sum_{j=1}^{\infty} b_j$ diverges then define $s_j = \sum_{\ell=1}^{j} b_{\ell}$. Discuss convergence or divergence for the series $\sum_{j=1}^{\infty} \frac{b_j}{s_j}$.

*10. Let $\sum_{j=1}^{\infty} b_j$ be a rearrangement of conditionally convergent series conditionally convergent series of real numbers. Let β be a real number. Prove that there is a rearrangement of the series that converges to β. (**Hint:** First observe that the positive terms of the given series must form a divergent series. Also, the negative terms form a divergent series. Now build the rearrangement by choosing finitely many positive terms whose sum "just exceeds" β. Then add on enough negative terms so that the sum is "just less than" β. Repeat this oscillatory procedure.)

*11. Do Exercise 10 in the case that β is $\pm\infty$.

*12. Let $\sum_{j=1}^{\infty} a_j$ be a conditionally convergent series of complex num-
bers conditionally convergent series of complex numbers. Let $\mathscr{S}$
be the set of all possible complex numbers to which the various
rearrangements could converge. What forms can $\mathscr{S}$ have? (**Hint:**
Experiment!)

3.4 Some Special Series

We begin with a series that defines a special constant of mathematical
analysis.

Definition 3.43: The series

$$\sum_{j=0}^{\infty} \frac{1}{j!},$$

where $j! \equiv j \cdot (j - 1) \cdot (j - 2) \cdots 1$ for $j \geq 1$ and $0! \equiv 1$, is convergent (by the Ratio
Test, for instance). Its sum is denoted by the symbol e in honor of the Swiss
mathematician Léonard Euler, who first studied it (see also Example 2.47,
where the number e is studied by way of a sequence). We shall see in
Proposition 3.44 that these two approaches to the number e are equivalent.

Like the number π, to be considered later in this book, the number e is one
which arises repeatedly in a number of contexts in mathematics. It has
many special properties. We first relate the series definition of e to the
sequence definition:

Proposition 3.44: *The limit*

$$\lim_{n \to \infty} \left(1 + \frac{1}{n} \right)^n$$

exists and equals e.

Proof: We need to compare the quantities

$$A_N \equiv \sum_{j=0}^{N} \frac{1}{j!} \quad \text{and} \quad B_N \equiv \left(1 + \frac{1}{N} \right)^N.$$

We use the binomial theorem to expand B_N:

$$B_N = 1 + \frac{N}{1} \cdot \frac{1}{N} + \frac{N \cdot (N-1)}{2 \cdot 1} \cdot \frac{1}{N^2} + \frac{N \cdot (N-1) \cdot (N-2)}{3 \cdot 2 \cdot 1} \cdot \frac{1}{N^3}$$

$$+ \cdots + \frac{N}{1} \cdot \frac{1}{N^{N-1}} + 1 \cdot \frac{1}{N^N}$$

$$= 1 + 1 + \frac{1}{2!} \cdot \frac{N-1}{N} + \frac{1}{3!} \cdot \frac{N-1}{N} \cdot \frac{N-2}{N} + \cdots$$

$$+ \frac{1}{(N-1)!} \cdot \frac{N-1}{N} \cdot \frac{N-2}{N} \cdots \frac{2}{N}$$

$$+ \frac{1}{N!} \cdot \frac{N-1}{N} \cdot \frac{N-2}{N} \cdots \frac{1}{N}$$

$$= 1 + 1 + \frac{1}{2!} \cdot \left(1 - \frac{1}{N}\right) + \frac{1}{3!} \cdot \left(1 - \frac{1}{N}\right) \cdot \left(1 - \frac{2}{N}\right) + \cdots$$

$$+ \frac{1}{(N-1)!} \cdot \left(1 - \frac{1}{N}\right) \cdot \left(1 - \frac{2}{N}\right) \cdots \left(1 - \frac{N-2}{N}\right)$$

$$+ \frac{1}{N!} \cdot \left(1 - \frac{1}{N}\right) \cdot \left(1 - \frac{2}{N}\right) \cdots \left(1 - \frac{N-1}{N}\right).$$

Notice that every summand that appears in this last equation is positive. Thus, for $0 \le M \le N$,

$$B_N \ge 1 + 1 + \frac{1}{2!} \cdot \left(1 - \frac{1}{N}\right) + \frac{1}{3!} \cdot \left(1 - \frac{1}{N}\right) \cdot \left(1 - \frac{2}{N}\right)$$

$$+ \cdots + \frac{1}{M!} \cdot \left(1 - \frac{1}{N}\right) \cdot \left(1 - \frac{2}{N}\right) \cdots \left(1 - \frac{M-1}{N}\right).$$

In this last inequality we hold M fixed and Let N tend to infinity. The result is that

$$\liminf_{N \Rightarrow \infty} B_N \ge 1 + 1 + \frac{1}{2!} + \frac{1}{3!} + \cdots + \frac{1}{M!} = A_M.$$

Now, as $M \to \infty$, the quantity A_M converges to e (by the definition of e). So we obtain

$$\liminf_{N \to \infty} B_N \ge e . \qquad (3.44.1)$$

On the other hand, our expansion for B_N allows us to observe that $B_N \le A_N$. Thus

$$\limsup_{N \to \infty} B_N \le e. \qquad (3.44.2)$$

Combining (3.44.1) and (3.44.2) we find that

$$e \le \lim_{N \to \infty} \inf B_N \le \lim_{N \to \infty} \sup B_N \le e$$

hence that $\lim_{N \to \infty} B_N$ exists and equals e. This is the desired result. □

Remark 3.45: The last proof illustrates the value of the concepts of lim inf and lim sup. For we do not know in advance that the limit of the expressions B_N exists, much less that the limit equals e. However, the lim inf and the lim sup always exist. So we estimate those instead, and find that they are equal and that they equal e. □

The next result tells us how rapidly the partial sums A_N of the series defining e converge to e. This is of theoretical interest, but will also be applied to determine the irrationality of e.

Proposition 3.46: *With A_N as above, we have that*

$$0 < e - A_N < \frac{1}{N \cdot N!}.$$

Proof: Observe that

$$
\begin{aligned}
e - A_N &= \frac{1}{(N+1)!} + \frac{1}{(N+2)!} + \frac{1}{(N+3)!} + \cdots \\
&= \frac{1}{(N+1)!} \cdot \left(1 + \frac{1}{N+2} + \frac{1}{(N+2)(N+3)} + \cdots \right) \\
&< \frac{1}{(N+1)!} \cdot \left(1 + \frac{1}{N+1} + \frac{1}{(N+1)^2} + \cdots \right).
\end{aligned}
$$

Now the expression in parantheses is a geometric series. It sums to $(N+1)/N$. Since $A_N < e$ we have

$$e - A_N = |e - A_N|$$

hence

$$|e - A_N| < \frac{1}{N \cdot N!},$$

proving thg result. □

Next we prove that e is an irrational number.

Theorem 3.47: *Euler's number e is irrational.*

Proof: Suppose to the contrary that e is rational. Then $e = p/q$ for some positive integers p and q. By the preceding proposition,

$$0 < e - A_q < \frac{1}{q \cdot q!}$$

or

$$0 < (q! \cdot (e - A_q)) < \frac{1}{q}. \tag{3.47.1}$$

Now

$$e - A_q = \frac{p}{q} - \left(1 + 1 + \frac{1}{2!} + \frac{1}{3!} + \cdots + \frac{1}{q!}\right)$$

hence

$$q! \cdot (e - A_q)$$

is an integer. But then equation (3.47.1) says that this integer lies between 0 and $1/q$. In particular, this integer lies strictly between 0 and 1. That, of course, is impossible. So e must be irrational.   □

It is a general principle of number theory that a real number that can be approximated *too rapidly* by rational numbers (the degree of rapidity being measured in terms of powers of the denominators of the rational nqmbers) must be irrational. Under suitable conditions an even strongertranccendental numbers conclusion holds: namely, the number in question turns out to be *transcendental*. A transcendental number is one which is not the solution of any polynomial equation with integer coefficients.

The subject of transcendental numbers is explored in the exercises. The exercises also contain a sketch of a proof that e is transcendental. Transcendental numbers are quite difficult to study. It is known that π and e are transcendental. But it is not known whether π^e or e^π is transcendental.

In Exercise 7 of Section 2.4, we briefly discuss Euler's number γ. Both this special number and also the more commonly encountered number π arise in many contexts in mathematics. It is unknown whether *gamma* is rational or irrational. The number π is known to be transcendental, but it is unknown whether $\pi + e$ (where e is Euler's number) is transcendental. In recent years, questions about the irrationality and transcendence of various numbers have become a matter of practical inter?. For these properties prove to be useful io making and breaking secret codes, and in encrypting information so that it is accessible to some users but not to others.

In Example 1.1 we prove that

$$S_N \equiv \sum_{j=1}^{N} j = \frac{N \cdot (N + 1)}{2}.$$

We conclude this section with a method for summing higher powers of j.

Suppose that we wish to calculate

$$S_{k,N} \equiv \sum_{j=1}^{N} j^k$$

for some positive integer k exceeding 1. We may proceed as follows: Write

$$(j + 1)^{k+1} - j^{k+1} = \left[j^{k+1} + (k + 1) \cdot j^k + \frac{(k + 1) \cdot k}{2} \cdot j^{k-1} \right.$$
$$+ \cdots + \frac{(k + 1) \cdot k}{2} \cdot j^2 + (k + 1) \cdot j + 1 \Big]$$
$$- j^{k+1}$$
$$= (k + 1) \cdot j^k + \frac{(k + 1) \cdot k}{2} \cdot j^{k-1} + \cdots$$
$$+ \frac{(k + 1) \cdot k}{2} \cdot j^2 + (k + 1) \cdot j + 1.$$

Summing from $j = 1$ to $j = N$ yields

$$\sum_{j=1}^{N} \{(j + 1)^{k+1} - j^{k+1}\} = (k + 1) \cdot S_{k,N} + \frac{(k + 1) \cdot k}{2} \cdot S_{k-1,N} + \cdots$$
$$+ \frac{(k + 1) \cdot k}{2} \cdot S_{2,N} + (k + 1) \cdot S_{1,N} + N,$$

The sum on the left collapses to $(N + 1)^{k+1} - 1$. We may solve for $S_{k,N}$ and obtain

$$S_{k,N} = \frac{1}{k+1} \cdot [(N+1)^{k+1} - 1 - N - \frac{(k+1) \cdot k}{2} \cdot S_{k-1,N}.$$
$$- \cdots - \frac{(k+1) \cdot k}{2} \cdot S_{2,N} - (k+1) \cdot S_{1,N}].$$

We have succeeded in expressing $S_{k,N}$ in terms of $S_{1,N}, S_{2,N}, \ldots, S_{k-1,N}$. Thus we may inductively obtain formulas for $S_{k,N}$, any k. It turns out that

$$S_{1,N} = \frac{N(N+1)}{2}$$

$$S_{2,N} = \frac{N(N+1)(2N+1)}{6}$$

$$S_{3,N} = \frac{N^2(N+1)^2}{4}$$

$$S_{4,N} = \frac{(N+1)N(2N+1)(3N^2 + 3N - 1)}{30}.$$

These formulas are treated in further detail in the exercises.

Exercises

1. Use mathematical induction to prove the formulas provided in the text for the sum of the first N perfect squares, the first N perfect cubes, and the first N perfect fourth powers.

2. A real number s is called *algebraic* if it satisfies a polynomial equation of the form

$$a_0 + a_1 x + a_2 x^2 + \cdots + a_m x^m = 0$$

   with the coefficients a_j being integers and $a_m \neq 0$. Prove that if we replace the word "integers" in this definition with "rational numbers," then the set of algebraic numbers remains the same. Prove that $n^{p/q}$ is algebraic for any positive integers n, p, q.
   A number which is not algebraic is called *transcendental*.

3. Discuss convergence of $\sum_j 1/[\ln j]^k$ for k a positive integer.

4. Discuss convergence of $\sum_j 1/p(j)$ for p a polynomial.

5. Discuss convergence of $\sum_j exp(p(j))$ for p a polynomial.

*6. Refer to Exercise **2** for terminology. Prove that the sum (or difference) of two algebraic numbers is algebraic.

7. Refer to Exercise **6**. It is not known whether $\pi + e$ or $\pi - e$ is transcendental. But one of them must be. Explain.

*8. Refer to Exercise **2** for terminology. transcendental numbers A number is called *transcendental* if it is not algebraic. Prove that the number of algebraic numbers is countable. Explain why this implies that the number mf transcendental numbers is uncountable. Thus most real numbers are transcendental; however, it is extremely difficult to verify that any particular real number is transcendental.

*9. Refer to Exercise **2** for terminology. Provide the details ofthe following sketch of a proof that Euler's number e is transcendental. [**Note:** In this argument we use some simple ideas of calculus. These ideas will be treated in rigorous detail later in the book.] Seeking a contradiction, we suppose that the number e satisfies a polynomial equation of the form

$$a_0 + a_1 x + \cdots + a_m x^m = 0$$

with integer coefficients a_j.

a. We may assume that $a_0 \neq 0$.

b. Let p be an odd prime that will be specified later. Define

$$g(x) = \frac{x^{p-1}(x-1)^p \cdots (x-m)^p}{(p-1)!}$$

and

$$G(x) = g(x) + g^{(1)}(x) + g^{(2)}(x) + \cdots g^{(mp+p-1)}(x).$$

(Here parenthetical exponents denote derivatives.) Verify that

$$|g(x)| < \frac{m^{mp+p-1}}{(p-1)!}$$

for a suitable range of x.

c. Check that

$$\frac{d}{dx}\{e^{-x}G(x)\} = -e^{-x}g(x)$$

and thus that

$$a_j \int_0^j e^{-x} g(x)\,dx = a_j G(0) - a_j e^{-j} G(j) \,. \qquad (*)$$

d. Multiply the last equation by e^j, sum from $j = 0$ to $j = m$, and use the polynomial equation that e satisfies to obtain that

$$\sum_{j=0}^m a_j e^j \int_0^j e^{-x} g(x)\,dx = -\sum_{j=0}^m \sum_{i=0}^{mp+p-1} a_j g^{(i)}(j). \qquad (**)$$

e. Check that $g^{(i)}(j)$ is an integer for all values of i and all j from 0 to m inclusive.

f. Referring to the last step, show that in fact $g^{(i)}(j)$ is an integer divisible by p *except* in the case that $j = 0$ and $i = p - 1$.

g. Check that

$$g^{(p-1)}(0) = (-1)^p(-2)^p \cdots (-m)^p.$$

Conclude that $g^{(p-1)}(0)$ is not divisible by p if $p > m$.

h. Check that if $p > |a_0|$ then the right side of equation $(**)$ consists of a sum of terms each of which is a multiple of p *except* for the term $- a_0 g^{(p-1)}(0)$. It follows that the sum on the right side of $(**)$ is a nonzero integer.

i. Use equation $(*)$ to check that, provided p is chosen sufficiently large, the left side of $(**)$ satisfies

$$\left| \sum_{j=0}^m a_j e^j \int_0^j e^{-x} g(x)\,dx \right| \le \left\{ \sum_{j=0}^m |a_j| \right\} e^m \frac{(m^{m+2})^{p-1}}{(p-1)!} < 1.$$

j. The last two steps contradict each other.
    This proof is from [NIV].

*10. What can you say about the convergence of $\sum_j [\sin j]^2 / j$?

*11. What can you say about the convergence of $\sum_j [\sin j]^k / j$ for k a positive integer?

12. Verify directly that $\sqrt{2} + \sqrt{3}$ is an algebraic number.

13. Let m and n be positive integers which are not perfect squares. Prove that $\sqrt{m} + \sqrt{n}$ is an algebraic number.

14. Prove that $\sqrt{2} + \sqrt[3]{5}$ is an algebraic number.

3.5 Operations on Series

Some operations on series, such as addition, subtraction, and scalar multiplication, are straightforward. Others, such as multiplication, entail subtleties. This section treats all these matters.

Proposition 3.48: *Let*

$$\sum_{j=1}^{\infty} a_j \quad \text{and} \quad \sum_{j=1}^{\infty} b_j$$

be convergent series of real or complex numbers; assume that the series sum to limits α and β respectively. Then

a. *The series $\sum_{j=1}^{\infty} (a_j + b_j)$ converges to the limit $\alpha + \beta$.*

b. *If c is a constant then the series $\sum_{j=1}^{\infty} c \cdot a_j$ converges to $c \cdot \alpha$.*

Proof: We shall prove assertion **(a)** and leave the easier assertion **(b)** as an exercise.

Pick $\epsilon > 0$. Choose an integer N_1 so large that $n > N_1$ implies that the partial sum $S_n \equiv \sum_{j=1}^{n} a_j$ satisfies $|S_n - \alpha| < \epsilon/2$. Choose N_2 so large that $n > N_2$ implies that the partial sum $T_n \equiv \sum_{j=1}^{n} b_j$ satisfies $|T_n - \beta| < \epsilon/2$. If U_n is the nth partial sum of the series $\sum_{j=1}^{\infty} (a_j + b_j)$ and if $n > N_0 \equiv \max(N_1, N_2)$ then

$$|U_n - (\alpha + \beta)| \leq |S_n - \alpha| + |T_n - \beta| < \frac{\epsilon}{2} + \frac{\epsilon}{2} = \epsilon.$$

Thus the sequence $\{U_n\}$ converges to $\alpha + \beta$. This proves part (a). The proof of (b) is similar. □

In order to keep our discussion of multiplication of series as straightforward as possible, we deal at first with absolutely convergent series. It is convenient in this discussion to begin our sums at $j = 0$ instead of $j = 1$. If we wish to multiply

$$\sum_{j=0}^{\infty} a_j \quad \text{and} \quad \sum_{j=0}^{\infty} b_j ,$$

then we need to specify what the partial sums of the product series should be. An obvious necessary condition that we wish to impose is that, if the first series converges to α and the second converges to β, then the product series, whatever we define it to be, should converge to $\alpha \cdot \beta$.

The naive method for defining the summands of the product series $\sum_j c_j$ is to let $c_j = a_j \cdot b_j$. However, a glance at the product of two partial sums of the given series shows that such a definition would be ignoring the distributivity of multiplication over addition.

Cauchy's idea was that the summands for the product series should be

$$c_m \equiv \sum_{j=0}^{m} a_j \cdot b_{m-j}.$$

This particular form for the summands can be easily motivated using power series considerations (which we shall provide in Section 9.1). For now we concentrate on verifying that this "Cauchy product" of two series really works.

Theorem 3.49: Let $\sum_{j=0}^{\infty} a_j$ and $\sum_{j=0}^{\infty} b_j$ be two absolutely convergent series which converge to limits α and β respectively. Define the series $\sum_{m=0}^{\infty} c_m$ with summands $c_m = \sum_{j=0}^{m} a_j \cdot b_{m-j}$. Then the series $\sum_{m=0}^{\infty} c_m$ converges absolutely to $\alpha \cdot \beta$.

Proof: Let A_n, B_n, and C_n be the partial sums of the three series in question. We calculate that

$$C_n = (a_0 b_0) + (a_0 b_1 + a_1 b_0) + (a_0 b_2 + a_1 b_1 + a_2 b_0)$$
$$+ \ldots + (a_0 b_n + a_1 b_{n-1} + \cdots + a_n b_0)$$
$$= a_0 \cdot B_n + a_1 \cdot B_{n-1} + a_2 \cdot B_{n-2} + \cdots + a_n \cdot B_0.$$

We set $\lambda_n = B_n - \beta$, each n, and rewrite the last line as

$$C_n = a_0 (\beta + \lambda_n) + a_1 (\beta + \lambda_{n-1}) + \cdots a_n (\beta + \lambda_0)$$
$$= A_n \cdot \beta + [a_0 \lambda_n + a_1 \cdot \lambda_{n-1} + \cdots + a_n \cdot \lambda_0].$$

Denote the expression in square brackets by the symbol ρ_n. Suppose that we could show that $\lim_{n \to \infty} \rho_n = 0$. Then we would have

$$\lim_{n \to \infty} C_n = \lim_{n \to \infty} (A_n \cdot \beta + \rho_n)$$
$$= (\lim_{n \to \infty} A_n) \cdot \beta + (\lim_{n \to \infty} \rho_n)$$
$$= \alpha \cdot \beta + 0$$
$$= \alpha \cdot \beta.$$

Thus it is enough to examine the limit of the expressions ρ_n.

Since $\sum_{j=1}^{\infty} a_j$ is absolutely convergent, we know that $A = \sum_{j=1}^{\infty} |a_j|$ is a finite number. Choose $\epsilon > 0$. Since $\sum_{j=1}^{\infty} b_j$ converges to β it follows that $\lambda_n \to 0$. Thus we may choose an integer $N > 0$ such that $n > N$ implies that $|\lambda_n| < \epsilon$. Thus, for $n = N + k$, $k > 0$, we may estimate

$$
\begin{aligned}
|\rho_{N+k}| &\leq |\lambda_0 a_{N+k} + \lambda_1 a_{N+k-1} + \cdots + \lambda_N a_k| \\
&\quad + |\lambda_{N+1} a_{k-1} + \lambda_{N+2} a_{k-2} + \cdots + \lambda_{N+k} a_0| \\
&\leq |\lambda_0 a_{N+k} + \lambda_1 a_{N+k-1} + \cdots ' + \lambda_N a_k| \\
&\quad + \max_{p \geq 1}\{|\lambda_{N+p}|\} \cdot (|a_{k-1}| + |a_{k-2}| + \cdots + |a_0|) \\
&\leq (N + 1)\max_{\ell \geq k}|a_\ell| \cdot \max 0 \leq j \leq N |\lambda_j| + \epsilon \cdot A.
\end{aligned}
$$

In this last estimate, we have used the fact (for the first term in absolute values) that (a) there are $N + 1$ summands, (b) the a terms all have index at least k, and (c) the λ terms have index between 0 and N. The second term (the "max" term) is easy to estimate because of our bound on λ_n.

With N fixed, we let $k \to \infty$ in the last inequality. Since $\max_{\ell \geq k}|a_\ell| \to 0$, we find that

$$
\limsup_{n \to \infty} |\rho_n| \leq \epsilon \cdot A.
$$

Since $\epsilon > 0$ was arbitrary, we conclude that

$$
\lim_{n \to \infty} |\rho_n| \to 0.
$$

This completes the proof. □

Notice that, in the proof of the theorem, we really only used the fact that one of the given series was absolutely convergent, not that both were absolutely convergent. Some hypothesis of this nature is necessary, as the following example shows.

Example 3.50: Consider the Cauchy product of the two conditionally convergent series

$$
\sum_{j=0}^{\infty} \frac{(-1)^j}{\sqrt{j+1}} \quad \text{and} \quad \sum_{j=0}^{\infty} \frac{(-1)^j}{\sqrt{j+1}}.
$$

Observe that

$$c_m = \frac{(-1)^0 (-1)^m}{\sqrt{1}\sqrt{m+1}} + \frac{(-1)^1 (-1)^{m-1}}{\sqrt{2}\sqrt{m}} + \cdots$$
$$+ \frac{(-1)^m (-1)^0}{\sqrt{m+1}\sqrt{1}}$$
$$= \sum_{j=0}^{m} (-1)^m \frac{1}{\sqrt{(j+1)\cdot(m+1-j)}}.$$

However, for $0 \le j \le m$,

$$(j+1)\cdot(m+1-j) \le (m+1)\cdot(m+1) = (m+1)^2.$$

Thus

$$|c_m| \ge \sum_{j=0}^{m} \frac{1}{m+1} = 1.$$

We thus see that the terms of the series $\sum_{m=0}^{\infty} c_m$ do not tend to zero, so the series cannot converge.

Exercises

1. Calculate the Cauchy product of the series $\sum_j 1/j^3$ and the series $\sum_j 1/j^4$.

2. Explain how you could discover the Cauchy product using multiplication of polynomials.

3. Discuss the concept of composition of power series.

4. Let $\sum_{j=1}^{\infty} a_j$ and $\sum_{j=1}^{\infty} b_j$ be convergent series of positive real numbers. Discuss division of these two series. Use the idea of the Cauchy product.

5. Let $\sum_{j=1}^{\infty} a_j$ and $\sum_{j=1}^{\infty} b_j$ be convergent series of positive real numbers. Discuss convergence of $\sum_{j=1}^{\infty} a_j b_j$.

6. If $\sum_j a_j$ is a convergent series of positive terms and if $\sum_j b_j$ is a convergent series of positive terms, then what can you say about $\sum_j (a_j/b_j)$?

7. Prove Proposition 3.48(b).

*8. Explain division of power series in the language of the Cauchy product.

*9. Discuss the concept of the exponential of a power series.

*10. Is there a way to calculate the square root of a power series?

*11. Is there a way to calculate the logarithm of a power series?

12. Let $\sum_j a_j x^j$ be a power series that converges on the interval $(-1, 1)$. Let $f(x)$ be a bounded, continuous function on $(-1, 1)$. Discuss convergence of $\sum_j f(x) a_j x^j$.

13. Let $\sum_j a_j x^j$ be a power series that converges on the interval $(-1, 1)$. Let $g(x)$ be a bounded, continuous function on $(-1, 1)$ that is bounded from 0 by some constant c. Discuss convergence of $\sum_j a_j x^j / g(x)$.

4

Basic Topology

4.1 Open and Closed Sets

To specify a topology on a set is to describe certain subsets that will play the role of neighborhoods. These sets are called *open sets*. Our purpose here is to be able to study the "shape" of a set without worrying about its rigid properties. People like to joke that a mathematiciandoes not know a coffee cup from a donut because they both have the same shape—a loop with a hole in the middle. See Figure 4.1. The purpose of this chapter is to make these ideas precise.

In what follows, we will use "interval notation": If $a \leq b$ are real numbers then we define

$$(a, b) = \{x \in \mathbb{R} : a < x < b\},$$
$$[a, b] = \{x \in \mathbb{R} : a \leq x \leq b\},$$
$$[a, b) = \{x \in \mathbb{R} : a \leq x < b\},$$
$$(a, b] = \{x \in \mathbb{R} : a < x \leq b\}.$$

Intervals of the form (a, b) are called *open*. Those of the form $[a, b]$ are called *closed*. The other two are called *half-open* or *half-closed*. See Figure 4.2.

Now we extend the terms "open" and "closed" more general sets.

Definition 4.1: A set $U \subseteq \mathbb{R}$ is called *open* if, for each $x \in U$, there is an $\epsilon > 0$ such that the interval $(x - \epsilon, x + \epsilon)$ is contained in U. See Figure 4.3.

Remark 4.2: The interval $(x - \epsilon, x + \epsilon)$ is frequently termed a *neighborhood* of x. □

Example 4.3: The set $U = \{x \in \mathbb{R} : |x - 3| < 2\}$ is open. To see this, choose a point $x \in U$. Let $\epsilon = 2 - |x - 3| > 0$. [Notice that we are choosing ϵ to be the distance of x to the boundary of U. Then we claim that the interval $I = (x - \epsilon, x + \epsilon) \subseteq U$.

FIGURE 4.1
A coffee cup and a donut.

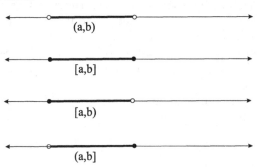

FIGURE 4.2
Intervals.

FIGURE 4.3
An open set.

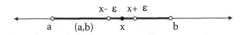

For, if $t \in I$, then

$$|t - 3| \leq |t - x| + |x - 3|$$
$$< \epsilon + |x - 3|$$
$$= (2 - |x - 3|) + |x - 3| = 2.$$

But this means that $t \in U$

We have shown that $t \in I$ implies $t \in U$. Therefore $I \subseteq U$. It follows from the definition that U is open.   □

Remark 4.4: The way to think about the definition of open set is that a set is open when none of its elements is at the "edge" of the set—each element is surrounded by other elements of the set, indeed a whole interval of them.

point at edge of set

FIGURE 4.4
A set that is not open.

See Figure 4.3 and contrast it with Figure 4.4. The remainder of this section will make these comments precise. □

Proposition 4.5: *If U_α are open sets, for α in some (possibly uncountable) index set A, then*

$$U = \bigcup_{\alpha \in A} U_\alpha$$

is open.

Proof: Let $x \in U$. By definition of union, the point x must lie in some U_α. But U_α is open. Therefore there is an interval $I = (x - \epsilon, \ x + \epsilon)$ such that $I \subseteq U_\alpha$. Therefore $I \subseteq U$. This proves that U is open. □

Proposition 4.6: *If $U_1, U_2, \ldots, U_k$ are open sets then the set*

$$V = \bigcap_{j=1}^{k} U_j$$

is also open.

Proof: Let $x \in V$. Then $x \in U_j$ for each j. Since each U_j is open there is for each j a positive number ϵ_j such that $I_j = (x - \epsilon_j, x + \epsilon_j)$ lies in U_j. Set $\epsilon = \min\{\epsilon_1, \ldots, \epsilon_k\}$. Then $\epsilon > 0$ and $(x - \epsilon, \ x + \epsilon) \subseteq U_j$ for every j. But that just means that $(x - \epsilon, \ x + \epsilon) \subseteq V$. Therefore V is open. □

Notice the difference between these two propositions: arbitrary unions of open sets are open. But, in order to guarantee that an intersection of open sets is still open, we had to assume that we were only intersecting finitely many such sets. If there were infinitely many sets then the minimum of the ϵ_j could be 0.

To understand this matter better, bear in mind the example of the open sets

$$U_j = \left(-\frac{1}{j}, \frac{1}{j}\right), \quad j = 1, 2, \ldots.$$

Each of the sets U_j is open, but the intersection of the sets U_j is the singleton $\{0\}$, which is *not* open.

The same analysis as in the first example shows that, if $a < b$, then the interval (a, b) is an open set. On the other hand, intervals of the form $(a, b]$ or $[a, b)$ or $[a, b]$ are *not* open. In the first instance, the point b is the center of no interval $(b - \epsilon, b + \epsilon)$ contained in $(a, b]$. Think about the other two intervals to understand why they are not open. We call intervals of the form (a, b) *open intervals*.

We are now in a position to give a complete description of all open sets.

Proposition 4.7: *Let $U \subseteq \mathbb{R}$ be a nonempty open set. Then there are either finitely many or countably many pairwise disjoint open intervals I_j such that*

$$U = \bigcup_{j=1}^{\infty} I_j.$$

See Figure 4.5.

Proof: Assume that U is an open subset of the real line. We define an equivalence relation on the set U. The resulting equivalence classes will be the open intervals I_j.

Let a and b be elements of U. We say that a is related to b if all real numbers between a and b are also elements of U. It is obvious that this relation is both reflexive and symmetric. For transitivity notice that if a is related to b and b is related to c then (assuming that a, b, c are distinct) one of the numbers a, b, c must lie between the other two. Assume for simplicity that $a < b < c$. Then all numbers between a and c lie in U, for all such numbers are either between a and b or between b and c or are b itself. Thus a is related to c. (The other possible orderings of a, b, c are left for you to consider.)

Thus we have an equivalence relation on the set U. Call the equivalence classes $\{U_\alpha\}_{\alpha \in A}$. We claim that each U_α is an open interval. In fact if a, b are elements of some U_α then all points between a and b are in U. But then a moment's thought shows that each of those "in between" points is related to both a and b. Therefore all points between a and b are elements of U_α. We conclude that U_α is an interval. Is it an *open* interval?

Let $x \in U_\alpha$. Then $x \in U$ so that there is an open interval $I = (x - \epsilon, x + \epsilon)$ contained in U. But x is related to all the elements of I; it follows that $I \subseteq U_\alpha$. Therefore U_α is open.

FIGURE 4.5
Structure of an open set.

We have exhibited the set U as a union of open intervals. These intervals are pairwise disjoint because they arise as the equivalence classes of an equivalence relation. Finally, each of these open intervals contains a (different) rational number (why?). Therefore there can be at most countably many of the intervals U_α.　　　　　　　　□

It is worth noting, and we shall learn more about this fact in Chapter 10, that there is no structure theorem for open sets (like the one that we just proved) in dimension 2 and higher. The geometry of Euclidean space gets *much* more complicated as the dimension increases. And real analysis in higher dimensions is a whole new subject that requires many new techniques.

Definition 4.8: A subset $F \subseteq \mathbb{R}$ is called *closed* if the complement $\mathbb{R} \setminus F$ is open. See Figure 4.6.

Example 4.9: The set $[0, 1]$ is closed. For its complement is

$$(-\infty, 0) \cup (1, \infty),$$

which is open.

Example 4.10: An interval of the form $[a, b] = \{x : a \le x \le b\}$ is closed. For its complement is $(-\infty, a) \cup (b, \infty)$, which is the union of two open intervals. The finite set $A = \{-4, -2, 5, 13\}$ is closed because its complement is

$$(-\infty, -4) \cup (-4, -2) \cup (-2, 5) \cup (5, 13) \cup (13, \infty),$$

which is open.
　　The set $B = \{1, 1/2, 1/3, 1/4, \ldots\} \cup \{0\}$ is closed, for its complement is the set

$$(-\infty, 0) \cup \left\{ \bigcup_{j=1}^{\infty} (1/(j + 1), 1/j) \right\} \cup (1, \infty),$$

which is open.

FIGURE 4.6
A closed set.

Verify for yourself that if the point 0 is omitted from the set B, then the set is no longer closed. □

Roughly speaking, a closed set is a set that contains all its limit points. An open set is just the opposite—open sets tend not to contain their limit points. The discussion below will make these ideas more precise.

Remark 4.11: A common mistake that students make is to suppose that every set is either open or closed. This is not true. For instance, the set $[0, 1) = \{x \in \mathbb{R} : 0 \leq x < 1\}$ is neither open nor closed. □

Proposition 4.12: *If E_α are closed sets, for α in some (possibly uncountable index set), then*

$$E = \underset{\alpha \in A}{\cap} E_\alpha$$

is closed.

Proof: This is just the contrapositive of Proposition 4.5 above: if U_α is the complement of E_α, each α, then U_α is open. Then $U = \cup U_\alpha$ is also open. But then

$$E = \cap E_\alpha = \cap{}^c(U_\alpha) = {}^c(\cup U_\alpha) = {}^c U$$

is closed. Here ${}^c S$ denotes the complement of a set S. □

The fact that the set B in the last example is closed, but that $B \setminus \{0\}$ is not, is placed in perspective by the next proposition.

Proposition 4.13: *Let S be a set of real numbers. Then S is closed if and only if every Cauchy sequence $\{(s_j\}$ of elements of S has a limit point which is also an element of S.*

Proof: First suppose that S is closed and let $\{s_j\}$ be a Cauchy sequence in S. We know. since the reals are complete, that there is an element $s \in \mathbb{R}$ such that $s_j \to s$. The point of this ha—f of the proof is to see that $s \in S$. If this statement were false then $s \in T = \mathbb{R} \setminus S$. But T must be open since it is the complement of a closed set. Thus there is an $\epsilon > 0$ such that the interval $I = (s - \epsilon, s + \epsilon) \subseteq T$. This means that no element of S lies in I. In particular, $|s - s_j| \geq \epsilon$ for every j. This contradicts the statement that $s_j \to s$. We conclude that $s \in S$.

Conversely, assume that every Cauchy sequence in S has its limit in S. If S were not closed then its complement would not be open. Hence there would be a point $t \in \mathbb{R} \setminus S$ with the property that no interval $(t - \epsilon, t + \epsilon)$ lies in $\mathbb{R} \setminus S$. In other words, $(t - \epsilon, t + \epsilon) \cap S \neq \emptyset$ for every $\epsilon > 0$. Thus, for $j = 1, 2, 3, \ldots$ we may choose a point $s_j \in (t - 1/j, t + 1/j) \cap S$. It follows

accumulation
point

FIGURE 4.7
The idea of an accumulation point.

that $\{s_j\}$ is a sequence of elements(of S that converges to $t \in \mathbb{R} \backslash S$. That contradicts our hypothesis. We conclude that S must be closed. $\quad\square$

Definition 4.14: Let S be a subset of $\mathbb{R}$. A point x is called an *accumulation point* of S if every neighborhood of x contains infinitely many distinct elements of S. See Figure 4.7. In particular, x is an accumulation point of S if it is the limit of a sequence of distinct elements in S.

The last proposition tells us that closed sets are characterized by the property that they contain all of their accumulation points.

Exercises

1. Let S be any set and let $\epsilon > 0$. Define $T = \{t \in \mathbb{R} : |t - s| < \epsilon$ for some $s \in S\}$. Prove that T is open.

2. Let S be any set and define $V = \{t \in \mathbb{R} : |t - s| \leq 1$ for somes $\in S\}$. Is V necessarily closed?

3. Let S be a set of real numbers. If S is not open then must it be closed? If S is lot closed then must it be open?

4. The *closure* of a set S is the intersection of all closed sets that contain S. Call a set S *robust* if it is the closure of its interior. Which sets of reals are robust?

5. Give an example of nonempty *closed sets* $X_1 \supseteq X_2 \supseteq \cdots$ such that $\cap_j X_j = \emptyset$.

6. Give an example of nonempty *closed* sets $X_1 \subset X_2 \cdots$ such that $\cup_j X_j$ is open.

7. Give an example of open sets $U_1 \supseteq U_2 \cdots$ such that $\cap_j U_j$ is closed and nonempty.

8. Exhibit a countable collection of open sets U_j such that each open set $O \subseteq \mathbb{R}$ can be written as a union of some of the sets U_j.

9. Let $S \subseteq \mathbb{R}$ be the rational numbers. Is S open? Is S closed?

10. Let S be an uncountable subset of $\mathbb{R}$. Prove that S must have infinitely many accumulation points. Must it have uncountably many?

*11. Let S be any set and define, for $x \in \mathbb{R}$,

$$\text{dis}(x, S) = \inf\{|x - s| : s \in S\}.$$

Prove that, if $x \notin \bar{S}$, then $\text{dis}(x, S) > 0$. If $x, y \in \mathbb{R}$ then prove that

$$|\text{dis}(x, S) - \text{dis}(y, S)| \le |x - y|.$$

*12. Every open subset of the real numbers is the countable pairwise disjoint union of open intervals. But there is no such structure theorem for closed sets. Explain why not.

13. The intersection of an open set and a closed set will, in general, be neither open nor closed. Explain why this is so.

14. The union of an open set and a closed set will, in general, be neither open nor closed. Explain why this is so.

4.2 Further Properties of Open and Closed Sets

Definition 4.15: Let $S \subseteq \mathbb{R}$ be a set. We call $b \in \mathbb{R}$ a *boundary point* of S if every nonempty neighborhood $(b - \epsilon, b + \epsilon)$ contains both points of S and points of $\mathbb{R} \backslash S$. See Figure 4.8. We denote the set of boundary points of S by ∂S.

A boundary point b might lie in S and might lie in the complement of S. The next example serves to illustrate the concept:

Example 4.16: Let S be the interval $(0, 1)$. Then no point of $(0, 1)$ is in the boundary of S since every point of $(0, 1)$ has a neighborhood that lies entirely inside $(0, 1)$. Also, no point of the complement of $T = [0, 1]$ lies in the boundary of S for a similar reason. Indeed, the only candidades for elements of the boundary of S are 0 and 1. See Figure 4.9. The point 0 *is* an element of the boundary since every neighborhood $(0 - \epsilon, 0 + \epsilon)$ contains the point $\epsilon/2 \in S$ and the point $-\epsilon/2 \in \mathbb{R} \backslash S$. A similar calculation shows that 1 lies in the boundary of S.

Now consider the set $T = [0, 1]$. Certainly there are no boundary points in the set $(0, 1)$, for the same reason as in the first paragraph. And there are no boundary points in $\mathbb{R} \backslash [0, 1]$, since the set is open. Thus the only candidates

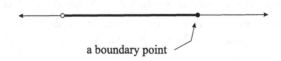

a boundary point

FIGURE 4.8
The idea of a boundary point.

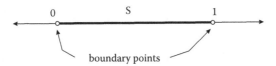

FIGURE 4.9
Boundary of the open unit interval.

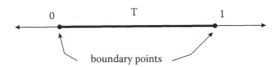

FIGURE 4.10
Boundary of the closed unit interval.

for elements of the boundary are 0 and 1. As in the first paragraph, these are both indeed boundary points for T. See Figure 4.10.

Notice that neither of the boundary points of S lie in S while both of the boundary points of T lie in T. □

The collection of all boundary points of a set S is called the *boundary* of S and is denoted by ∂S.

Example 4.17: The boundary of the set $\mathbb{Q}$ is the entire real line. For if x is any element of $\mathbb{R}$ then every interval $(x - \epsilon, x + \epsilon)$ contains both rational numbers and irrational numbers. □

The union of a set S with its boundary is called the *closure* of S, denoted by $\bar{S}$. The next example illustrates the concept.

Example 4.18: Let S be the set of rational numbers in the interval $[0, 1]$. Then the closure $\bar{S}$ of S is the entire interval $[0, 1]$.

Let T be the open interval $(0, 1)$. Then the closure $\bar{T}$ of T is the closed interval $[0, 1]$. □

Definition 4.19: Let $S \subseteq \mathbb{R}$. A point $s \in S$ is called an *interior point* of S if there is an $\epsilon > 0$ such that the interval $(s - \epsilon, s + \epsilon)$ lies in S. See Figure 4.11. We call the set of all interior points the *interior* of S, and we denote this set by $\overset{\circ}{S}$.

Definition 4.20: A point $t \in S$ is called an *isolated point* of S if there is an $\epsilon > 0$ such that the intersection of the interval $(t - \epsilon, t + \epsilon)$ with S is just the singleton $\{t\}$. See Figure 4.12.

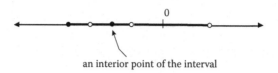

an interior point of the interval

FIGURE 4.11
The idea of an interior point.

isolated points

FIGURE 4.12
The idea of an isolated point.

By the definitions given here, an isolated point t of a set $S \subseteq \mathbb{R}$ is a boundary point. For any interval $(t - \epsilon, t + \epsilon)$ contains a point of S (namely, t itself) and points of $\mathbb{R} \setminus S$ (since t is isolated).

Proposition 4.21: *LetS $\subseteq \mathbb{R}$. Then each point of S is either an interior point or a boundary point of S.*

Proof: Fix $s \in S$. If s is not an interior point then no open interval centered at s contains only elements of s. Thus any interval centered at s contains an element of S (namely, s itself) and also contains points of $\mathbb{R} \setminus S$. Thus s is a boundary point of S. □

Example 4.22: Let $S = [0, 1]$. Then the interior points of S are the elements of $(0, 1)$. The boundary points of S are the points 0 and 1. The set S has no isolated points.

Let $T = \{1, 1/2, 1/3, ...\} \cup \{0\}$. Then the points 1, 1/2, 1/3, ... are isolated points of T. The point 0 is an accumulation point of T. Every element of T is a boundary point, and there are no others. □

Remark 4.23: Observe that the interior points of a set S are *elements* of S—by their very definition. Also isolated points of S are elements of S. However, a boundary point of S may or may not be an eLemenu of S.

If x is an accumulation point of S then every open neighborhood of x contains infinitely many elements of S. Hence x is either a boundary point of S or an interior point of S; it *cannot* be an isolated point of S. □

Proposition 4.24: *Let S be a subset of the real numbers. Then the boundary of S equals the boundary of $\mathbb{R} \setminus S$.*

Proof: If x is in the boundary of S, then any neighborhood of x contains points of S and points of ${}^cS$. Thus every neighborhood of x contains points of ${}^cS$ and points of S. So x is in the boundary of ${}^cS$. □

The next theorem allows us to use the concept of boundary to distinguish open sets from closed sets.

Theorem 4.25: *A closed set contains all of its boundary points. An open set contains none of its boundary points.*

Proof: Let S be closed and let x be an element of its boundary. If every neighborhood of x contains points of S *other than x itself* then x is an accumulation point of S hence $x \in S$. If not every neighborhood of x contains points of S other than x itself, then there is an $\epsilon > 0$ such that $\{(x - \epsilon, x) \cup (x, x + \epsilon)\} \cap S = \emptyset$. The only way that x can be an element of ∂S in this circumstance is if $x \in S$. That is what we wished to prove.

For the other half of the theorem notice that if T is open then ${}^cT$ is closed. But then ${}^cT$ will contain all its boundary points, which are the same as the boundary points of T itself (why is this true?). Thus T can contain none of its boundary points. □

Proposition 4.26: *Every nonisolated boundary point of a set S is an accumulation point of the set S.*

Proof: This proof is treated in the exercises. □

Definition 4.27: A subset S of the real numbers is called *bounded* if there is a positive number M such that $|s| \le M$ for every element s of S. See Figure 4.13.

The next result is one of the great theorems of nineteenth century analysis. It is essentially a restatement of the Bolzano–Weierstrass theorem of Section 2.2.

Theorem 4.28: (Bolzano–Weierstrass) *Every bounded, infinite subset of $\mathbb{R}$ has an accumulation point.*

Proof: Let S be a bounded, infinite set of real numbers. Let $\{a_j\}$ be a sequence of distinct elements of S. By Theorem 2.24, there is a subsequence $\{a_{j_k}\}$ that converges to a limit α. Then α is an accumulation point of S. □

FIGURE 4.13
A bounded set.

Corollary 4.29: *Let $S \subseteq \mathbb{R}$ be a nonempty, closed, and bounded set. If $\{a_j\}$ is any sequence in S, then there is a Cauchy subsequence $\{a_{j_k}\}$ that converges to an element of S.*

Proof: Merely combine the Bolzano–Weierstrass theorem with Proposition 4.13 of the last section.   □

Example 4.30: Consider the set $\{\sin j\}$. This set of real numbers is bounded by 1. By the Bolzano–Weierstrass theorem, it therefore has an accumulation point. So there is a sequence $\{\sin j_k\}$ that converges to some limit point p, even though it would be difficult to say precisely what that sequence is. □

****

KARL THEODOR WILHELM WEIERSTRASS

Karl Theodor Wilhelm Weierstrass (1815–1897) was a German mathematician. He was born in Ostenfelde, part of Ennigerloh, Province of Westphalia. Despite leaving university without a degree, he studied mathematics and trained as a school teacher, eventually teaching mathematics, physics, botany and gymnastics. He later received an honorary doctorate and became professor of mathematics in Berlin.

Among many other contributions, Weierstrass formalized the definition of the continuity of a function, proved the intermediate value theorem and the Bolzano–Weierstrass theorem, and used the latter to study the properties of continuous functions on closed bounded intervals. Of course he produced the Weierstrass nowhere differentiable function and proved the Weierstrass approximation theorem.

Weierstrass was the son of Wilhelm Weierstrass, a government official, and Theodora Vonderforst. His interest in mathematics began while he was a gymnasium student at the Theodorianum in Paderborn. He was sent to the University of Bonn upon graduation to prepare for a government position. His studies were to be in the fields of law, economics, and finance, and this conflicted with his hopes to study mathematics. He resolved the conflict by paying little heed to his planned course of study but continuing private study in mathematics. The outcome was that he left the university without a degree. He then studied mathematics at the Münster Academy (which was even then famous for mathematics) and his father was able to obtain a place for him in a teacher training school in Münster.

In 1843 he taught in Deutsch Krone in West Prussia and since 1848 he taught at the Lyceum Hosianum in Braunsberg. Besides mathematics he also taught physics, botany, and gymnastics.

After 1850 Weierstrass suffered from a long period of illness, but was able to publish mathematical articles that brought him fame and distinction. The University of Königsberg conferred an honorary doctor's degree on him on 31 March 1854. In 1864 he became professor at the Friedrich-Wilhelms-Universität Berlin, which later became the Humboldt Universität zu Berlin.

In 1870, at the age of fifty-five, Weierstrass met Sofia Kovalevsky whom he tutored privately after failing to secure her admission to the University. They had a fruitful intellectual, but troubled personal, relationship. The misinterpretation of this relationship and Kovalevsky's early death in 1891 was said to have contributed to Weierstrass' later ill health. He was immobile for the last three years of his life, and died in Berlin from pneumonia.

****

Exercises

1. Let S be any set of real numbers. Prove that $S \subseteq \bar{S}$. Prove that $\bar{S}$ is a closed set. Prove that $\bar{S} \setminus \overset{\circ}{S}$ is the boundary of S.

2. What is the interior of the set $S = \{1, 1/2, 1/3, \ldots\} \cup \{0\}$? What is the boundary of the set?

3. The union of infinitely many closed sets need not be closed. It need not be open either. Give examples to illustrate the possibilities.

4. The intersection of infinitely many open sets need not be open. It need not be closed either. Give examples to illustrate the possibilities.

5. Let S be any set of real numbers. Prove that $\overset{\circ}{S}$ is open. Prove that S is open if and only if S equals its interior.

6. Prove Proposition 4.26.

7. Let $S \subseteq \mathbb{R}$ be the rational numbers. What is the interior of S? What is the boundary of S? What is the closure of S?

*8. Give an example of a one-to-one, onto, continuous function f with a continuous inverse from the halfline $(0, \infty)$ to the full line $(-\infty, \infty)$.

*9. Give an example of a closed set in the plane (refer to Chapter 10) whose projection on the x-axis is not closed.

*10. Show that the projection of an open set in the plane (refer to Chapter 10) into the x-axis must be open.

11. What is the interior of the Cantor ternary set? What is the boundary of this set?

12. What is the boundary of the set of irrational numbers?

13. For each positive rational number q consider the interval $(q - q/2, q + q/2)$.

What is the union of all these intervals?

4.3 Compact Sets

Compact sets are sets (usually infinite) which share many of the most important properties of finite sets. They play an important role in real analysis.

Definition 4.31: A set $S \subseteq \mathbb{R}$ is called *compact* if every sequence in S has a subsequence that converges *to an element of S.*

Theorem 4.32: (Heine–Borel) *A set $S \subseteq \mathbb{R}$ is compact if and only if it is closed and bounded.*

Proof: That a closed, bounded set has the property of compactness is the content of Corollary 4.29 and Proposition 4.13.

Now let S be a set that is compact. If S is not bounded, then there is an element s_1 of S that has absolute value larger than 1. Also there must be an element s_2 of S that has absolute value larger than 2. Continuing, we find elements $s_j \in S$ satisfying

$$|s_j| > j$$

for each j. But then no subsequence of the sequence $\{sj\}$ can be Cauchy. This contradiction shows that S must be bounded.

If S is compact but S is not closed, then there is a point x which is the limit of a sequence $\{s_j\} \subseteq S$ but which is not itself in S. But every sequence into S is, by definition of "compact," supposed to have a subsequence converging an element of S. For the sequence $\{s_j\}$ that we are considering, x is the only candidate for the limit of a subsequence. Thus it must be that $x \in S$. That contradiction establishes that S is closed. □

In the abstract theory of topology (where there is no notion of distance), sequences cannot be used to characterize topological properties. Therefore a different definition of compactness is used. For interest's sake, and for future use, we now show that the definition of compactness that we have been discussing is equivalent to the one used in abstract topology theory. First we need a new definition.

an open cover of the set S

FIGURE 4.14
Open covers and compactness.

Definition 4.33: Let S be a subset of the real numbers. A collection of open of sets $\{O_\alpha\}_{\alpha \in A}$ (each O_α is an open set of real numbers) is called an *open covering* of S if

$$\bigcup_{\alpha \in A} O_\alpha \supseteq S.$$

See Figure 4.14.

Example 4.34: The collection $C = \{(1/j, 1)\}_{j=1}^{\infty}$ is an open covering of the interval $I = (0, 1)$. No finite subcollection of the elements of C covers I.

The collection $\mathcal{D} = \{(1/j, 1)\}_{j=1}^{\infty} \cup \{(-1/5, 1/5), (4/5, 6/5)\}$ is an open cov-ering of the interval $J = [0, 1]$. However, not all the elements $\mathcal{D}$ are actually needed to cover J. In fact

$$(-1/5, 1/5), \quad (1/6, 1), \quad (4/5, 6/5)$$

cover the interval J. □

It is the distinction displayed in this example that distinguishes compact sets from the point of view of topology. To understand the point, we need another definition:

Definition 4.35: If C is an open covering of a set S and if $\mathcal{D}$ is another open covering of S such that each element of $\mathcal{D}$ is also an element of C then we call $\mathcal{D}$ a *subcovering* of C.

We call $\mathcal{D}$ a *finite subcovering* if $\mathcal{D}$ has just finitely many elements

Example 4.36: The collection of intervals

$$C = \{(j - 1, j + 1)\}_{j=1}^{\infty}$$

is an open covering of the set $S = [5, 9]$. The collection

$$\mathcal{D} = \{(j - 1, j + 1)\}_{j=5}^{\infty}$$

is a subcovering.

However, the collection

$$\mathcal{E} = \{(4, 6), (5, 7), (6, 8), (7, 9), (8, 10)\}$$

is a *finite* subcovering.   □

Theorem 4.37: *A set $S \subseteq \mathbb{R}$ is compact if and only if every open covering $C = \{O_\alpha\}_{\alpha \in A}$ of S has a finite subcovering.*

Proof: Assume that S is a compact set and let $C = \{O_\alpha\}_{\alpha \in A}$ be an open covering of S.

By Theorem 4.29, S is closed and bounded. Therefore it holds that $a = \inf S$ is a finite real number, and an element of S. Likewise, $b = \sup S$ is a finite real number and an element of S. Write $I = [a, b]$. The case $a = b$ is trivial so we assume that $a < b$.

Set

$$\mathcal{A} = \{x \in I : C \text{ contains a finite subcover that covers } S \cap [a, x]\}.$$

Then $\mathcal{A}$ is nonempty since $a \in \mathcal{A}$. Let $t = \sup \mathcal{A}$. Then some element O_0 of C contains t. Let s be an element of O_0 to the left of t. Then, by the definition of t, s is an element of $\mathcal{A}$. So there is a finite subcovering C' of C that covers $S \cap [a, s]$. But then $\mathcal{D} = C' \cup \{O_0\}$ covers $S \cap [a, t]$, showing that $t = \sup \mathcal{A}$ lies in $\mathcal{A}$. But in fact $\mathcal{D}$ even covers points to the right of t. Thus t cannot be the supremum of $\mathcal{A}$ unless $t = b$.

We have learned that t must be the point b itself and that therefore $b \in \mathcal{A}$. But that says that $S \cap [a, b] = S$ can be covered by finitely many of the elements of C. That is what we wished to prove.

For the converse, assume that every open covering of S has a finite subcovering. Let $\{a_j\}$ be a sequence in S. Assume, seeking a contradiction, that the sequence has no subsequence that converges to an element of S. This must mean that for every $s \in S$ there is an $\epsilon_s > 0$ such that no element of the sequence satisfies $0 < |a_j - s| < \epsilon_s$. Let $I_s = (s - \epsilon_s, s + \epsilon_s)$. The collection $C = \{I_s\}$ is then an open covering of the set S. By hypothesis, there exists a finite subcov-ering $I_{s1}, \ldots I_{sk}$ of open intervals that cover S. But then $S \subseteq \cup_{j=1}^{k} I_{sj}$ contains no element of the sequence $\{a_j\}$, and that is a contradiction.   □

Example 4.38: If $A \subseteq B$ and both sets are nonempty then $A \cap B = A \neq \emptyset$. A similar assertion holds when intersecting *finitely many* nonempty sets $A_1 \supseteq A_2 \supseteq \cdots \supseteq A_k$; it holds in this circumstance that $\cap_{j=1}^{k} A_j = A_k$.

However, it is possible to have infinitely many nonempty nested sets with null intersection. An example is the sets $I_j = (0, 1/j)$. We see that $I_1 \supseteq I_2 \supseteq I_3 \supseteq \cdots$ yet

$$\bigcap_{j=1}^{\infty} I_j = \varnothing.$$

By contrast, if we take $K_j = [0, 1/j]$ then

$$\bigcap_{j=1}^{\infty} K_j = \{0\}.$$

The next proposition shows that compact sets have the intuitively appealing property of the K_js rather than the unsettling property of the I_js. □

Proposition 4.39: *Let*

$$K_1 \supseteq K_2 \supseteq \cdots \supseteq K_j \supseteq \ldots$$

be nonempty compact sets of real numbers. Set

$$\mathcal{K} = \bigcap_{j=1}^{\infty} K_j.$$

Then $\mathcal{K}$ is compact and $\mathcal{K} \neq \varnothing$.

Proof: Each K_j is closed and bounded hence $\mathcal{K}$ is closed and bounded. Thus $\mathcal{K}$ is compact. Let $x_j \in K_j$, each j. Then $\{x_j\} \subseteq K_1$. By compactness, there is a convergent subsequence $\{x_{jk}\}$ with limit $x_0 \in K_1$. However, $\{x_{jk}\}_{k=2}^{\infty} \subseteq K_2$. Thus $x_0 \in K_2$. Similar reasoning shows that $x_0 \in K_m$ for all $m = 1, 2, \ldots$. In conclusion, $x_0 \in \cap_j \mathcal{K}_j = \mathcal{K}$. □

Exercises

1. Prove that the intersection of a compact set and a closed set is compact.

2. Let K be a compact set and let U be an open set that contains K. Prove that there is an $\epsilon > 0$ such that, if $k \in K$, then the interval $(k - \epsilon, k + \epsilon)$ is contained in U.

3. Let K be compact and L closed, and assume that the two sets are disjoint. Show that there is a positive distance between the two sets.

4. Let K be a compact set. Let $\delta > 0$. Prove that there is a finite collection of intervals of radius δ that covers K.

5. Let K be a compact set. Let $\mathcal{U} = \{U_j\}_{j=1}^{k}$ be a finite open covering of K. Show that there is a $\delta > 0$ so that, if x is any point of K, then the disc or interval of center x and radius δ lies entirely in one of the U_j.

6. Assume that we have intervals $[a_1, b_1] \supseteq [a_2, b_2] \supseteq \cdots$, each of positive length, and that $\lim_{j \to \infty} |a_j - b_j| = 0$. Prove that there is a point x such that $x \in [a_j, b_j]$ for every j.

7. Let $K \subseteq \mathbb{R}$ be a compact set. Let $\epsilon > 0$. Define

$$\hat{K} = \{t \in \mathbb{R} : |t - k| \le \epsilon \text{ for some } k \in K\}.$$

Is $\hat{K}$ compact? Why or why not?

8. If K in $\mathbb{R}$ is compact then show that $^c K$ is not compact.

9. Prove that the intersection of any number of compact sets is compact. The analogous statement for unions is false.

10. Let $U \subset \mathbb{R}$ be any open set. Show that there exist compact sets $K_1 \subset K_2 \subset \cdots$ so that $\cup_j K_j = U$.

11. Produce an open set U in the real line so that U may not be written as the decreasing intersection of compact sets.

12. Let K be a compact set and E a closed set. Assume that these sets are disjoint. Define

$$d = \inf\{|k - e| : k \in K, e \in E\}.$$

Prove that d is a positive number.

13. Prove that the result of Exercise 12 is false for two closed sets.

14. Prove that the result of Exercise 12 is false for two open sets.

4.4 The Cantor Set

In this section we describe the construction of a remarkable subset of $\mathbb{R}$ with many pathological properties. It only begins to suggest the richness of the structure of the real number system.

We begin with the unit interval $S_0 = [0, 1]$. We extract from S_0 its open middle third; thus $S_1 = S_0 \setminus (1/3, 2/3)$. Observe that S_1 consists of two closed intervals of equal length $1/3$. See Figure 4.15.

Now we construct S_2 from S_1 by extracting from each of its two intervals the middle third: $S_2 = [0, 1/9] \cup [2/9, 1/3] \cup [6/9, 7/9] \cup [8/9, 1]$. Figure 4.16 shows S_2.

Continuing in this fashion, we construct S_{j+1} from S_j by extracting the middle third from each of its component subintervals. We define the Cantor set C to be

FIGURE 4.15
Construction of the Cantor set.

FIGURE 4.16
Second step in the construction of the Cantor set.

$$C = \bigcap_{j=1}^{\infty} S_j$$

Notice that each of the sets S_j is closed and bounded, hence compact. By Proposition 4.39 of the last section, C is therefore not empty (one can also note that the endpoints of the removed intervals are all in the Cantor set). The set C is closed and bounded, hence compact.

Proposition 4.40: *The Cantor set C has zero length, in the sense that the complementary set $[0, 1]\backslash C$ has length 1.*

Proof: In the construction of S_1, we removed from the unit interval one interval of length 3^{-1}. In constructing S_2, we further removed two intervals of length 3^{-2}. In constructing S_j, we removed 2^{j-1} intervals of length 3^{-j}. Thus the total length of the intervals removed from the unit interval is

$$\sum_{j=1}^{\infty} 2^{j-1} \cdot 3^{-j}.$$

This last equals

$$\frac{1}{3} \sum_{j=1}^{\infty} \left(\frac{2}{3}\right)^j.$$

The geometric series sums easily and we find that the total length of the intervals removed is

$$\frac{1}{3}\left(\frac{1}{1 - 2/3}\right) = 1.$$

Thus the Cantor set has length zero because its complement in the unit interval has length one. □

Proposition 4.41: *The Cantor set is uncountable.*

Proof: We assign to each element of the Cantor set a "label" consisting of a sequence of 0s and 1s that identifies its location in the set.

Fix an element x in the Cantor set. Then x is in S_1. If x is in the left half of S_1, then the first digit in the "label" of x is 0; otherwise it is 1. See Figure 4.17.

Likewise $x \in S_2$. By the first part of this argument, it is either in the left half S_2^0 of S_2 (when the first digit in the label is 0) or the right half S_2^1 of S_2 (when the first digit of the label is 1). Whichever of these is correct, that half will consist of two intervals of length 3^{-2}. If x is in the leftmost of these two intervals then the second digit of the "label" of x is 0. Otherwise the second digit is 1. Refer to Figure 4.18.

Continuing in this fashion, we may assign to x an infinite sequence of 0s and 1s.

Conversely, if a, b, c, ... is a sequence of 0s and 1s, then we may locate a unique corresponding element y of the Cantor set. If the first digit is a zero then y is in the left half of S_1; otherwise y is in the right half of S_1. Likewise the second digit locates y within S_2, and so forth.

Thus we have a one-to-one correspondence between the Cantor set and the collection of all infinite sequences of zeroes and ones. [Notice that we are in effect thinking of the point assigned to a sequence $c_1 c_2 c_3$ ... of 0s and 1s as the limit of the points assigned to c_1, $c_1 c_2$, $c_1 c_2 c_3$, ... Thus we are using the fact that C is closed.] However, as we can learn in the Appendix at the end of the book, the set of all infinite sequences of zeroes and ones is un-countable. Thus we see that the Cantor set is uncountable.   □

FIGURE 4.17
The first digit of the label of a point in the Cantor set.

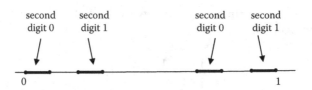

FIGURE 4.18
The second digit of the label of a point in the Cantor set.

The Cantor set is quite thin (it has zero length) but it is large in the sense that it has uncountably many elements. Also it is compact. The next result reveals a surprising, and not generally well known, property of this "thin" set.

Theorem 4.42: *Let C be the Cantor set and define*

$$S = \{x + y : x \in C, y \in C\}.$$

Then $S = [0, 2]$.

Proof: We sketch the proof here and treat the details in the exercises.

Since $C \subseteq [0, 1]$ it is clear that $S \subseteq [0, 2]$. For the reverse inclusion, fix an element $t \in [0, 2]$. Our job is to find two elements c and d in C such that $c + d = t$.

First observe that $\{x + y : x \in S_1, y \in S_1\} = [0, 2]$. Therefore there exist $x_1 \in S_1$ and $y_1 \in S_1$ such that $x_1 + y_1 = t$.

Similarly, $\{x + y : x \in S_2, y \in S_2\} = [0, 2]$. Therefore there exist $x_2 \in S_2$ and $y_2 \in S_2$ such that $x_2 + y_2 = t$.

Continuing in this fashion we may find for each j numbers x_j and y_j such that $x_j, y_j \in S_j$ and $x_j + y_j = t$. Of course $\{x_j\} \subseteq C$ and $\{y_j\} \subseteq C$ hence there are subsequences $\{x_{jk}\}$ and $\{y_{jk}\}$ which converge to real numbers c and d respectively. Since C is compact, we can be sure that $c \in C$ and $d \in C$. But the operation of addition respects limits, thus we may pass to the limit as $k \to \infty$ in the equation

$$x_{jk} + y_{jk} = t$$

to obtain

$$c + d = t.$$

Therefore $[0, 2] \subseteq \{x + y : x \in C\}$. This completes the proof. □

In the exercises at the end of the section we shall explore constructions of other Cantor sets, some of which have zero length and some of which have positive length. The Cantor set that we have discussed in detail in the present section is sometimes distinguished with the name "the Cantor ternary set." We shall also consider in the exercises other ways to construct the Cantor ternary set.

Observe that, whereas any open set is the countable or finite disjoint union of open intervals, the existence of the Cantor set shows us that there is no such structure theorem for closed sets. That is to say, we cannot hope to write an arbitrary closed set as the disjoint union of closed intervals.

[However, de Morgan's Law shows that an arbitrary closed set can be written as the countable intersection of sets, each of which is the union of disjoint closed intervals.] In fact closed intervals are atypically simple when considered as examples of closed sets.

Exercises

1. Construct a Cantor-like set by removing the middle *fifth* from the unit interval, removing the middle fifth of each of the remaining intervals, and so on. What is the length of the set that you construct in this fashion?
   Is it uncountable? Is it different from the Cantor set constructed in the text?

2. Refer to Exercise 1. Construct a Cantor set by removing, at the jth step, a middle subinterval of length 3^{-2j+1} from each existing interval. The Cantor-like set that results should have positive length. What is that length? Does this Cantor set have the other properties of the Cantor set constructed in the text?

3. Prove that it is not the case that there is a positive distance between two disjoint open sets.

4. Let $0 < \lambda < 1$. Imitate the construction of the Cantor set to produce a subset of the unit interval whose complement has length λ.

5. How many endpoints of the intervals in the S_j are there in the Cantor set? How many non-endpoints?

6. What is the interior of the Cantor set? What is the boundary?

7. Fix the sequence $a_j = 3^{-j}, j = 1, 2, \ldots$ Consider the set S of all sums

$$\sum_{j=1}^{\infty} \mu_j a_j,$$

   where each μ_j is one of the numbers 0 or 2. Show that S is the Cantor set. If s is an element of S, $s = \sum \mu_j a_j$, and if $\mu_j = 0$ for all j sufficiently large, then show that s is an endpoint of one of the intervals in one of the sets S_j that were used to construct the Cantor set in the text.

8. Let us examine the proof that $\{x + y : x \in C, y \in C\}$ equals $[0, 2]$ more carefully.

   a. Prove for each j that $\{x + y : x \in S_j, y \in S_j\}$ equals the interval $[0, 2]$.

   b. For $t \in [0, 1]$, explain how the subsequences $\{x_{jk}\}$ and $\{y_{jk}\}$ in S_j can be chosen to satisfy $x_{jk} + y_{jk} = t$ and so that $x_{jk} \to x_0 \in C$ and $y_{jk} \to y_0 \in C$. Observe that it is important for the proof that the index j_k be the same for both subsequences.

    c. Formulate a suitable statement concerning the assertion that the binary operation of addition "respects limits" as required in the argument in the text. Prove this statement and explain how it allows us to pass to the limit in the equation $x_{jk} + y_{jk} = t$.

9. Use the characterization of the Cantor set from Exercise **8** to give a new proof of the fact that $\{x + y : x \in C, y \in C\}$ equals the interval $[0, 2]$.

10. How many points in the Cantor set have finite ternary expansions? How many have infinite ternary expansions?

*11. Discuss which sequences a_j of positive numbers could be used as in Exercise **7** to construct sets which are like the Cantor set.

*12. Describe how to produce a two-dimensional Cantor-like set in the plane.

13. When we construct the Cantor set we remove open intervals from the interval $[0, 1]$. Suppose we instead attempted to construct a Cantor-like set by removing closed intervals from the interval $(0, 1)$? Why is this program doomed to failure?

**14. Let C be the Cantor ternary set and construct a new set

$$E = \{c - d : c \in C, d \in C\}.$$

Can you give an explicit description of E? [**Hint:** Begin by subtracting S_1 from S_1. Then subtract S_2 from S_2. This should give you some idea of what is going on.

4.5 Connected and Disconnected Sets

Definition 4.43: Let S be a set of real numbers. We say that S is *disconnected* if it is possible to find a pair of open sets U and V such that

$$U \cap S \neq \varnothing, \, V \cap S \neq \varnothing,$$

$$(U \cap S) \cap (V \cap S) = \varnothing,$$

and

$$S = (U \cap S) \cup (V \cap S).$$

a disconnected set

FIGURE 4.19
The idea of disconnected.

See Figure 4.19. If no such U and V exist then we call S *connected*.

Example 4.44: The set $T = \{x \in \mathbb{R} : |x| < 1, x \neq 0\}$ is disconnected. Take $U = \{x : x < 0\}$ and $V = \{x : x > 0\}$. Then

$$U \cap T = \{x : -1 < x < 0\} \neq \emptyset$$

and

$$V \cap T = \{x : -0 < x < 1\} \neq \emptyset$$

Also $(U \cap T) \cap (V \cap T) = \emptyset$. Clearly $T = (U \cap T) \cup (V \cap T)$, hence T is disconnected. □

It is clear that a disconnected set has the property that there are disjoint open sets that, in effect, *disconnect* the set. The next example looks at the contrapositive.

Example 4.45: The set $X = [-1, 1]$ is connected. To see this, suppose to the contrary that there exist open sets U and V such that $U \cap X \neq \emptyset, V \cap X \neq \emptyset, (U \cap X) \cap (V \cap X) = \emptyset$, and

$$X = (U \cap X) \cup (V \cap X).$$

Choose $a \in U \cap X$ and $b \in V \cap X$. Set

$$\alpha = \sup(U \cap [a, b]\}).$$

Now $[a, b] \subseteq X$ hence $U \cap [a, b]$ is disjoint from V. Thus $\alpha \leq b$. But ^cV is closed hence $\alpha \notin V$. It follows that $\alpha < b$.

If $\alpha \in U$ then, because U is open, there exists an $\tilde{\alpha} \in U$ such that $\alpha < \tilde{\alpha} < b$. This would mean that we chose α incorrectly. Hence $\alpha \notin U$. But $\alpha \notin U$ and $\alpha \notin V$ means $\alpha \notin X$. On the other hand, α is the supremum of a subset of X (since $a \in X$, $b \in X$, and X is an interval). Since X is a closed interval, we conclude that $\alpha \in X$. This contradiction shows that X must be connected. □

With small modifications, the discussion in the last example demonstrates that any closed interval is connected (Exercise 1). See Figure 4.20. Also

FIGURE 4.20
A closed interval is connected.

(see Exercise 2), we may similarly see that any open interval or half-open interval is connected. In fact the converse is true as well:

Theorem 4.46: *A subset S of $\mathbb{R}$ is connected if and only if S is an interval.*

Proof: If S is not an interval then there exist $a \in S, b \in S$ and a point t between a and b such that $t \notin S$. Define $U = \{x \in \mathbb{R} : x < t\}$ and $V = \{x \in \mathbb{R} : t < x\}$. Then U and V are open and disjoint, $U \cap S \neq \varnothing$, $V \cap S \neq \varnothing$, and

$$S = (U \cap S) \cup (V \cap S)$$

Thus S is disconnected.

   If S is an interval then we prove that it is connected using the methodology of Example 4.45. □

   The Cantor set is not connected; indeed it is disconnected in a special sense.

   Call a set S *totally disconnected* if, for each distinct $x \in S, y \in S$, there exist disjoint open sets U and V such that $x \in U, y \in V$, and $S = (U \cap S) \cup (V \cap S)$.

Proposition 4.47: *The Cantor set is totally disconnected.*

Proof: Let $x, y \in C$ be distinct and assume that $x < y$. Set $\delta = |x-y|$. Choose j so large that $3^{-j} < \delta$. Then $x, y \in S_j$, but x and y cannot both be in the same interval of S_j (since the intervals will have length equal to 3^{-j}). It follows that there is a point t between x and y that is not an element of S_j, hence not an element of C. Set $U = \{s : s < t\}$ and $V = \{s : s > t\}$. Then $x \in U \cap C$ hence $U \cap C \neq \varnothing$; likewise $V \cap C \neq \varnothing$. Also $(U \cap C) \cap (V \cap C) = \varnothing$. Finally $C = (C \cap U) \cup (C \cap V)$. Thus C is totally disconnected. □

Exercises
   1. Imitate Example 4.45 in the text to prove that any closed interval is connected.
   2. Imitate Example 4.45 in the text to prove that any open interval or halfopen interval is connected.
   3. Give an example of a totally disconnected set $S \subseteq [0, 1]$ such that $\bar{S} = [0, 1]$.

4. Write the real line as the union of two totally disconnected sets.

5. Let $S \subseteq \mathbb{R}$ be a set. Let $s, t \in S$. We say that s and t are in the same *connected component* of S if the entire interval $[s, t]$ lies in S. What are the connected components of the Cantor set? Is it possible to have a set S with countably many connected components? With uncountably many connected components?

6. If A is connected and B is connected then will $A \cap B$ be connected?

7. If A is connected and B is connected then will $A \cup B$ be connected?

8. What can you say about the sum of two disconnected sets? Give some examples.

9. If A is connected and B is disconnected then what can you say about $A \cap B$?

10. If sets U_j form the basis of a topology on a space X (that is to say, each open set in X can be written as a union of some of the U_j) and if each U_j is connected, then what can you say about X?

*11. If A is connected and B is connected then does it follow that $A \times B$ is connected?

12. Give an example of two disconnected sets E and F so that $E \cap F$ is connected.

13. Give an example of two disconnected sets E and F so that $E \cup F$ is connected.

14. Give an example of two disconnected sets E and F so that $E \backslash F$ is connected.

4.6 Perfect Sets

Definition 4.48: A set $S \subseteq \mathbb{R}$ is called *perfect* if it is closed and if every point of S is an accumulation point of S.

The property of being perfect is a rather special one: it means that the set has no isolated points.

Example 4.49: Consider the set $S = [0, 2]$. This set is perfect. Because **(i)** it is closed, **(ii)** any interior point is clearly an accumulation point, **(iii)** 0 is the limit of $\{1/j\}$ so is an accumulation point, and **(iv)** 2 is the limit of $\{2 - 1/j\}$ so is an accumulation point. □

Clearly a closed interval $[a, b]$ is perfect. After all, a point x in the interior of the interval is surrounded by an entire open interval $(x - \epsilon, x + \epsilon)$ of elements of the interval; moreover a is the limit of elements from the right and b is the limit of elements from the left.

Example 4.50: The Cantor set, *a totally disconnected set*, is perfect. It is definitely closed. Now fix $x \in C$. Then $x \in S_1$. Thus x is in one of the two intervals composing S_1. One (or perhaps both) of the endpoints of that interval does not equal x. Call that endpoint a_1. Likewise $x \in S_2$. Therefore x lies in one of the intervals of S_2. Choose an endpoint a_2 of that interval which does not equal x. Continuing in this fashion, we construct a sequence $\{a_j\}$. Notice that *each of the elements of this sequence lies in the Cantor set* (why?). Finally, $|x - a_j| \leq 3^{-j}$ for each j. Therefore x is the limit of the sequence. We have thus proved that the Cantor set is perfect. □

The fundamental theorem about perfect sets tells us that such a set must be rather large. We have

Theorem 4.51: *A nonempty perfect set must be uncountable.*

Proof: Let S be a nonempty perfect set. Since S has accumulation points, it cannot be finite. Therefore it is either countable or uncountable.

Seeking a contradiction, we suppose that S is countable. Write $S = \{s_1, s_2, \ldots\}$. Set $U_1 = (s_1 - 1, s_1 + 1)$. Then U_1 is a neighborhood of s_1. Now s_1 is a limit point of S so there must be infinitely many elements of S lying in U_1.

We select a bounded open interval U_2 such that $\bar{U}_2 \subseteq U_1$, $\bar{U}_2$ does not contain s_1, and U_2 does contain some element of S.

Continuing in this fashion, assume that $s_1, \ldots, s_j$ have been selected and choose a bounded interval U_{j+1} such that (i) $\bar{U}_{j+1} \subseteq U_j$, (ii) $s_j \notin \bar{U}_{j+1}$, and (iii) U_{j+1} contains some element of S.

Observe that each set $V_j = \bar{U}_j \cap S$ is closed and bounded, hence compact. Also each V_j is nonempty by construction but V_j does not contain s_{j-1}. It follows that $V = \cap_j V_j$ cannot contain s_1 (since V_2 does not), cannot contain s_2 (since V_3 does not), indeed cannot contain any element of S. Hence V, being a subset of S, is empty. But V is the decreasing intersection of nonempty compact sets, hence cannot be empty!

This contradiction shows that S cannot be countable. So it must be uncountable. □

Corollary 4.52: *If $a < b$ then the closed interval $[a, b]$ is uncountable.*

Proof: The interval $[a, b]$ is perfect. □

We also have a new way of seeing that the Cantor set is uncountable, since it is perfect:

Corollary 4.53: *The Cantor set is uncountable.*

Exercises

1. Let $U_1 \subseteq U_2 \cdots$ be open sets and assume that each of these sets has bounded, nonempty complement. Is it true that $\cup_j U_j \neq \mathbb{R}$?

2. Let $X_1, X_2 \ldots$ each be perfect sets and suppose that $X_1 \supseteq X_2 \supseteq \ldots$. Set $X = \cap_j X_j$. Is X perfect?

3. Is the product of perfect sets perfect?

4. If $A \cap B$ is perfect, then what may we conclude about A and B?

5. If $A \cup B$ is perfect, then what may we conclude about A and B?

6. Call a set imperfect if its complement is perfect. Which sets are imperfect? Can you specify a connected imperfect set?

7. What can you say about the interior of a perfect set?

8. What can you say about the boundary of a perfect set?

*9. Let $S_1, S_2, \ldots$ be closed sets and assume that $\cup_j S_j = \mathbb{R}$. Prove that at least one of the sets S_j has nonempty interior. (**Hint:** Use an idea from the proof that perfect sets are uncountable.)

*10. Let S be a nonempty set of real numbers. A point x is called a *condensation point* of S if every neighborhood of x contains uncountably many points of S. Prove that the set of condensation points of S is closed. Is it necessarily nonempty? Is it nonempty when S is uncountable?
   If T is an uncountable set, then show that the set of its condensation points is perfect.

*11. Prove that any closed set can be written as the union of a perfect set and a countable set. (**Hint:** Refer to Exercise **10** above.)

12. Prove that every open set U contains a perfect set.

13. Prove that every closed set E is contained in a perfect set.

14. Prove that every open set W is contained in a perfect set.

5

Limits and Continuity of Functions

5.1 Definition and Basic Properties of the Limit of a Function

In this chapter, we are going to treat some topics that you have seen before in your calculus class. However, we shall use the deep properties of the real numbers that we have developed in this text to obtain important new insights. Therefore, you should *not* think of this chapter as review. Look at the concepts introduced here with the power of your new understanding of analysis.

Definition 5.1: Let $E \subseteq \mathbb{R}$ be a set and let f be a real-valued function with domain E. Fix a point $P \in \mathbb{R}$ that is an accumulation point of E. Let ℓ be a real number. We say that

$$\lim_{E \ni x \to P} f(x) = \ell$$

if, for each $\epsilon > 0$, there is a $\delta > 0$ such that, when $x \in E$ and $0 < |x - P| < \delta$, then

$$|f(x) - \ell| < \epsilon.$$

In other words, we say that the limit as x tends to P of f is equal to ℓ.

The definition makes precise the notion that we can force $f(x)$ to be just as close as we please to ℓ by making x sufficiently close to P. Notice that the definition puts the condition $0 < |x - P| < \delta$ on x, so that x is not allowed to take the value P. In other words, we do not look at $x = P$, but rather at x *near* to P.

Also, observe that we only consider the limit of f at a point P that is not isolated. In the exercises you will be asked to discuss why it would be nonsensical to use the earlier definition to study the limit at an isolated point.

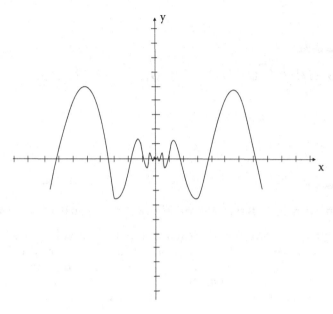

FIGURE 5.1
The limit of an oscillatory function.

Example 5.2: Let $E = \mathbb{R} \setminus \{0\}$ and

$$f(x) = x \cdot \sin(1/x) \text{ if } x \in E.$$

See Figure 5.1. Then, $\lim_{x \to 0} f(x) = 0$. To see this, let $\epsilon > 0$. Choose $\delta = \epsilon$. If $0 < |x - 0| < \delta$, then

$$|f(x) - 0| = |x \cdot \sin(1/x)| \le |x| < \delta = \epsilon,$$

as desired. Thus, the limit exists and equals 0.   □

Example 5.3: Let $E = \mathbb{R}$ and

$$g(x) = \begin{cases} 1 & \text{if } x \text{ is rational} \\ 0 & \text{if } x \text{ is irrational.} \end{cases}$$

Then, $\lim_{x \to P} g(x)$ does not exist for any point P of E.

To see this, fix $P \in \mathbb{R}$. Seeking a contradiction, assume that there is a limiting value ℓ for g at P. If this is so, then we take $\epsilon = 1/2$ and we can find a $\delta > 0$ such that $0 < |x - P| < \delta$ implies

$$|g(x) - \ell| < \epsilon = \frac{1}{2}. \tag{5.3.1}$$

If we take x to be rational, then (5.3.1) says that

$$|1 - \ell| < \frac{1}{2}, \tag{5.3.2}$$

while if we take x irrational, then (5.3.1) says that

$$|0 - \ell| < \frac{1}{2}. \tag{5.3.3}$$

But then, the triangle inequality gives that

$$\begin{aligned} |1 - 0| &= |(1 - \ell) + (\ell - 0)| \\ &\leq |1 - \ell| + |\ell - 0|, \end{aligned}$$

which by (5.3.2) and (5.3.3) is

$$< 1.$$

This contradiction, that $1 < 1$, allows us to conclude that the limit does not exist at P. □

Proposition 5.4: *Let f be a function with domain E, and let P be an accumulation point of E. If $\lim_{x \to P} f(x) = \ell$ and $\lim_{x \to P} f(x) = m$, then $\ell = m$.*

Proof: Let $\epsilon > 0$. Choose $\delta_1 > 0$ such that, if $x \in E$ and $0 < |x - P| < \delta_1$, then $|f(x) - \ell| < \epsilon/2$. Similarly choose $\delta_2 > 0$ such that, if $x \in E$ and $0 < |x - P| < \delta_2$, then $|f(x) - m| < \epsilon/2$. Define δ to be the minimum of δ_1 and δ_2. If $x \in E$ and $0 < |x - P| < \delta$, then the triangle inequality tells us that

$$\begin{aligned} |\ell - m| &= |(\ell - f(x)) + (f(x) - m)| \\ &\leq |(\ell - f(x))| + |f(x) - m| \\ &< \frac{\epsilon}{2} + \frac{\epsilon}{2} \\ &= \epsilon \end{aligned}$$

Since $|\ell - m| < \epsilon$ for every positive ϵ we conclude that $\ell = m$. That is the desired result. □

The point of the last proposition is that if a limit is calculated by two different methods, then the same answer will result. While of primarily philosophical interest now, this will be important information later when we establish the existence of certain limits.

This is a good time to observe that the limits

$$\lim_{x \to P} f(x)$$

and

$$\lim_{h \to 0} f(P + h)$$

are equal in the sense that if one limit exists, then so does the other and they both have the same value.

To facilitate checking that certain limits exist, we now record some elementary properties of the limit. This requires that we first recall how functions are combined.

Suppose that f and g are each functions which have domain E. We define the *sum* or *difference* of f and g to be the function

$$(f \pm g)(x) = f(x) \pm g(x),$$

the *product* of f and g to be the function

$$(f \cdot g)(x) = f(x) \cdot g(x),$$

and the *quotient* of f and g to be

$$\left(\frac{f}{g} \right)(x) = \frac{f(x)}{g(x)}.$$

Notice that the quotient is only defined at points x for which $g(x) \neq 0$. Now, we have:

Theorem 5.5: (Elementary Properties of Limits of Functions) *Let f and g be functions with domain E and fix a point P that is an accumulation point of E. Assume that*

  i. $\lim\limits_{x \to P} f(x) = \ell$
  ii. $\lim\limits_{x \to P} g(x) = m.$

Then,

  a. $\lim\limits_{x \to P} (f \pm g)(x) = \ell \pm m$
  b. $\lim\limits_{x \to P} (f \cdot g)(x) = \ell \cdot m$
  c. $\lim\limits_{x \to P} (f/g)(x) = \ell/m$ *provided $m \neq 0$.*

Proof: We prove part **(b)**. Parts **(a)** and **(c)** are treated in the exercises.

Let $\epsilon > 0$. We may also assume that $\epsilon < 1$. Choose $\delta_1 > 0$ such that, if $x \in E$ and $0 < |x - P| < \delta_1$, then

$$|f(x) - \ell| < \frac{\epsilon}{2(|m| + 1)}.$$

Choose $\delta_2 > 0$ such that if $x \in E$ and $0 < |x - P| < \delta_2$, then

$$|g(x) - m| < \frac{\epsilon}{2(|\ell| + 1)}.$$

(Notice that this last inequality implies that $|g(x)| < |m| + |\epsilon|$.) Let δ be the minimum of δ_1 and δ_2. If $x \in E$ and $0 < |x - P| < \delta$, then

$$
\begin{aligned}
|f(x) \cdot g(x) - \ell \cdot m| &= |(f(x) - \ell) \cdot g(x) + (g(x) - m) \cdot \ell| \\
&\leq |(f(x) - \ell) \cdot g(x)| + |(g(x) - m) \cdot \ell| \\
&< \left(\frac{\epsilon}{2(|m| + 1)}\right) \cdot |g(x)| + \left(\frac{\epsilon}{2(|\ell| + 1)}\right) \cdot |\ell| \\
&\leq \left(\frac{\epsilon}{2(|m| + 1)}\right) \cdot (|m| + |\epsilon|) + \frac{\epsilon}{2} \\
&< \frac{\epsilon}{2} + \frac{\epsilon}{2} \\
&= \epsilon.
\end{aligned}
$$

$\square$

Example 5.6: It is a simple matter to check that, if $f(x) = x$, then

$$\lim_{x \to P} f(x) = P$$

for every real P. (Indeed, for $\epsilon > 0$ we may take $\delta = \epsilon$.) Also, if $g(x) \equiv \alpha$ is the constant function taking value α, then

$$\lim_{x \to P} g(x) = \alpha.$$

It then follows from parts **(a)** and **(b)** of the theorem that, if $f(x)$ is any polynomial function, then

$$\lim_{x \to P} f(x) = f(P).$$

Moreover, if $r(x)$ is any *rational function* (quotient of polynomials), then we may also use part **(c)** of the theorem to conclude that

$$\lim_{x \to P} r(x) = r(P)$$

for all points P at which the rational function $r(x)$ is defined.   □

Example 5.7: If x is a small, positive real number, then $0 < \sin x < x$. This is true because $\sin x$ is the nearest distance from the point $(\cos x, \ \sin x)$ to the x-axis while x is the distance from that point to the x-axis along an arc. If $\epsilon > 0$, then we set $\delta = \epsilon$. We conclude that if $0 < |x - 0| < \delta$, then

$$| \sin x - 0| < |x| < \delta = \epsilon.$$

Since $\sin(-x) = - \sin x$, the same result holds when x is a negative number with small absolute value. Therefore

$$\lim_{x \to 0} \sin x = 0.$$

Since

$$\cos x = \sqrt{1 - \sin^2 x} \text{ for all } x \in [-\pi/2, \pi/2],$$

we may conclude from the preceding theorem that

$$\lim_{x \to 0} \cos x = 1.$$

Now, fix any real number P. We have

$$\begin{aligned}
\lim_{x \to P} \sin x &= \lim_{h \to 0} \sin(P + h) \\
&= \lim_{h \to 0} (\sin P \cos h + \cos P \sin h) \\
&= \sin P \cdot 1 + \cos P \cdot 0 \\
&= \sin P.
\end{aligned}$$

We of course have used parts **(a)** and **(b)** of the theorem to commute the limit process with addition and multiplication. A similar argument shows that

$$\lim_{x \to P} \cos x = \cos P.$$

                                                                                            □

Remark 5.8: In the last example, we have used the definition of the sine function and the cosine function that you learned in calculus. In Chapter 8, when we learn about series of functions, we will learn a more rigorous method for treating the trigonometric functions. □

We conclude by giving a characterization of the limit of a function using sequences.

Proposition 5.9: *Let f be a function with domain E and P be an accumulation point of E. Then,*

$$\lim_{x \to P} f(x) = \ell \qquad (5.9.1)$$

if and only if, for any sequence $\{a_j\} \subseteq E \setminus \{P\}$ satisfying $\lim_{j \to \infty} a_j = P$, it holds that

$$\lim_{j \to \infty} f(a_j) = \ell. \qquad (5.9.2)$$

Proof: Assume that condition (5.9.1) fails. Then, there is an $\epsilon > 0$ such that for no $\delta > 0$ is it the case that when $0 < |x - P| < \delta$, then $|f(x) - \ell| < \epsilon$. Thus, for each $\delta = 1/j$, we may choose a number $a_j \in E \setminus \{P\}$ with $0 < |a_j - P| < 1/j$ and $|f(a_j) - \ell| \geq \epsilon$. But then, condition (5.9.2) fails for this sequence $\{a_j\}$.

If condition (5.9.2) fails, then there is some sequence $\{a_j\}$ such that $\lim_{j \to \infty} a_j = P$ but $\lim_{j \to \infty} f(a_j) \neq \ell$. This means that there is an $\epsilon > 0$ such that for infinitely many a_j it holds that $|f(a_j) - \ell| \geq \epsilon$. But then, no matter how small $\delta > 0$, there will be an a_j satisfying $0 < |a_j - P| < \delta$ (since $a_j \to P$) and $|f(a_j) - \ell| \geq \epsilon$. Thus, (5.9.1) fails.

Exercises

1. Let f and g be functions on a set $A = (a, c) \cup (c, b)$ and assume that $f(x) \leq g(x)$ for all $x \in A$. Assuming that both limits exist, show that

$$\lim_{x \to c} f(x) \leq \lim_{x \to c} g(x).$$

Does the conclusion improve if we assume that $f(x) < g(x)$ for all $x \in A$?

2. Explain why it makes no sense to consider the limit of a function at an isolated point of the domain of the function.

3. Give a definition of limit using the concept of open set.

4. If $\lim_{x \to c} f(x) = \ell > 0$, then prove that there is a $\delta > 0$ so small that $|x - c| < \delta$ guarantees that $f(x) > \ell/2$.

5. Give an example of a function with domain $\mathbb{R}$ such that $\lim_{x \to c} f(x)$ exists at every point c but f is discontinuous at a dense set of points.

6. Prove that $\lim_{x \to P} f(x) = \lim_{h \to 0} f(P + h)$ whenever both expressions make sense.

7. Prove parts **(a)** and **(c)** of Theorem 5.5.

8. Give an example of a function $f : \mathbb{R} \to \mathbb{R}$ so that $\lim_{x \to c}$ does not exist for any $c \in \mathbb{R}$.

9. Discuss the limiting properties at the origin of the functions

$$f(x) = \begin{cases} x \sin(1/x) & \text{if } x \neq 0 \\ 0 & \text{if } x = 0. \end{cases}$$

and

$$g(x) = \begin{cases} \sin(1/x) & \text{if } x \neq 0 \\ 0 & \text{if } x = 0. \end{cases}$$

10. Show that, if f is an increasing or decreasing function, then f has a limit at "most" points. What does the word "most" mean in this context?

*11. Give an example of a function $f : \mathbb{R} \to \mathbb{R}$ so that $\lim_{x \to c} f(x)$ exists when c is irrational but does not exist when c is rational.

*12. Give a definition of limit using the concept of distance.

*13. Express the concept of limit of a function at a point using the idea of sequences from Chapter 2.

14. Give examples of functions f and g so that $\lim_{x \to c} f(x)$ does not exist and $\lim_{x \to c} g(x)$ does not exist and but $\lim_{x \to c} f(x) \cdot g(x)$ *does exist and*

15. Give examples of functions f and g so that $\lim_{x \to c} f(x)$ does not exist and $\lim_{x \to c} g(x)$ does not exist and but $\lim_{x \to c} f(x)/g(x)$ *does exist.*

5.2 Continuous Functions

Definition 5.10: Let $E \subseteq \mathbb{R}$ be a set and let f be a real-valued function with domain E. Fix a point P which is in E and is also an accumulation point of E. We say that f is *continuous* at P if

$$\lim_{x \to P} f(x) = f(P).$$

We learned from the penultimate example of Section 5.1 that polynomial functions are continuous at every real x. So are the transcendental functions $\sin x$ and $\cos x$ (see Example 5.7). A rational function is continuous at every point of its domain.

Example 5.11: The function

$$h(x) = \begin{cases} \sin(1/x) & \text{if } x \neq 0 \\ 1 & \text{if } x = 0 \end{cases}$$

is discontinuous at 0. See Figure 5.2. The reason is that

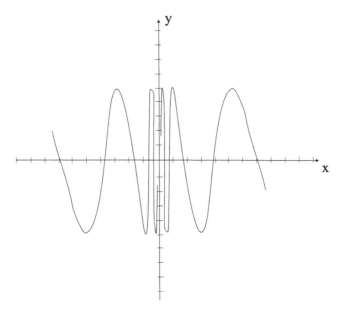

FIGURE 5.2
A function discontinuous at 0.

$$\lim_{x \to 0} h(x)$$

does not exist. (Details of this assertion are left for you: notice that $h(1/(j\pi)) = 0$ while $h(2/[(4j + 1)\pi]) = 1$ for $j = 1, 2, \ldots$.)
   The function

$$k(x) = \begin{cases} x \cdot \sin(1/x) & \text{if } x \neq 0 \\ 1 & \text{if } x = 0 \end{cases}$$

is also discontinuous at $x = 0$. This time the limit $\lim_{x \to 0} k(x)$ exists (see Example 5.2), but the limit does not agree with $k(0)$.
   However, the function

$$m(x) = \begin{cases} x \cdot \sin(1/x) & \text{if } x \neq 0 \\ 0 & \text{if } x = 0 \end{cases}$$

is continuous at $x = 0$ because the limit at 0 exists and agrees with the value of the function there. See Figure 5.3.   □

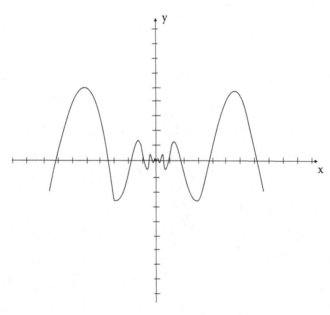

FIGURE 5.3
A function continuous at 0.

The arithmetic operations $+$, $-$, $\times$, and $\div$ preserve continuity (so long as we avoid division by zero). We now formulate this assertion as a theorem.

Theorem 5.12: *Let f and g be functions with domain E and let P be a point of E which is also an accumulation point of E. If f and g are continuous at P, then so are $f \pm g$, $f \cdot g$, and (provided $g(P) \neq 0$) f/g.*

Proof: Apply Theorem 5.5 of Section 5.1. □

Continuous functions may also be characterized using sequences:

Proposition 5.13: *Let f be a function with domain E and fix $P \in E$ which is an accumulation point of E. The function f is continuous at P if and only if, for every sequence $\{a_j\} \subseteq E$ satisfying $\lim_{j \to \infty} a_j = P$, it holds that*

$$\lim_{j \to \infty} f(a_j) = f(P).$$

Proof: Apply Proposition 5.9 of Section 5.1. □

Recall that, if g is a function with domain D and range E, and if f is a function with domain E and range F, then the *composition* of f and g is

$$f \circ g(x) = f(g(x)).$$

See Figure 5.4.

Proposition 5.14: *Let g have domain D and range E and let f have domain E and range F. Let $P \in D$. Suppose that P is an accumulation point of D and $g(P)$ is an accumulation point of E. Assume that g is continuous at P and that f is continuous at $g(P)$. Then, $f \circ g$ is continuous at P.*

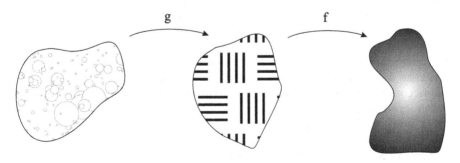

FIGURE 5.4
Composition of functions.

Proof: Let $\{a_j\}$ be any sequence in D such that $\lim_{j\to\infty} a_j = P$. Then,

$$\lim_{j\to\infty} f\circ g\,(a_j) = \lim_{j\to\infty} f\,(g\,(a_j)) = f\left(\lim_{j\to\infty} g\,(a_j)\right)$$

$$= f\left(g\left(\lim_{j\to\infty} a_j\right)\right) = f\,(g\,(P)) = f\circ g\,(P).$$

Now, apply Proposition 5.9. □

Example 5.15: It is not the case that if

$$\lim_{x\to P} g\,(x) = \ell$$

and

$$\lim_{t\to\ell} f\,(t) = m$$

then,

$$\lim_{x\to P} f\circ g\,(x) = m.$$

A counterexample is given by the functions

$$g\,(x) = 0$$

$$f\,(x) = \begin{cases} 2 & \text{if } x \neq 0 \\ 5 & \text{if } x = 0. \end{cases}$$

Notice that $\lim_{x\to 0} g\,(x) = 0$, $\lim_{t\to 0} f\,(t) = 2$, yet $\lim_{x\to 0} f\circ g\,(x) = 5$.

The additional hypothesis that f be continuous at ℓ is necessary in order to guarantee that the limit of the composition will behave as expected. □

Next we explore the topological approach to the concept of continuity. Whereas the analytic approach that we have been discussing so far considers continuity one point at a time, the topological approach considers all points simultaneously. Let us call a function continuous if it is continuous at every point of its domain.

Definition 5.16: Let f be a function with domain E and let W be any set of real numbers. We define

$$f^{-1}(W) = \{x \in E: f(x) \in W\}.$$

We sometimes refer to $f^{-1}(W)$ as the *inverse image* of W under f.

Theorem 5.17: *Let f be a function with domain E. The function f is continuous if and only if the inverse image of any open set under f is the intersection of E with an open set.*

In particular, if E is open, then f is continuous if and only if the inverse image of every open set under f is open.

Proof: Assume that f is continuous. Let O be any open set in $\mathbb{R}$ and let $P \in f^{-1}(O)$. Then, by definition, $f(P) \in O$. Since O is open, there is an $\epsilon > 0$ such that the interval $(f(P) - \epsilon, f(P) + \epsilon)$ lies in O. By the continuity of f we may select a $\delta > 0$ such that if $x \in E$ and $|x - P| < \delta$, then $|f(x) - f(P)| < \epsilon$. In other words, if $x \in E$ and $|x - P| < \delta$, then $f(x) \in O$ or $x \in f^{-1}(O)$. Thus, we have found an open interval $I = (P - \delta, P + \delta)$ about P whose intersection with E is contained in $f^{-1}(O)$. So $f^{-1}(O)$ is the intersection of E with an open set.

Conversely, suppose that for any open set $O \subseteq \mathbb{R}$ we have that $f^{-1}(O)$ is the intersection of E with an open set. Fix $P \in E$. Choose $\epsilon > 0$. Then, the interval $(f(P) - \epsilon, f(P) + \epsilon)$ is an open set. By hypothesis the set $f^{-1}((f(P) - \epsilon, f(P) + \epsilon))$ is the intersection of E with an open set. This set contains the point P. Thus, there is a $\delta > 0$ such that

$$E \cap (P - \delta, P + \delta) \subseteq f^{-1}((f(P) - \epsilon, f(P) + \epsilon)).$$

But that just says that

$$f(E \cap (P - \delta, P + \delta)) \subseteq (f(P) - \epsilon, f(P) + \epsilon).$$

In other words, if $|x - P| < \delta$ and $x \in E$, then $|f(x) - f(P)| < \epsilon$. But that means that f is continuous at P. $\qquad\square$

Example 5.18: Since any open subset of the real numbers is a countable or finite disjoint union of intervals, therefore—in order to check that the inverse image under a function f of every open set is open—it is enough to check that the inverse image of any open interval is open. This is frequently easy to do.

For example, if $f(x) = x^2$, then the inverse image of an open interval (a, b) is $(-\sqrt{b}, -\sqrt{a}) \cup (\sqrt{a}, \sqrt{b})$ if $a > 0$, is $(-\sqrt{b}, \sqrt{b})$ if $a \le 0, b \ge 0$, and is $\emptyset$ if $a < b < 0$. Thus, the function f is continuous.

Note that, by contrast, it is somewhat tedious to give an $\epsilon - \delta$ proof of the continuity of $f(x) = x^2$. $\qquad\square$

Corollary 5.19: *Let f be a function with domain E. The function f is continuous if and only if the inverse image of any closed set F under f is the intersection of E with some closed set. continuity and closed sets.*

In particular, if E is closed, then f is continuous if and only if the inverse image of any closed set F under f is closed.

Proof: It is enough to prove that

$$f^{-1}(^cF) = {}^c(f^{-1}(F)).$$

We leave this assertion as an exercise for you.

Exercises

1. Define the function

$$g(x) = \begin{cases} 0 & \text{if } x \text{ is irrational} \\ x & \text{if } x \text{ is rational} \end{cases}.$$

   At which points x is g continuous? At which points is it discontinuous?

2. Let f be a continuous function whose domain contains an open interval (a, b). What form can $f((a, b))$ have? (**Hint:** There are just four possibilities.)

3. Explain why it would be foolish to define the concept of continuity at an isolated point.

*4. Let f be a continuous function on the open interval (a,b). Under what circumstances can f be extended to a continuous function on $[a, b]$?

5. Define an onto, continuous function from $\mathbb{R}^2$ to $\mathbb{R}$.

6. Define continuity using the notion of closed set.

7. Define the function $f(x)$ to equal 0 if x is irrational and to equal b if $x = a/b$ is a rational number in lowest terms. At which points is f continuous? At which points discontinuous?

8. The image of a compact set under a continuous function is compact (see the next section). But the image of a closed set need not be closed. Explain. The *inverse image* of a compact set under a continuous function need not be compact. Explain.

9. Give a careful proof of Corollary 5.19.

*10. See the next section for terminology. In particular, a function f on a set E is uniformly continuous if, given $\epsilon > 0$, there is a $\delta > 0$ such that $|f(s) - f(t)| < \epsilon$ whenever $|s - t| < \delta$.

Let $0 < \alpha \le 1$. A function f with domain E is said to satisfy aLipschitz condition *Lipschitz condition* of order α if there is a constant $C > 0$ such that, for any $s, t \in E$, it holds that $|f(s) - f(t)| \le C \cdot |s - t|^\alpha$. Prove that such a function must be uniformly continuous.

*11. See Exercise 10 for terminology. Is the composition of uniformly continuous functions uniformly continuous?

12. Refer to Exercise 10 for terminology. Show that the sum of two uniformly continuous functions is uniformly continuous.

13. Refer to Exercise 10 for terminology. Show that the product of two uniformly continuous functions is uniformly continuous.

14. Refer to Exercise 10 for terminology. Under what circumstances is the quotient of two uniformly continuous functions uniformly continuous?

15. Let f be a continuous function on the interval $[0, 1]$. Let $\epsilon > 0$. Show that there is a piecewise linear function φ on $[0,1]$ such that $|f(x) - \varphi(x)| < \epsilon$ for every $x \in [0, 1]$. (**Hint:** A piecewise linear function is a continuous function whose graph consists of finitely many line segments.)

16. Give an example of a sequence of continuous functions on $\mathbb{R}$ such that $\lim_{j \to \infty} f_j(x)$ exists for every x but the limit function is discontinuous.

*17. Refer to Exercise 16. How big can the set of discontinuities of the limit function be?

5.3 Topological Properties and Continuity

Recall that in Chapter 4 we learned a characterization of compact sets in terms of open covers. In Section 5.2 of the present chapter we learned a characterization of continuous functions in terms of inverse images of open sets. Thus, it is not surprising that compact sets and continuous functions interact in a natural way. We explore this interaction in the present section.

Definition 5.20: Let f be aimage of a function function with domain E and let L be a subset of E. We define

$$f(L) = \{f(x): x \in L\}.$$

The set $f(L)$ is called the *image* of L under f. See Figure 5.5.

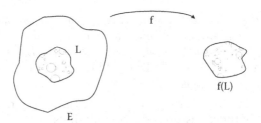

FIGURE 5.5
The image of the set L under the function F.

Theorem 5.21: *The image of a compact set under a continuous function is also compact.*

Proof: Let f be a continuous function with domain E and let K be a subset of E that is compact. Our job is to show that $f(K)$ is compact.

Let $C = \{O_\alpha\}$ be an open covering of $f(K)$. Since f is continuous we know that, for each α, the set $f^{-1}(O_\alpha)$ is the intersection of E with an open set $\mathcal{U}_\alpha$. Let $\hat{C} = \{\mathcal{U}_\alpha\}_{\alpha \in A}$. Since C covers $f(K)$ it follows that $\hat{C}$ covers K. But K is compact; therefore (Theorem 4.37) there is a finite subcovering

$$\{\mathcal{U}_{\alpha_1}, \mathcal{U}_{\alpha_2}, \dots \mathcal{U}_{\alpha_m}\}$$

of K. But then, it follows that $f(\mathcal{U}_{\alpha_1} \cap E), \dots, f(\mathcal{U}_{\alpha_m} \cap E)$ covers $f(K)$, hence

$$O_{\alpha_1}, O_{\alpha_2}, \dots, O_{\alpha_m}$$

covers $f(K)$.

We have taken an arbitrary open cover C for $f(K)$ and extracted from it a finite subcovering. It follows that $f(K)$ is compact. $\qquad \square$

It is not the case that the continuous image of a closed set is closed. For instance, take $f(x) = 1/(1 + x^2)$ and $E = \mathbb{R}$: the set E is closed and f is continuous but $f(E) = (0, 1]$ is not closed.

It is also not the case that the continuous image of a bounded set is bounded. As an example, take $f(x) = 1/x$ and $E = (0, 1)$. Then, E is bounded and f continuous but $f(E) = (1, \infty)$ is unbounded.

However, the combined properties of closedness *and* boundedness (that is, compactness) are preserved. That is the content of the preceding theorem.

Corollary 5.22: *Let f be a continuous, real-valued function with compact domain $K \subseteq \mathbb{R}$. Then, there is a number L such that*

$$|f(x)| \le L$$

for all $x \in K$.

Proof: We know from the theorem that $f(K)$ is compact. By Theorem 4.29, we conclude that $f(K)$ is bounded. Thus, there is a number L such that $|t| \le L$ for all $t \in f(K)$. But that is just the assertion that we wish to prove. □

In fact we can prove an important strengthening of the corollary. Since $f(K)$ is compact, it contains its supremum M and its infimum m. Therefore, there must be a number $C \in K$ such that $f(C) = M$ and a number $c \in K$ such that $f(c) = m$. In other words, $f(c) \le f(x) \le f(C)$ for all $x \in K$. We summarize:

Theorem 5.23: *Let f be a continuous function on a compact set $K \subseteq \mathbb{R}$. Then, there exist numbers c and C in K such that $f(c) \le f(x) \le f(C)$ for all $x \in K$. We call c an absolute minimum for f on K and C an absolute maximum for f on K. We call $f(c)$ the absolute minimum value for f on K and $f(C)$ the absolute maximum value for f on K.*

Notice that, in the last theorem, the location of the absolute maximum and absolute minimum need not be unique. For instance, the function $\sin x$ on the compact interval $[0, 4\pi]$ has an absolute minimum at $3\pi/2$ and $7\pi/2$. It has an absolute maximum at $\pi/2$ and at $5\pi/2$.

Now, we define a refined type of continuity called "uniform continuity." We shall learn that this new notion of continuous function arises naturally for a continuous function on a compact set. It will also play an important role in our later studies, especially in the context of the integral.

Definition 5.24: Let f be a function with domain $E \subseteq \mathbb{R}$. We say that f is *uniformly continuous* on E if, for each $\epsilon > 0$, there is a $\delta > 0$ such that, whenever $s, t \in E$ and $|s - t| < \delta$, then $|f(s) - f(t)| < \epsilon$.

Observe that "uniform continuity" differs from "continuity" in that it treats all points of the domain simultaneously: the $\delta > 0$ that is chosen is independent of the points $s, t \in E$. This difference is highlighted by the next two examples.

Example 5.25: Suppose that a function $f \colon \mathbb{R} \to \mathbb{R}$ satisfies the condition

$$|f(s) - f(t)| \le C \cdot |s - t|, \tag{5.25.1}$$

where C is some positive constant. This is called a *Lipschitz condition*, and it arises frequently in analysis. Let $\epsilon > 0$ and set $\delta = \epsilon/C$. If $|x - y| < \delta$, then, by (5.25.1),

$$|f(x) - f(y)| \le C \cdot |x - y| < C \cdot \delta = C \cdot \frac{\varepsilon}{C} = \varepsilon.$$

It follows that f is uniformly continuous.   □

Example 5.26: Consider the function $f(x) = x^2$. Fix a point $P \in \mathbb{R}$, $P > 0$, and let $\varepsilon > 0$. To guarantee that $|f(x) - f(P)| < \varepsilon$ we must have (for $x > 0$)

$$|x^2 - P^2| < \varepsilon$$

or

$$|x - P| < \frac{\varepsilon}{x + P}.$$

Since x will range over a neighborhood of P, we see that the required δ in the definition of continuity cannot be larger than $\varepsilon/(2P)$. In fact the choice $|x - P| < \delta = \varepsilon/(2P + 1)$ will do the job.

Put in slightly different words, let $\varepsilon = 1$. Then, $|f(j + 1/j) - f(j)| > \varepsilon = 1$ for any j. Thus, for this ε, we may not take δ to be $1/j$ for any j. So no uniform δ exists.

Thus, the choice of δ depends not only on ε (which we have come to expect) but also on P. In particular, f is not uniformly continuous on $\mathbb{R}$. This is a quantitative reflection of the fact that the graph of f becomes ever steeper as the variable x moves to the right.

Notice that the same calculation shows that the function f with restricted domain $[a, b]$, $0 < a < b < \infty$, is uniformly continuous. That is because, when the function is restricted to $[a, b]$, its slope does not become arbitrarily large. See Figure 5.6.   □

Now, the main result about uniform continuity is the following:

Theorem 5.27: *Let f be a continuous function with compact domain K. Then, f is uniformly continuous on K.*

Proof: Pick $\varepsilon > 0$. By the definition of continuity there is for each point $x \in K$ a number $\delta_x > 0$ such that if $|x - t| < \delta_x$, then $|f(t) - f(x)| < \varepsilon/2$. The intervals $I_x = (x - \delta_x/2, x + \delta_x/2)$ form an open covering of K. Since K is compact, we may therefore (by Theorem 4.34) extract a finite subcovering

$$I_{x_1}, \ldots I_{x_m}.$$

Now, let $\delta = \min\{\delta_{x_1}/2, \ldots, \delta_{x_m}/2\} > 0$. If $s, t \in K$ and $|s - t| < \delta$, then $s \in I_{x_j}$ for some $1 \le j \le m$. It follows that

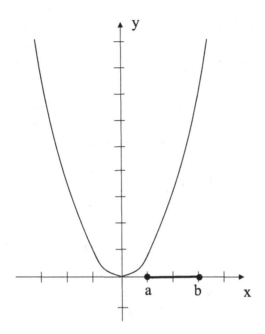

FIGURE 5.6
Uniform continuity on the interval $[a, b]$.

$$|s - x_j| < \delta_{x_j}/2$$

and

$$|t - x_j| \le |t - s| + |s - x_j| < \delta + \delta_{x_j}/2 \le \delta_{x_j}/2 + \delta_{x_j}/2 = \delta_{x_j}.$$

We know that

$$|f(s) - f(t)| \le |f(s) - f(x_j)| + |f(x_j) - f(t)|.$$

But since each of s and t is within δ_{x_j} of x_j we may conclude that the last line is less than

$$\frac{\epsilon}{2} + \frac{\epsilon}{2} = \epsilon.$$

Notice that our choice of δ does not depend on s and t (indeed, we chose δ *before* we chose s and t). We conclude that f is uniformly continuous. □

Remark 5.28: Where in the proof did the compactness play a role? We defined δ to be the minimum of $\delta_{x_1}, \ldots \delta_{x_m}$. To guarantee that δ be *positive* it is crucial that we be taking the minimum of *finitely many* positive numbers. So we needed a *finite* subcovering.   □

Example 5.29: The function $f(x) = \sin(1/x)$ is continuous on the domain $E = (0, \infty)$ since it is the composition of continuous functions (refer again to Figure 5.2). However, it is not uniformly continuous since

$$\left| f\left(\frac{1}{2j\pi}\right) - f\left(\frac{1}{\frac{(4j+1)\pi}{2}}\right) \right| = 1$$

for $j = 1, 2, \ldots$. Thus, even though the arguments are becoming arbitrarily close together, the images of these arguments remain bounded apart. We conclude that f cannot be uniformly continuous. See Figure 5.2.

However, if f is considered as a function on any interval of the form $[a, b]$, $0 < a < b < \infty$, then the preceding theorem tells us that the function f is uniformly continuous.   □

As an exercise, you should check that

$$g(x) = \begin{cases} x \ \sin(1/x) & \text{if } x \neq 0 \\ 0 & \text{if } x = 0 \end{cases}$$

is uniformly continuous on any interval of the form $[-N, N]$. See Figure 5.3. Next we show that continuous functions preserve connectedness.

Theorem 5.30: *Let f be a continuous function with domain an open interval continuous images of connected sets I. Suppose that L is a connected subset of I. Then, $f(L)$ is connected.*

Proof: Suppose to the contrary that there are open sets U and V such that

$$U \cap f(L) \neq \emptyset, V \cap f(L) \neq \emptyset,$$

$$(U \cap f(L)) \cap (V \cap f(L)) = \emptyset,$$

and

$$f(L) = (U \cap f(L)) \cup (V \cap f(L)).$$

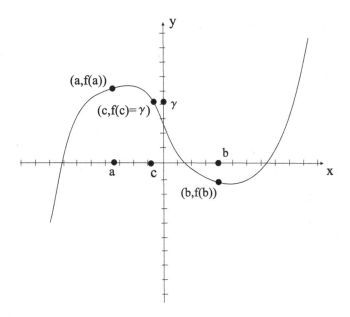

FIGURE 5.7
The Intermediate Value Theorem.

Since f is continuous, $f^{-1}(U)$ and $f^{-1}(V)$ are open. They each have nonempty intersection with L since $U \cap f(L)$ and $V \cap f(L)$ are nonempty. By the definition of f^{-1}, they are disjoint. And since $U \cup V$ contains $f(L)$ it follows, by definition, that $f^{-1}(U) \cup f^{-1}(V)$ contains L. But this shows that L is disconnected, and that is a contradiction. □

Corollary 5.31: (The Intermediate Value Theorem) *Let f be a continuous function whose domain contains the interval $[a, b]$. Let γ be a number that lies between $f(a)$ and $f(b)$). Then, there is a number c between a and b such that $f(c) = \gamma$. Refer to* Figure 5.7.

Proof: The set $[a, b]$ is connected. Therefore, $f([a, b])$ is connected. But $f([a, b])$ contains the points $f(a)$ and $f(b)$. By connectivity, $f([a, b])$ must contain the interval that has $f(a)$ and $f(b)$ as endpoints. In particular, $f([a, b])$ must contain any number γ that lies between $f(a)$ and $f(b)$. But this just says that there is a number c lying between a and b such that $f(c) = \gamma$. That is the desired conclusion.

Example 5.32: Let f be a continuous function with domain the interval $[0, 1]$ and range the interval $[0, 1]$. We claim that f has a fixed point, that is, a point p such that $f(p) = p$.

To see this, suppose not. Consider the function $g(x) = f(x) - x$. Since f has no fixed point, $f(0) > 0$ so $g(0) > 0$. Also, $f(1) < 1$ so that $g(1) < 0$. By the Intermediate Value Theorem, it follows that there is a point p at which $g(p) = 0$. But that means that $f(p) = p$. □

Exercises

1. If f is continuous on $[0, 1]$ and if $f(x)$ is positive for each rational x, then does it follow that f is positive at all x?

2. Give an example of a continuous function f and a connected set E such that $f^{-1}(E)$ is not connected. Is there a condition you can add that will force $f^{-1}(E)$ to be connected?

3. Give an example of a continuous function f and an open set U so that $f(U)$ is not open.

4. Let S be any subset of $\mathbb{R}$. Define the function

$$f(x) = \inf\{|x - s| : s \in S\}.$$

(We think of $f(x)$ as the distance of x to S.) Prove that f is uniformly continuous.

5. Let f be any function whose domain and range is the entire real line. If A and B are disjoint sets does it follow that $f(A)$ and $f(B)$ are disjoint sets? If C and D are disjoint sets does it follow that $f^{-1}(C)$ and $f^{-1}(D)$ are disjoint?

6. Let f be any function whose domain is the entire real line. If A and B are sets, then is $f(A \cup B) = f(A) \cup f(B)$? If C and D are sets, then is $f^{-1}(C \cup D) = f^{-1}(C) \cup f^{-1}(D)$? What is the answer to these questions if we replace $\cup$ by $\cap$?

7. We know that the continuous image of a connected set (i.e., an interval) is also a connected set (another interval). Suppose now that A is the union of k disjoint intervals and that f is a continuous function. What can you say about the set $f(A)$?

8. A function f with domain A and range B is called a *homeomorphism* if it homeomorphism is one-to-one, onto, continuous, and has a continuous inverse. If such an f exists, then we say that A and B are *homeomorphic*. Which sets of reals are homeomorphic to the open unit interval $(0, 1)$? Which sets of reals are homeomorphic to the closed unit interval $[0, 1]$?

9. Let f be a continuous function with domain $[0, 1]$ and range $[0, 1]$. We know from Example 5.32 that there exists a point $P \in [0, 1]$ such that $f(P) = P$. Prove that this result is false if the domain and range of the function are both $(0, 1)$.

10. Let f be a continuous function and let $\{a_j\}$ be a Cauchy sequence in the domain of f. Does it follow that $\{f(a_j)\}$ is a Cauchy sequence? What if we assume instead that f is uniformly continuous?

11. Let E and F be disjoint closed sets of real numbers. Prove that there is a continuous function f with domain the real numbers such that $\{x: f(x) = 0\} = E$ and $\{x: f(x) = 1\} = F$.

12. If K and L are sets, then define

$$K + L = \{k + \ell: k \in K \text{ and } \ell \in L\}.$$

If K and L are compact, then prove that $K + L$ is compact. If K and L are merely closed, does it follow that $K + L$ is closed?

13. A function f from an interval (a, b) to an interval (c, d) is called *proper* if, for any compact set $K \subseteq (c, d)$, it holds that $f^{-1}(K)$ is compact. Prove that if f is proper, then either

$$\lim_{x \to a^+} f(x) = c \text{ or } \lim_{x \to a^+} f(x) = d.$$

Likewise prove that either

$$\lim_{x \to b^-} f(x) = c \text{ or } \lim_{x \to b^-} f(x) = d.$$

*14. Let E be any closed set of real numbers. Prove that there is a continuous function f with domain $\mathbb{R}$ such that $\{x: f(x) = 0\} = E$.

*15. Prove that the function $f(x) = \cos x$ can be written, on the interval $(0, 2\pi)$, as the difference of two increasing functions.

16. A function is said to be *proper* if the inverse image of any compact set is compact. Give an equivalent definition of proper using the language of sequences. Give an example of a function $f: \mathbb{R} \to \mathbb{R}$ that is proper. Give another example of a function $g: \mathbb{R} \to \mathbb{R}$ that is not proper.

17. Let $f: \mathbb{R} \to \mathbb{R}$ be a function. If f^2 is continuous and f^3 is continuous, then does it follow that f is continuous?

5.4 Classifying Discontinuities and Monotonicity

We begin by refining our notion of limit:

Definition 5.33: Fix $P \in \mathbb{R}$. Let f be a function with domain E. Suppose that P is a limit point of $E \cap [P - 1, P)$. We say that f has *left limit* ℓ at P, and write

$$\lim_{x \to P^-} f(x) = \ell$$

if, for every $\epsilon > 0$, there is a $\delta > 0$ such that, whenever $x \in E$ and $P - \delta < x < P$, then it holds that

$$|f(x) - \ell| < \epsilon.$$

Now, suppose that P is a limit point of $E \cap (P, P + 1]$. We say that f has *right limit* m at P, and write

$$\lim_{x \to P^+} f(x) = m$$

if, for every $\epsilon > 0$, there is a $\delta > 0$ such that, whenever $x \in E$ and $P < x < P + \delta$, then it holds that

$$|f(x) - m| < \epsilon.$$

This definition simply formalizes the notion of either letting x tend to P from the left only or from the right only.   □

Example 5.34: Let

$$f(x) = \begin{cases} 0 & \text{if} \quad x \le 0 \\ \sin(1/x) & \text{if} \quad x > 0. \end{cases}$$

Then, $\lim_{x \to 0^-} f(x) = 0$ and $\lim_{x \to 0^+} f(x)$ does not exist.

Definition 5.35: Fix $P \in \mathbb{R}$. Let f be a function with domain E. Suppose that P is a limit point of $E \cap [P - 1, P)$ and that P is an element of E. We say that f is *left continuous* at P if

$$\lim_{x \to P^-} f(x) = f(P).$$

Likewise, in case P is a limit point of $E \cap (P, P + 1]$ and is also an element of E, we say that f is *right continuous* at P if

$$\lim_{x \to P^+} f(x) = f(P).$$

Example 5.36: Define

$$f(x) = \begin{cases} 1 & \text{if } x < 0 \\ 0 & \text{if } x = 0 \\ x\sin(1/x) & \text{if } x > 0. \end{cases}$$

Then, f is right continuous at 0 but f is not left continuous at 0. □

Let f be a function with domain E. Let P in E and assume that f is discontinuous at P. There are two ways in which this discontinuity can occur:

I. If $\lim_{x \to P^-} f(x)$ and $\lim_{x \to P^+} f(x)$ both exist but either do not equal each other or do not equal $f(P)$, then we say that f has a *discontinuity of the first kind* (or sometimes a *simple discontinuity*) at P.

II. If either $\lim_{x \to P^-}$ does not exist or $\lim_{x \to P^+}$ does not exist, then we say that f has a *discontinuity of the second kind* at P.

Refer to Figure 5.8.

Example 5.37: Define

$$f(x) = \begin{cases} \sin(1/x) & \text{if } x \neq 0 \\ 0 & \text{if } x = 0 \end{cases}$$

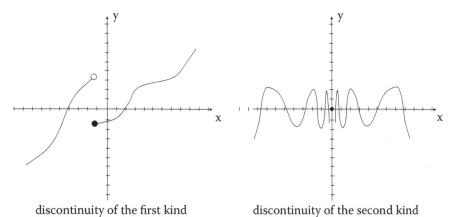

discontinuity of the first kind discontinuity of the second kind

FIGURE 5.8
Discontinuities of the first and second kind.

$$g(x) = \begin{cases} 1 & \text{if } x > 0 \\ 0 & \text{if } x = 0 \\ -1 & \text{if } x < 0 \end{cases}$$

$$h(x) = \begin{cases} 1 & \text{if } x \text{ is irrational} \\ 0 & \text{if } x \text{ is rational} \end{cases}$$

Then, f has a discontinuity of the second kind at 0 while g has a discontinuity of the first kind at 0. The function h has a discontinuity of the second kind at every point.   □

Definition 5.38: Let f be a function whose domain contains an open interval (a, b). We say that f is *increasing* on (a, b) if, whenever $a < s < t < b$, it holds that $f(s) \leq f(t)$. We say that f is *decreasing* on (a, b) if, whenever $a < s < t < b$, it holds that $f(s) \geq f(t)$. See Figure 5.9.

If a function is either increasing or decreasing, then we call it *monotone* or *monotonic*. Compare with the definition of monotonic sequences in Section 2.1.

As with sequences, the word "monotonic" is superfluous in many contexts. But its use is traditional and occasionally convenient.

Proposition 5.39: *Let f be a monotonic function on an open interval (a,b). Then, all of the discontinuities of f are of the first kind.*

Proof: It is enough to show that, for each $P \in (a, b)$, the limits

$$\lim_{x \to P^-} f(x)$$

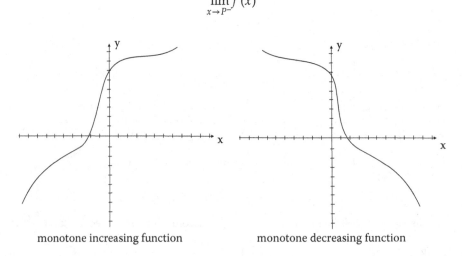

monotone increasing function   monotone decreasing function

FIGURE 5.9
Increasing and decreasing functions.

and

$$\lim_{x \to P^+} f(x)$$

exist.

Let us first assume that f is monotonically increasing. Fix $P \in (a, b)$. If $a < s < P$, then $f(s) \le f(P)$. Therefore, $S = \{f(s): a < s < P\}$ is bounded above. Let M be the least upper bound of S. Pick $\epsilon > 0$. By definition of least upper bound there must be an $f(s) \in S$ such that $|f(s) - M| < \epsilon$. Let $\delta = |P - s|$. If $P - \delta < t < P$, then $s < t < P$ and $f(s) \le f(t) \le M$ or $|f(t) - M| < \epsilon$. Thus, $\lim_{x \to P^-} f(x)$ exists and equals M.

If we set m equal to the infimum of the set $T = \{f(t): P < t < b\}$, then a similar argument shows that $\lim_{x \to P^+} f(x)$ exists and equals m. That completes the proof.

The argument in case f is monotonically decreasing is just the same, and we omit the details. ☐

Corollary 5.40: *Let f be a monotonic function on an interval (a, b). Then, f has at most countably many discontinuities.*

Proof: Assume for simplicity that f is monotonically increasing. If P is a discontinuity, then the proposition tells us that

$$\lim_{x \to P^-} f(x) < \lim_{x \to P^+} f(x).$$

Therefore there is a rational number q_P between $\lim_{x \to P^-} f(x)$ and $\lim_{x \to P^+} f(x)$. Notice that different discontinuities will have different rational numbers associated to them because if $\hat{P}$ is another discontinuity and, say, $\hat{P} < P$, then

$$\lim_{x \to \hat{P}^-} f(x) < q_{\hat{P}} < \lim_{x \to \hat{P}^+} f(x) \le \lim_{x \to P^-} f(x) < q_P < \lim_{x \to P^+} f(x).$$

Thus, we have exhibited a one-to-one function from the set of discontinuities of f into the set of rational numbers. It follows that the set of discontinuities is countable.

The argument in case f is monotonically decreasing is just the same, and we omit the details. ☐

A continuous function f has the property that the inverse image under f of any open set is open. However, it is not in general true that the *image* under f

of any open set is open. A counterexample is the function $f(x) = x^2$ and the open set $O = (-1, 1)$ whose image under f is $[0, 1)$.

Example 5.41: Consider the greatest integer function $f(x) = [x]$. This means that $f(x)$ equals the greatest integer which is less than or equal to x. Then, f is monotone increasing, and its discontinuities are of the first kind and are at the integers. This example illustrates Proposition 5.39 and Corollary 5.40.   □

Definition 5.42: Suppose that f is a function on (a,b) such that $a < s < t < b$ implies $f(s) < f(t)$. Such a function isstrictly increasingstrictly decreasing called *strictly increasing* (*strictly decreasing* functions are defined similarly). We refer to such functions as *strictly monotone*.

It is clear that a strictly increasing (resp. decreasing) function is one-to-one, hence has an inverse. Now, we prove:

Theorem 5.43: *Let f be a strictly monotone, continuous function with domain $[a,b]$. Then, f^{-1}exists and is continuous.*

Proof: Assume without loss of generality that f is strictly monotone *increasing*. Let us extend f to the entire real line by defining

$$f(x) = \begin{cases} (x - a) + f(a) & \text{if } x < a \\ \text{as given} & \text{if } a \le x \le b \\ (x - b) + f(b) & \text{if } x > b. \end{cases}$$

See Figure 5.10. Then, it is easy to see that this extended version of f is still continuous and is strictly monotone increasing on all of $\mathbb{R}$.

That f^{-1} exists has already been discussed. The extended function f takes any open interval (c, d) to the open interval $(f(c), f(d))$. Since any open set is a union of open intervals, we see that f takes any open set to an open set. In other words, $[f^{-1}]^{-1}$ takes open sets to open sets. But this just says that f^{-1} is continuous.

Since the inverse of the extended function f is continuous, then so is the inverse of the original function f. That completes the proof.

Example 5.44: Consider the function

$$f(x) = e^x.$$

It is strictly inreasing on the entire real line, so it has an inverse. Its inverse is in fact the natural logarithm function $\ln x$.

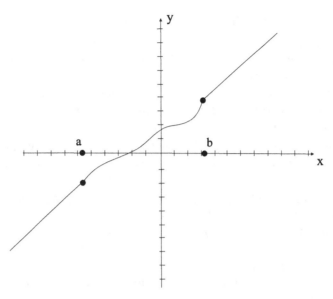

FIGURE 5.10
A strictly monotonically increasing function.

Now, take a look at the function

$$g(x) = \begin{cases} x & \text{if } x \le -1 \\ -1 & \text{if } -1 < x < 1 \\ x - 2 & \text{if } 1 \le x. \end{cases}$$

$\square$

This is a monotone increasing function, but it is *not* strictly increasing. And it has no inverse because it is not one-to-one.

Exercises

1. Let A be any left-to-right ordered, countable subset of the reals. Assume that A has no accumulation points. In particular, $A = \{a_j\}$ and $a_j \to \infty$. Construct an increasing function whose set of points of discontinuity is precisely the set A. Explain why this is, in general, impossible for an uncountable set A.

2. Give an example of two functions, discontinuous at $x = 0$, whose sum *is* continuous at $x = 0$. Give an example of two such functions whose product is continuous at $x = 0$. How does the problem change if we replace "product" by "quotient"?

3. Let f be a function with domain $\mathbb{R}$. If $f^2(x) = f(x) \cdot f(x)$ is continuous, then does it follow that f is continuous? If

$f^3(x) = f(x) \cdot f(x) \cdot f(x)$ is continuous, then does it follow that f is continuous?

4. Fix an interval (a, b). Is the collection of increasing functions on (a, b) closed under $+$, $-$, $\times$ or $\div$?

5. Let f be a continuous function whose domain contains a closed, bounded interval $[a, b]$. What topological properties does $f([a, b])$ possess? Is this set necessarily an interval?

6. Refer to Exercise **8** of Section 5.3 for terminology. Show that there is no homeomorphism from the real line to the interval $[0, 1)$.

7. Let f be a function with domain $\mathbb{R}$. Prove that the set of discontinuities of the first kind for f is countable. (**Hint:** If the left and right limits at a point disagree, then you can slip a rational number between them.)

8. Let $a_1 < a_2 < \cdots$ with the a_j increasing to infinity and the a_j having no finite accumulation points. Give an example of a function with a discontinuity of the second kind at each a_j and no other discontinuities.

9. Let $I \subseteq \mathbb{R}$ be an open interval and $f: I \to \mathbb{R}$ a function. We say that f is *convex* if whenever $\alpha, \beta \in I$ and $0 \leq t \leq 1$, then

$$f((1 - t)\alpha + t\beta) \leq (1 - t)f(\alpha) + tf(\beta).$$

Prove that a convex function must be continuous. What does this definition of convex function have to do with the notion of "concave up" that you learned in calculus?

*10. Refer to Exercise **9** for terminology. What can you say about differentiability of a convex function?

*11. TRUE or FALSE: If f is a continuous function with domain and range the real numbers and which is both one-to-one and onto, then f must be either increasing or decreasing. Does your answer change if we assume that f is continuously differentiable (see the next chapter for this terminology)?

12. Refer to Exercise 9. Is the sum of two convex functions convex? Is the product of two convex functions convex?

13. What kind of discontinuity does the function $f(x) = \sin(1/x)$ have at the origin?

14. Discuss discontinuities for the trigonometric functions tangent, cotangent, secant, and cosecant. What kinds of discontinuities do they have?

6

Differentiation of Functions

6.1 The Concept of Derivative

Let f be a function with domain an open interval I. If $x \in I$, then the quantity

$$\frac{f(t) - f(x)}{t - x}$$

measures the slope of the chord of the graph of f that connects the points $(x, f(x))$ and $(t, f(t))$. See Figure 6.1. If we let $t \to x$, then the limit of the quantity represented by this "Newton quotient" should represent the slope of the graph *at the point* x. These considerations motivate the definition of the following derivative:

Definition 6.1: If f is a function with domain an open interval I and if $x \in I$, then the limit

$$\lim_{t \to x} \frac{f(t) - f(x)}{t - x},$$

when it exists, is called the *derivative* of f at x. See Figure 6.2. If the derivative of f at x exists, then we say that f is *differentiable* at x. If f is differentiable at every $x \in I$, then we say that f is *differentiable on I*.

We write the derivative of f at x either as

$$f'(x) \quad \text{or} \quad \frac{d}{dx}f \quad \text{or} \quad \frac{df}{dx} \quad \text{or} \quad \dot{f}.$$

We begin our discussion of the derivative by establishing some basic properties and relating the notion of derivative to continuity.

DOI: 10.1201/9781003222682-7

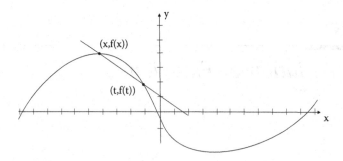

FIGURE 6.1
The Newton quotient.

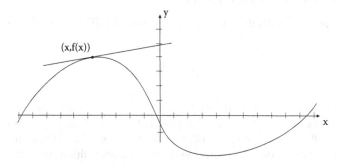

FIGURE 6.2
The derivative.

Lemma 6.2: *If f is differentiable at a point x, then f is continuous at x. In particular, $\lim_{t \to x} f(t) = f(x)$.*

Proof: We use Theorem 5.5**(b)** about limits to see that

$$\lim_{t \to x} (f(t) - f(x)) = \lim_{t \to x} \left((t - x) \cdot \frac{f(t) - f(x)}{t - x} \right)$$
$$= \lim_{t \to x} (t - x) \cdot \lim_{t \to x} \frac{f(t) - f(x)}{t - x}$$
$$= 0 \cdot f'(x)$$
$$= 0.$$

Therefore $\lim_{t \to x} f(t) = f(x)$ and f is continuous at x.   □

Example 6.3: All differentiable functions are continuous: differentiability is a stronger property than continuity. Observe that the function $f(x) = |x|$ is continuous at every x but is not differentiable at 0. So continuity does not imply differentiability. Details appear in Example 6.5.   □

Theorem 6.4: *Assume that f and g are functions with domain an open interval I and that f and g are differentiable at $x \in I$. Then, $f \pm g, f \cdot g$, and f/g are differentiable at x (for f/g we assume that $g(x) \neq 0$). Moreover*

a. $(f \pm g)''(x) = f'(x) \pm g'(x)$;

b. $(f \cdot g)'(x) = f'(x) \cdot g(x) + f(x) \cdot g'(x)$;

c. $\left(\dfrac{f}{g}\right)'(x) = \dfrac{g(x) \cdot f'(x) - f(x) \cdot g'(x)}{g^2(x)}$.

Proof: Assertion **(a)** is easy and we leave it as an exercise for you. For **(b)**, we write

$$\lim_{t \to x} \frac{(f \cdot g)(t) - (f \cdot g)(x)}{t - x} = \lim_{t \to x} \left(\frac{\dfrac{(f(t) - f(x)) \cdot g(t)}{t - x}}{+ \dfrac{(g(t) - g(x)) \cdot f(x)}{t - x}} \right)$$

$$= \lim_{t \to x} \left(\frac{(f(t) - f(x)) \cdot g(t)}{t - x} \right)$$

$$+ \lim_{t \to x} \left(\frac{(g(t) - g(x)) \cdot f(x)}{t - x} \right)$$

$$= \lim_{t \to x} \left(\frac{(f(t) - f(x))}{t - x} \right) \cdot \left[\lim_{t \to x} g(t) \right]$$

$$+ \lim_{t \to x} \left(\frac{(g(t) - g(x))}{t - x} \right) \cdot \left[\lim_{t \to x} f(x) \right],$$

where we have used Theorem 5.5 about limits. Now the first limit is the derivative of f at x, while the third limit is the derivative of g at x. Also notice that the limit of $g(t)$ equals $g(x)$ by the lemma. The result is that the last line equals

$$f'(x) \cdot g(x) + g'(x) \cdot f(x),$$

as desired.

To prove **(c)**, write

$$\lim_{t \to x} \frac{(f/g)(t) - (f/g)(x)}{t - x} = \lim_{t \to x} \frac{1}{g(t) \cdot g(x)} \left(\frac{\dfrac{f(t) - f(x)}{t - x} \cdot g(x)}{- \dfrac{g(t) - g(x)}{t - x} \cdot f(x)} \right).$$

The proof is now completed by using Theorem 5.5 about limits to evaluate the individual limits in this expression.   □

Example 6.5: That $f(x) = x$ is differentiable follows from

$$\lim_{t \to x} \frac{t - x}{t - x} = 1.$$

Any constant function is differentiable (with derivative identically zero) by a similar argument. It follows from the theorem that any polynomial function is differentiable.

On the other hand, the continuous function $f(x) = |x|$ is *not* differentiable at the point $x = 0$. This is so because

$$\lim_{t \to 0^-} \frac{|t| - |0|}{t - x} = \lim_{t \to 0^-} \frac{-t - 0}{t - 0} = -1$$

while

$$\lim_{t \to 0^+} \frac{|t| - |0|}{t - x} = \lim_{t \to 0^+} \frac{t - 0}{t - 0} = 1.$$

So the required limit does not exist.   □

Since the subject of differential calculus is concerned with learning uses of the derivative, it concentrates on functions which *are* differentiable. One comes away from the subject with the impression that most functions are differentiable except at a few isolated points—as is the case with the function $f(x) = |x|$. Indeed this was what the mathematicians of the nineteenth century thought. Therefore it came as a shock when Karl Weierstrass produced a continuous function that is not differentiable at *any point*. In a sense that can be made precise, *most* continuous functions are of this nature: their graphs "wiggle" so much that they cannot have a tangent line at any point. Now we turn to an elegant variant of the example of Weierstrass that is due to B. L. van der Waerden (1903–1996).

Theorem 6.6: *Define a function ψ with domain $\mathbb{R}$ by the rule*

$$\psi(x) = \begin{cases} x - n & \text{if } n \leq x < n + 1 \text{ and } n \text{ is even} \\ n + 1 - x & \text{if } n \leq x < n + 1 \text{ and } n \text{ is odd} \end{cases}$$

for every integer n. The graph of this function is exhibited in Figure 6.3. *Then, the function*

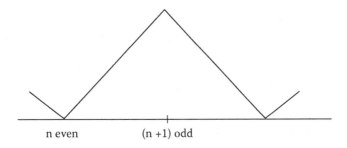

FIGURE 6.3
The van der Waerden example.

$$f(x) = \sum_{j=1}^{\infty} \left(\frac{3}{4}\right)^{j} \psi(4^{j}x)$$

is continuous at every real x and differentiable at no real x.

Proof: Since we have not yet discussed series of functions, we take a moment to understand the definition of f. Fix a real x. Notice that $0 \le \psi(x) \le 1$ for every x. Then, the series becomes a series of numbers, and the jth summand does not exceed $(3/4)^{j}$ in absolute value. Thus, the series converges absolutely; therefore it converges. So it is clear that the displayed formula defines a function of x.

Step I: f is continuous. To see that f is continuous, pick an $\epsilon > 0$. Choose N so large that

$$\sum_{j=N+1}^{\infty} \left(\frac{3}{4}\right)^{j} < \frac{\epsilon}{4}$$

(we can of course do this because the series $\sum \left(\frac{3}{4}\right)^{j}$ converges). Now fix x. Observe that, since ψ is continuous and the graph of ψ is composed of segments of slope 1, we have

$$|\psi(s) - \psi(t)| \le |s - t|$$

for all s and t. Moreover $|\psi(s) - \psi(t)| \le 1$ for all s, t.
For $j = 1, 2, \ldots, N$, pick $\delta_j > 0$ so that, when $|t - x| < \delta_j$, then

$$|\psi(4^{j}t) - \psi(4^{j}x)| < \frac{\epsilon}{8}.$$

Let δ be the minimum of $\delta 1, \ldots \delta N$.

Now, if $|t - x| < \delta$, then

$$|f(t) - f(x)| = \left| \sum_{j=1}^{N} \left(\frac{3}{4} \right)^j \cdot (\psi(4^j t) - \psi(4^j x)) \right.$$

$$\left. + \sum_{j=N+1}^{\infty} \left(\frac{3}{4} \right)^j \cdot (\psi(4^j t) - \psi(4^j x)) \right|$$

$$\leq \sum_{j=1}^{N} \left(\frac{3}{4} \right)^j | (\psi(4^j t) - \psi(4^j x)) |$$

$$+ \sum_{j=N+1}^{\infty} \left(\frac{3}{4} \right)^j |\psi(4^j t) - \psi(4^j x)|$$

$$\leq \sum_{j=1}^{N} \left(\frac{3}{4} \right)^j \cdot \frac{\epsilon}{8} + \sum_{j=N+1}^{\infty} \left(\frac{3}{4} \right)^j.$$

Here, we have used the choice of δ to estimate the summands in the first sum. The first sum is thus less than $\epsilon/2$ (just notice that $\sum_{j=1}^{\infty} (3/4)^j < 4$). The second sum is less than $\epsilon/2$ by the choice of N. Altogether then

$$|f(t) - f(x)| < \epsilon$$

whenever $|t - x| < \delta$. Therefore f is continuous, indeed uniformly so.

Step II: f is nowhere differentiable. Fix x. For $\ell = 1, 2, \ldots$ define $t_\ell = x \pm 4^{-\ell}/2$. We will say whether the sign is plus or minus in a moment (this will depend on the position of x relative to the integers). Then,

$$\left| \frac{f(t_\ell) - f(x)}{t_\ell - x} \right| = \left| \frac{1}{t_\ell - x} \left[\sum_{j=1}^{\ell} \left(\frac{3}{4} \right)^j (\psi(4^j t_\ell) - \psi(4^j x)) \right. \right.$$

$$\left. \left. + \sum_{j=\ell+1}^{\infty} \left(\frac{3}{4} \right)^j (\psi(4^j t_\ell) - \psi(4^j x)) \right] \right|. \tag{6.6.1}$$

Notice that, when $j \geq \ell+1$, then $4^j t_\ell$ and $4^j x$ differ by an even integer. Since ψ has period 2, we find that each of the summands in the second sum is 0. Next we turn to the first sum.

We choose the sign—plus or minus—in the definition of t_ℓ so that there is no integer lying between $4^\ell t_\ell$ and $4^\ell x$. We can do this because the two numbers differ by $1/2$. But then, the ℓth summand has magnitude

$$(3/4)^\ell \cdot |4^\ell t_\ell - 4^\ell x| = 3^\ell |t_\ell - x|.$$

On the other hand, the first $\ell - 1$ summands add up to not more than

$$\sum_{j=1}^{\ell-1} \left(\frac{3}{4}\right)^j \cdot |4^j t_\ell - 4^j x| = \sum_{j=1}^{\ell-1} 3^j \cdot 4^{-\ell}/2 \le \frac{3^\ell - 1}{3 - 1} \cdot 4^{-\ell}/2 \le 3^\ell \cdot 4^{-\ell-1}.$$

It follows that

$$\left| \frac{f(t_\ell) - f(x)}{t_\ell - x} \right| = \frac{1}{|t_\ell - x|} \cdot \left| \sum_{j=1}^{\ell} \left(\frac{3}{4}\right)^j (\psi(4^j t_\ell) - \psi(4^j x)) \right|$$

$$= \frac{1}{|t_\ell - x|} \cdot \left| \sum_{j=1}^{\ell-1} \left(\frac{3}{4}\right)^j (\psi(4^j t_\ell) - \psi(4^j x)) + \left(\frac{3}{4}\right)^\ell (\psi(4^\ell t_\ell) - \psi(4^\ell x)) \right|$$

$$\ge \frac{1}{|t_\ell - x|} \cdot \left| \left(\frac{3}{4}\right)^\ell \psi(4^\ell t_\ell) - \left(\frac{3}{4}\right)^\ell \psi(4^\ell x) \right|$$

$$- \frac{1}{|t_\ell - x|} \left| \sum_{j=1}^{\ell-1} \left(\frac{3}{4}\right)^j (\psi(4^j t_\ell) - \psi(4^j x)) \right|$$

$$\ge 3^\ell - \frac{1}{(4^{-\ell}/2)} \cdot 3^\ell \cdot 4^{-\ell-1}$$

$$\ge 3^{\ell-1}.$$

Thus, $t_\ell \to x$ but the Newton quotients blow up as $\ell \to \infty$. Therefore the limit

$$\lim_{t \to x} \frac{f(t) - f(x)}{t - x}$$

cannot exist. The function f is not differentiable at x. □

The proof of the last theorem was long, but the idea is simple: the function f is built by piling oscillations on top of oscillations. When the ℓth oscillation is added, it is made very small in size so that it does not cancel the previous oscillations. But it is made very steep so that it will cause the derivative to become large.

The practical meaning of Weierstrass's example is that we should realize that differentiability is a very strong and special property of functions. Most continuous functions are not differentiable at any point (see Section 11.3 for a rigorous statement and proof). When we are proving theorems about continuous functions, we should *not* think of them in terms of properties of differentiable functions.

Next we turn to the Chain Rule.

Theorem 6.7: *Let g be a differentiable function on an open interval I and let f be a differentiable function on an open interval that contains the range of g. Then, f ∘ g is differentiable on the interval I and*

$$(f \circ g)'(x) = f'(g(x)) \cdot g'(x)$$

for each x ∈ I.

Proof: We use the notation Δt to stand for an increment in the variable t. Let us use the symbol $V(r)$ to stand for any expression which tends to 0 as $\Delta r \to 0$. Fix $x \in I$. Set $r = g(x)$. By hypothesis,

$$\lim_{\Delta r \to 0} \frac{f(r + \Delta r) - f(r)}{\Delta r} = f'(r)$$

or

$$\frac{f(r + \Delta r) - f(r)}{\Delta r} - f'(r) = V(r)$$

or

$$f(r + \Delta r) = f(r) + \Delta r \cdot f'(r) + \Delta r \cdot V(r). \qquad (6.7.1)$$

Notice that equation (6.7.1) is valid even when $\Delta r = 0$. Since Δr in equation (6.7.1) can be any small quantity, we set

$$\Delta r = \Delta x \cdot [g'(x) + V(x)].$$

Substituting this expression into (6.7.1) and using the fact that $r = g(x)$ yields

$$f(g(x) + \Delta x [g'(x) + V(x)]) =$$
$$f(r) + (\Delta x \cdot [g'(x) + V(x)]) \cdot f'(r) + (\Delta x \cdot [g'(x) + V(x)]) \cdot V(r) \quad (6.7.2)$$
$$= f(g(x)) + \Delta x \cdot f'(g(x)) \cdot g'(x) + \Delta x \cdot V(x).$$

Just as we derived (6.7.1), we may also obtain

$$g(x + \Delta x) = g(x) + \Delta x \cdot g'(x) + \Delta x \cdot V(x)$$
$$= g(x) + \Delta x [g'(x) + V(x)].$$

We may substitute this equality into the left side of (6.7.2) to obtain

$$f(g(x + \Delta x)) = f(g(x)) + \Delta x \cdot f'(g(x)) \cdot g'(x) + \Delta x \cdot V(x).$$

With some algebra this can be rewritten as

$$\frac{f(g(x + \Delta x)) - f(g(x))}{\Delta x} - f'(g(x)) \cdot g'(x) = V(x).$$

But this just says that

$$\lim_{\Delta x \to 0} \frac{(f \circ g)(x + \Delta x) - (f \circ g)(x)}{\Delta x} = f'(g(x)) \cdot g'(x).$$

That is, $(f \circ g)'(x)$ exists and equals $f'(g(x)) \cdot g'(x)$, as desired. □

Example 6.8: The derivative of

$$f(x) = \sin(x^3 - x^2)$$

is

$$f'(x) = [\cos(x^3 - x^2)] \cdot (3x^2 - 2x).$$

□

****

ISAAC NEWTON

Sir Isaac Newton PRS (1642–1726) was an English mathematician, physicist, astronomer, theologian, and author. He is widely recognised as one of the greatest mathematicians and most influential scientists of all time. His book *Philosophia Naturalis Principia Mathematica*, first published in 1687, established classical mechanics.

In *Principia*, Newton formulated the laws of motion and universal gravitation that formed the dominant scientific viewpoint until it was superseded by the theory of relativity. Newton used his mathematical description of gravity to derive Kepler's laws of planetary motion, account for tides, the trajectories of comets, the precession of the equinoxes and other phenomena, eradicating doubt about the Solar System's heliocentricity. Newton's inference that the Earth is an oblate spheroid was later confirmed by the geodetic measurements of Maupertuis, La Condamine, and others, convincing most European scientists of the superiority of Newtonian mechanics over earlier systems.

Newton built the first practical reflecting telescope and developed a sophisticated theory of colour based on the observation that a prism separates white light into the colours of the visible spectrum. His work on light was collected in his highly influential book Opticks, published in 1704. In addition to his work on calculus, as a mathematician Newton contributed to the study of power series, generalised the binomial theorem to non-integer exponents, developed a method for approximating the roots of a function, and classified most of the cubic plane curves.

Newton was a fellow of Trinity College and the second Lucasian Professor of Mathematics at the University of Cambridge. Newton dedicated much of his time to the study of alchemy and biblical chronology, but most of his work in those areas remained unpublished until long after his death.

Isaac Newton was born on Christmas Day, 25 December 1642. His father, also named Isaac Newton, had died three months before. When Newton was three, his mother remarried and went to live with her new husband, the Reverend Barnabas Smith, leaving her son in the care of his maternal grandmother, Margery Ayscough (née Blythe).

In June 1661, Newton was admitted to Trinity College, Cambridge, on the recommendation of his uncle Rev William Ayscough, who had studied there.

He started as a subsizar—paying his way by performing valet's duties—until he was awarded a scholarship in 1664, guaranteeing him four more years until he could get his MA. In 1665, he discovered the generalised binomial theorem and began to develop a mathematical theory that later became calculus. Soon after Newton had obtained his BA degree in August 1665, the university temporarily closed as a precaution against the Great Plague.

In April 1667, he returned to Cambridge and in October was elected as a fellow of Trinity.

His studies had impressed the Lucasian professor Isaac Barrow, who was more anxious to develop his own religious and administrative potential (he became master of Trinity two years later); in 1669 Newton succeeded him, only one year after receiving his MA. He was elected a Fellow of the Royal Society (FRS) in 1672.

Newton later became involved in a dispute with Leibniz over priority in the development of calculus (the Leibniz–Newton calculus controversy). Most modern historians believe that Newton and Leibniz developed calculus independently, although with very different mathematical notations. Occasionally it has been suggested that Newton published almost nothing about it until 1693, and did not give a full account until 1704, while Leibniz began publishing a full account of his methods in 1684.

Because of this, the *Principia* has been called "a book dense with the theory and application of the infinitesimal calculus" in modern times and in Newton's time "nearly all of it is of this calculus."

Newton is generally credited with the generalised binomial theorem, valid for any exponent. He discovered Newton's identities, Newton's method, classified cubic plane curves (polynomials of degree three in two variables), made substantial contributions to the theory of finite differences, and was the first to use fractional indices and to employ coordinate geometry to derive solutions to Diophantine equations.

He was appointed Lucasian Professor of Mathematics in 1669, on Barrow's recommendation. During that time, any Fellow of a college at Cambridge or Oxford was required to take holy orders and become an ordained Anglican priest. However, the terms of the Lucasian professorship required that the holder not be active in the church—presumably so as to have more time for science. Newton argued that this should exempt him from the ordination requirement, and Charles II, whose permission was needed, accepted this argument.

Facsimile of a 1682 letter from Isaac Newton to Dr William Briggs, commenting on Briggs' A New Theory of Vision. Newton argued that light is composed of particles or corpuscles, which were refracted by accelerating into a denser medium. He verged on soundlike waves to explain the repeated pattern of reflection and transmission by thin films (Opticks Bk.II, Props. 12), but still retained his theory of 'fits' that disposed corpuscles to be reflected or transmitted. However, later physicists favoured a purely wavelike explanation of light to account for the interference patterns and the general phenomenon of diffraction. Today's quantum mechanics, photons, and the idea of waveparticle duality bear only a minor resemblance to Newton's understanding of light.

In 1704, Newton published *Opticks*, in which he expounded his corpuscular theory of light. He considered light to be made up of extremely subtle corpuscles, that ordinary matter was made of grosser corpuscles and speculated that through a kind of alchemical transmutation.

In 1679, Newton returned to his work on celestial mechanics by considering gravitation and its effect on the orbits of planets with reference to Kepler's laws of planetary motion. Newton communicated his results to Edmond Halley and to the Royal Society in De motu corporum in gyrum, a tract written on about nine sheets which was copied into the Royal Society's Register Book in December 1684. This tract contained the nucleus that Newton developed and expanded to form the *Principia*.

The *Principia* was published on 5 July 1687 with encouragement and financial help from Edmond Halley. In this work, Newton stated the three universal laws of motion. Together, these laws describe the relationship between any object, the forces acting upon it and the resulting motion, laying the foundation for classical mechanics. They contributed to many advances during the Industrial Revolution which soon followed and were not improved upon for more than 200 years.

Newton's postulate of an invisible force able to act over vast distances led to him being criticised for introducing "occult agencies" into science. Later, in the second edition of the *Principia* (1713), Newton firmly rejected such criticisms in a concluding General Scholium, writing that it was enough that the phenomena implied a gravitational attraction, as they did; but they did not so far indicate its cause, and it was both unnecessary and improper to frame hypotheses of things that were not implied by the phenomena.

Newton was also a member of the Parliament of England for Cambridge University in 1689 and 1701, but according to some accounts his only comments were to complain about a cold draught in the chamber and request that the window be closed. He was, however, noted by Cambridge diarist Abraham de la Pryme to have rebuked students who were frightening locals by claiming that a house was haunted.

Newton moved to London to take up the post of warden of the Royal Mint in 1696, a position that he had obtained through the patronage of Charles Montagu, 1st Earl of Halifax, then Chancellor of the Exchequer. He took charge of England's great recoining, trod on the toes of Lord Lucas, Governor of the Tower, and secured the job of deputy comptroller of the temporary Chester branch for Edmond Halley.

Although it was claimed that he was once engaged, Newton never married. The French writer and philosopher Voltaire, who was in London at the time of Newton's funeral, said that he "was never sensible to any passion, was not subject to the common frailties of mankind, nor had any commerce with women—a circumstance which was assured me by the physician and surgeon who attended him in his last moments." This now-widespread belief that he died a virgin has been commented on by writers as diverse as mathematician Charles Hutton, economist John Maynard Keynes, and physicist Carl Sagan.

Newton had a close friendship with the Swiss mathematician Nicolas Fatio de Duillier, whom he met in London around 1689—some of their correspondence has survived. Their relationship came to an abrupt and unexplained end in 1693, and at the same time Newton suffered a nervous breakdown which included sending wild accusatory letters to his friends Samuel Pepys and John Locke—his note to the latter included the charge that Locke "endeavoured to embroil me with woemen."

****

Exercises

1. For which positive integers k is it true that if $f^k = f \cdot f \cdots f$ is differentiable at x, then f is differentiable at x?

2. Let f be a function that has domain an interval I and takes values in the complex numbers. Then, we may write $f(x) = u(x) + iv(x)$ with u and v each being real-valued functions. We say that f is differentiable at a point $x \in I$ if both u and v are. Formulate an alternative definition of differentiability of f at a point x which

makes no reference to u and v (but instead defines the derivative directly in terms of f) and prove that your new definition is equivalent to the definition in terms of u and v.

3. Let $f(x)$ equal 0 if x is irrational; let $f(x)$ equal $1/q$ if x is a rational number that can be expressed in lowest terms as p/q. Is f differentiable at any x?

4. Assume that f is a continuous function on $(-1, 1)$ and that f is differentiable on $(-1, 0) \cup (0, 1)$. If the limit $\lim_{x \to 0} f'(x)$ exists, then is f differentiable at $x = 0$?

5. Formulate notions of "left differentiable" and "right differentiable" for functions defined on suitable half-open intervals. Also formulate definitions of "left continuous" and "right continuous." If you have done things correctly, then you should be able to prove that a left differentiable (right differentiable) function is left continuous (right continuous).

6. Define

$$f(x) = \begin{cases} x^{3/2} \cdot \sin(1/x) & \text{if } x \neq 0 \\ 0 & \text{if } x = 0. \end{cases}$$

Prove that f is differentiable at every point, but that the derivative function f' is discontinuous at 0.

7. Refer to Exercise **6**. Is the discontinuity at the origin of the first kind or the second kind?

8. Prove part (a) of Theorem 6.4.

9. Refer to Exercise 2. Verify the properties of the derivative presented in Theorem 6.4 in the new context of complex-valued functions.

*10. Let $E \subseteq \mathbb{R}$ be a closed set. Fix a nonnegative integer k. Show that there is a function f in $C^k(\mathbb{R})$ (that is, a k-times continuously differentiable function) such that $E = \{x : f(x) = 0\}$.

*11. Prove that the nowhere differentiable function constructed in Theorem 6.6 is in Lip_α for all $\alpha < 1$.

*12. Prove that the Weierstrass nowhere differentiable function f constructed in Theorem 6.6 satisfies

$$\frac{|f(x+h) + f(x-h) - 2f(x)|}{|h|} \leq C|h|$$

for all nonzero h but f is *not* Lipschitz-1.

13. Fill in the details for this alternative proof of the product rule for the following derivatives:

   a. Prove that $(f^2)' = 2f \cdot f'$.

   b. Apply the result of part **(a)** to the function $f + g$.

   c. Use the result of part **(a)** to cancel terms in the formula from part **(b)** to obtain the product rule.

14. With suitable hypotheses, the quotient rule for differentiation can be derived from the product rule. Explain.

15. Prove rigorously that the function $f(x) = \sin x$ is differentiable at every point.

16. Prove rigorously that the function $g(x) = \log|x|$ is differentiable at all $x \neq 0$.

6.2 The Mean Value Theorem and Applications

We begin this section with some remarks about local maxima and minima of functions.

Definition 6.9: Let f be a function with domain (a, b). A point $C \in (a, b)$ is called a *local maximum* for f (we also say that f has a local maximum at C) if there is a $\delta > 0$ such that $f(t) \leq f(C)$ for all $t \in (C - \delta, C + \delta)$. A point $c \in (a, b)$ is called a *local minimum* for f (we also say that f has a local minimum at c) if there is a $\delta > 0$ such that $f(t) \geq f(c)$ for all $t \in (c - \delta, c + \delta)$. See Figure 6.4.

   Local minima (plural of minimum) and local maxima (plural of maximum) are referred to collectively as *local extrema*.

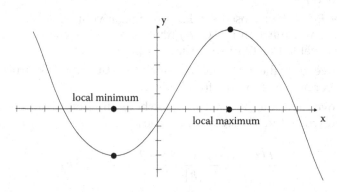

FIGURE 6.4
Some extrema.

Proposition 6.10: (Fermat) *If f is a function with domain (a, b), if f has a local extremum at $x \in (a, b)$, and if f is differentiable at x, then $f'(x) = 0$.*

Proof: Suppose that *f* has a local minimum at *x*. Then, there is a $\delta > 0$ such that if $x - \delta < t < x$, then $f(t) \geq f(x)$. Then,

$$\frac{f(t) - f(x)}{t - x} \leq 0.$$

Letting $t \to x$, it follows that $f'(x) \leq 0$. Similarly, if $x < t < x + \delta$ for suitable δ, then

$$\frac{f(t) - f(x)}{t - x} \geq 0.$$

It follows that $f'(x) \geq 0$. We must conclude that $f'(x) = 0$.

A similar argument applies if *f* has a local maximum at *x*. The proof is therefore complete. □

Example 6.11: Consider the function $f(x) = \sin x$ on the interval $[\pi/3, 11\pi/3]$. Surely $f'(x) = \cos x$, and we see that *f'* vanishes at the points $\pi/2, 3\pi/2, 5\pi/2, 7\pi/2$ of the interval. By Fermat's theorem, these are candidates to be local maxima or minima of *f*. And, indeed, $\pi/2, 5\pi/2$ are local maxima and $3\pi/2, 7\pi/2$ are local minima. □

Before going on to mean value theorems, we provide a striking application of the following proposition:

Theorem 6.12: (Darboux's Theorem) *Let f be a differentiable function on an open interval I. Pick points $s < t$ in I and suppose that $f'(s) < \rho < f'(t)$. Then, there is a point u between s and t such that $f'(u) = \rho$.*

Proof: Consider the function $g(x) = f(x) - \rho x$. Then, $g'(s) < 0$ and $g'(t) > 0$. Assume for simplicity that $s < t$. The sign of the derivative at *s* shows that $g(s) < g(s)$ for *s* greater than *s* and near *s*. The sign of the derivative at *t* implies that $g(t) < g(t)$ for *t* less than *t* and near *t*. Thus the minimum of the continuous function *g* on the compact interval $[s, t]$ must occur at some point *u* in the interior (s, t). The preceding proposition guarantees that $g'(u) = 0$, or $f'(u) = \rho$ as claimed. □

Example 6.13: If *f'* were a continuous function, then the theorem would just be a special instance of the Intermediate Value Property of continuous functions (see Corollary 5.31). But derivatives need not be continuous, as the example

$$f(x) = \begin{cases} x^{3/2} \cdot \sin(1/x) & \text{if} \quad x \neq 0 \\ 0 & \text{if} \quad x = 0 \end{cases}$$

illustrates. Check for yourself that $f'(0)$ exists and vanishes but $\lim_{x \to 0} f'(x)$ does not exist. This example illustrates the significance of the theorem. □

Since the theorem says that f' will always satisfy the Intermediate Value Property (even when it is not continuous), its discontinuities cannot be of the first kind. In other words:

Proposition 6.14: *If f is a differentiable function on an open interval I, then the discontinuities of f' are all of the second kind.*

Next, we turn to the simplest form of the Mean Value Theorem.

Theorem 6.15: (Rolle's Theorem) *Let f be a continuous function on the closed interval [a, b] which is differentiable on (a,b). If f(a) = f(b) = 0, then there is a point $\xi \in (a, b)$ such that $f'(\xi) = 0$. See Figure 6.5.*

Proof: If f is a constant function, then any point ξ in the interval will do. So assume that f is nonconstant.

Theorem 5.23 guarantees that f will have both a maximum and a minimum in $[a, b]$. If one of these occurs in (a, b), then Proposition 6.10 guarantees that f' will vanish at that point and we are done. If both occur at the endpoints, then all the values of f lie between 0 and 0. In other words f is constant, contradicting our assumption.   □

Example 6.16: Of course the point ξ in Rolle's theorem need not be unique. If $f(x) = x^3 - x^2 - 2x$ on the interval $[-1, 2]$, then $f(-1) = f(2) = 0$ and $f'(x) = 3x^2 -$

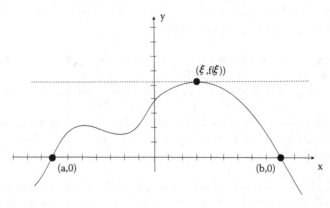

FIGURE 6.5
Rolle's theorem.

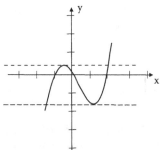

FIGURE 6.6
An example of Rolle's theorem.

$2x - 2$ vanishes at *two* points (namely, $(2 + \sqrt{28})/6$ and $(2 - \sqrt{28})/6$) of the interval $(-1, 2)$. Refer to Figure 6.6.   □

If you rotate the graph of a function satisfying the hypotheses of Rolle's theorem, the result suggests that, for any continuous function f on an interval $[a, b]$, differentiable on (a,b), we should be able to relate the slope of the chord connecting $(a, f(a))$ and $(b, f(b))$ with the value of f' at some interior point. That is the content of the following standard Mean Value Theorem:

Theorem 6.17: (The Mean Value Theorem) *Let f be a continuous function on the closed interval $[a, b]$ that is differentiable on (a, b). There exists a point $\xi \in (a, b)$ such that*

$$\frac{f(b) - f(a)}{b - a} = f'(\xi).$$

See Figure 6.7.

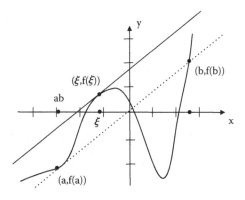

FIGURE 6.7
The Mean Value Theorem.

Proof: Our scheme is to implement the remarks preceding the theorem: we "rotate" the picture to reduce to the case of Rolle's theorem. More precisely, define

$$g(x) = f(x) - \left[f(a) + \frac{f(b) - f(a)}{b - a} \cdot (x - a) \right] \quad \text{if} \quad x \in [a, b].$$

By direct verification, g is continuous on $[a, b]$ and differentiable on (a, b) (after all, g is obtained from f by elementary arithmetic operations). Also $g(a) = g(b) = 0$. Thus, we may apply Rolle's theorem to g and we find that there is a $\xi \in (a, b)$ such that $g'(\xi) = 0$. Remembering that x is the variable, we differentiate the formula for g to find that

$$0 = g'(\xi) = \left[f'(x) - \frac{f(b) - f(a)}{b - a} \right]\Bigg|_{x = \xi}$$

$$= \left[f'(\xi) - \frac{f(b) - f(a)}{b - a} \right].$$

As a result,

$$f'(\xi) = \frac{f(b) - f(a)}{b - a}.$$

□

Corollary 6.18: *If f is a differentiable function on the open interval I and if $f'(x) = 0$ for all $x \in I$, then f is a constant function.*

Proof: If s and t are any two elements of I, then the theorem tells us that

$$f(s) - f(t) = f'(\xi) \cdot (s - t)$$

for some ξ between s and t. But, by hypothesis, $f'(\xi) = 0$. We conclude that $f(s) = f(t)$. But, since s and t were chosen arbitrarily, we must conclude that f is constant.

□

Corollary 6.19: *If f is differentiable on an open interval I and $f'(x) \geq 0$ for all $x \in I$, then f is increasing on I; that is, if $s < t$ are elements of I, then $f(s) \leq f(t)$.*
If f is differentiable on an open interval I and $f'(x) \leq 0$ for all $x \in I$, then f is decreasing on I; that is, if $s < t$ are elements of I, then $f(s) \geq f(t)$.

Proof: Similar to the preceding corollary.

□

Example 6.20: Let us verify that, if f is a differentiable function on $\mathbb{R}$, and if $|f'(x)| \leq 1$ for all x, then $|f(s) - f(t)| \leq |s - t|$ for all real s and t.

In fact, for $s \neq t$ there is a ξ between s and t such that

$$\frac{f(s) - f(t)}{s - t} = f'(\xi).$$

But $|f'(\xi)| \leq 1$ by hypothesis; hence,

$$\left| \frac{f(s) - f(t)}{s - t} \right| = |f'(\xi)| \leq 1$$

or

$$|f(s) - f(t)| \leq |s - t|.$$

$\square$

Example 6.21: Let us verify that

$$\lim_{x \to +\infty} (\sqrt{x + 5} - \sqrt{x}) = 0.$$

Here, the limit operation means that, for any $\epsilon > 0$, there is an $N > 0$ such that $x > N$ implies that the expression in parentheses has absolute value less than ϵ.

Define $f(x) = \sqrt{x}$ for $x > 0$. Then, the expression in parentheses is just $f(x + 5) - f(x)$. By the Mean Value Theorem this equals

$$f'(\xi) \cdot 5$$

for some $x < \xi < x + 5$. But this last expression is

$$\frac{1}{2} \cdot \xi^{-1/2} \cdot 5.$$

By the bounds on ξ, this is

$$\leq \frac{5}{2} x^{-1/2}.$$

Clearly, as $x \to +\infty$, this expression tends to zero.

$\square$

A powerful tool in analysis is a generalization of the usual Mean Value Theorem that is due to A. L. Cauchy (1789–1857).

Theorem 6.22: (Cauchy's Mean Value Theorem) *Let f and g be continuous functions on the interval $[a,b]$ which are both differentiable on the interval (a, b), $a < b$. Assume that $g' \neq 0$ on the interval and that $g(a) \neq g(b)$. Then, there is a point $\xi \in (a, b)$ such that*

$$\frac{f(b) - f(a)}{g(b) - g(a)} = \frac{f'(\xi)}{g'(\xi)}.$$

Proof: Apply the usual Mean Value Theorem to the function

$$h(x) = g(x) \cdot \{f(b) - f(a)\} - f(x) \cdot \{g(b) - g(a)\}.$$

□

Clearly the usual Mean Value Theorem (Theorem 6.17) is obtained from Cauchy's by taking $g(x)$ to be the function x. We conclude this section by illustrating a typical application of the result.

Example 6.23: Let f be a differentiable function on an interval I such that f' is differentiable at a point $x \in I$. Then,

$$\lim_{h \to 0+} \frac{f(x + h) + f(x - h) - 2f(x)}{h^2} = (f')'(x) \equiv f''(x).$$

To see this, fix x and define $\mathcal{F}(h) = f(x + h) + f(x - h) - 2f(x)$ and $\mathcal{G}(h) = h^2$. Then,

$$\frac{f(x + h) + f(x - h) - 2f(x)}{h^2} = \frac{\mathcal{F}(h) - \mathcal{F}(0)}{\mathcal{G}(h) - \mathcal{G}(0)}.$$

According to Cauchy's Mean Value Theorem, there is a ξ between 0 and h such that the last line equals

$$\frac{\mathcal{F}'(\xi)}{\mathcal{G}'(\xi)}.$$

Writing this last expression out gives

$$\frac{f'(x + \xi) - f'(x - \xi)}{2\xi} = \frac{1}{2} \cdot \frac{f'(x + \xi) - f'(x)}{\xi} + \frac{1}{2} \cdot \frac{f'(x - \xi) - f'(x)}{-\xi},$$

and the last line tends as $h \to 0$, by the definition of the derivative, to the quantity $(f')'(x)$. □

It is a fact that the standard proof of l'Hôpital's Rule (Guillaume François Antoine de l'Hôpital, Marquis de St.-Mesme, 1661–1704) is obtained by way of Cauchy's Mean Value Theorem. This line of reasoning is explored in the next section.

Exercises

1. Let f be a function that is continuous on $[0, \infty)$ and differentiable on $(0, \infty)$. If $f(0) = 0$ and $|f'(x)| \le |f(x)|$ for all $x > 0$, then prove that $|f(x)| \le e^x$ for all x. [This result is often called Gronwall's inequality.]

2. Let f be a continuous function on $[a, b]$ that is differentiable on (a, b). Assume that $f(a) = m$ and that $|f'(x)| \le K$ for all $x \in (a, b)$. What bound can you then put on the magnitude of $f(b)$?

3. Let f be a differentiable function on an open interval I and assume that f has no local minima nor local maxima on I. Prove that f is either increasing or decreasing on I.

4. Let $0 < \alpha \le 1$. Prove that there is a constant $C_\alpha > 0$ such that, for $0 < x < 1$, it holds that

$$|\ln x| \le C_\alpha \cdot x^{-\alpha}.$$

Prove that the constant cannot be taken to be independent of α.

5. Let f be a function that is twice continuously differentiable on $[0, \infty)$ and assume that $f''(x) \ge c > 0$ for all x. Prove that f is not bounded from above.

6. Let f be differentiable on an interval I and $f'(x) > 0$ for all $x \in I$. Does it follow that $(f^2)' > 0$ for all $x \in I$? What additional hypothesis on f will make the conclusion true?

7. Answer Exercise **6** with the exponent 2 replaced by any positive integer exponent.

8. Use the Mean Value Theorem to say something about the behavior at $+\infty$ of the function $g(x) = \sqrt[3]{x^4 + 1} - x^{4/3}$.

9. Use the Mean Value Theorem to say something about the behavior at $+\infty$ of the function $f(x) = \sqrt{x + 1} - \sqrt{x}$.

10. Refer to Exercise **9**. What can you say about the asymptotics at $+\infty$ of $\sqrt{x + 1} / \sqrt{x}$?

11. Supply the details of the proof of Theorem 6.22.

12. Give an example of a function f for which the limit in Example 6.23 exists at some x but for which f is not twice differentiable at x.

13. Use the Mean Value Theorem to prove that $|\sin x| \leq |x|$ for all x.

14. Apply the Mean Value Theorem to the Fundamental Theorem of Calculus. What conclusion can you draw?

15. Use the Mean Value Theorem to prove that $\ln x \leq x$ for $x \geq 1$.

6.3 More on the Theory of Differentiation

l'Hôpital's Rule (actually due to his teacher J. Bernoulli (1667–1748)) is a useful device for calculating limits, and a nice application of the Cauchy Mean Value Theorem. Here, we present a special case of the theorem.

Theorem 6.24: *Suppose that f and g are differentiable functions on an open interval I and that $p \in I$. If $\lim_{x \to p} f(x) = \lim_{x \to p} g(x) = 0$ and if*

$$\lim_{x \to p} \frac{f'(x)}{g'(x)} \tag{6.24.1}$$

exists and equals a real number ℓ, then

$$\lim_{x \to p} \frac{f(x)}{g(x)} = \ell.$$

Proof: Fix a real number $a > \ell$. By (6.24.1) there is a number $q > p$ such that, if $p < x < q$, then

$$\frac{f'(x)}{g'(x)} < a. \tag{6.24.2}$$

But now, if $p < s < t < q$, then

$$\frac{f(t) - f(s)}{g(t) - g(s)} = \frac{f'(x)}{g'(x)}$$

for some $s < x < t$ (by Cauchy's Mean Value Theorem). It follows then from (6.24.2) that

$$\frac{f(t) - f(s)}{g(t) - g(s)} < a.$$

Now let $s \to p$ and invoke the hypothesis about the zero limit of f and g at p to conclude that

$$\frac{f(t)}{g(t)} \leq a$$

when $p < t < q$. Since a is an arbitrary number to the right of ℓ we conclude that

$$\limsup_{t \to p^+} \frac{f(t)}{g(t)} \leq \ell.$$

Similar arguments show that

$$\liminf_{t \to p^+} \frac{f(t)}{g(t)} \geq \ell;$$

$$\limsup_{t \to p^-} \frac{f(t)}{g(t)} \leq \ell;$$

$$\liminf_{t \to p^-} \frac{f(t)}{g(t)} \geq \ell.$$

We conclude that the desired limit exists and equals ℓ. ☐

A closely related result, with a similar proof, is as follows:

Theorem 6.25: *Suppose that f and g are differentiable functions on an open interval I and that $p \in I$. If $\lim_{x \to p} f(x) = \lim_{x \to p} g(x) = \pm\infty$ and if*

$$\lim_{x \to p} \frac{f'(x)}{g'(x)} \tag{6.25.1}$$

exists and equals a real number ℓ, then

$$\lim_{x \to p} \frac{f(x)}{g(x)} = \ell.$$

Example 6.26: Let

$$f(x) = |\ln|x||^{(x^2)}.$$

We wish to determine $\lim_{x \to 0} f(x)$. To do so, we define

$$F(x) = \ln f(x) = x^2 \ln|\ln|x|| = \frac{\ln|\ln|x||}{1/x^2}.$$

Notice that both the numerator and the denominator tend to $\pm\infty$ as $x \to 0$. So the hypotheses of l'Hôpital's rule are satisfied and the limit is

$$\lim_{x \to 0} \frac{\ln|\ln|x||}{1/x^2} = \lim_{x \to 0} \frac{(\ln|\ln|x|)'}{(1/x^2)'} = \lim_{x \to 0} \frac{1/[x\ \ln|x|]}{-2/x^3} = \lim_{x \to 0} \frac{-x^2}{2\ \ln|x|} = 0.$$

Since $\lim_{x \to 0} F(x) = 0$ we may calculate that the original limit has value $\lim_{x \to 0} f(x) = 1$. $\qquad\qquad\square$

Proposition 6.27: *Let f be an invertible function on an interval (a,b) with nonzero derivative at a point $x \in (a, b)$. Let $X = f(x)$. Then, $(f^{-1})'$ (X) exists and equals $1/f'(x)$.*

Proof: Observe that, for $T \neq X$,

$$\frac{f^{-1}(T) - f^{-1}(X)}{T - X} = \frac{1}{\frac{f(t)-f(x)}{t-x}}, \qquad\qquad (6.27.1)$$

where $T = f(t)$. Since $f'(x) \neq 0$, the difference quotients for f in the denominator are bounded from zero; hence, the limit of the formula in (6.27.1) exists. This proves that f^{-1} is differentiable at X and that the derivative at that point equals $1/f'(x)$. $\qquad\qquad\square$

Example 6.28: We know that the function $f(x) = x^k$, k a positive integer, is one-to-one and differentiable on the interval $(0,1)$. Moreover the derivative $k \cdot x^{k-1}$ never vanishes on that interval. Therefore the proposition applies and we find for $X \in (0,1) = f((0,1))$ that

$$(f^{-1})'(X) = \frac{1}{f'(x)} = \frac{1}{f'(X^{1/k})}$$

$$= \frac{1}{k \cdot X^{1-1/k}} = \frac{1}{k} \cdot X^{1/k-1}.$$

In other words,

$$(X^{1/k})' = \frac{1}{k} X^{1/k-1}.$$

$\qquad\qquad\square$

We conclude this section by saying a few words about higher derivatives. If f is a differentiable function on an open interval I, then we may ask whether the function f' is differentiable. If it is, then we denote its derivative by

$$f'' \quad \text{or} \quad f^{(2)} \quad \text{or} \quad \frac{d^2}{dx^2}f \quad \text{or} \quad \frac{d^2f}{dx^2},$$

and call it the second derivative of f. Likewise the derivative of the $(k-1)$th derivative, if it exists, is called the kth derivative and is denoted

$$f''\cdots' \quad \text{or} \quad f^{(k)} \quad \text{or} \quad \frac{d^k}{dx^k}f \quad \text{or} \quad \frac{d^kf}{dx^k}.$$

Observe that we cannot even consider whether $f^{(k)}$ exists at a point unless $f^{(k-1)}$ exists in a *neighborhood* of that point.

If f is k times differentiable on an open interval I and if each of the derivatives $f^{(1)}, f^{(2)}, \ldots, f^{(k)}$ is continuous on I, then we say that the function f is k *times continuously differentiable* on I. We write $f \in C^k(I)$. Obviously there is some redundancy in this definition since the continuity of $f^{(j-1)}$ follows from the existence of $f^{(j)}$. Thus, only the continuity of the last derivative $f^{(k)}$ need be checked. Continuously differentiable functions are useful tools in analysis. We denote the class of k times continuously differentiable functions on I by $C^k(I)$.

Example 6.29: For $k = 1, 2, \ldots$ the function

$$f_k(x) = \begin{cases} x^{k+1} & \text{if } x \geq 0 \\ -x^{k+1} & \text{if } x < 0 \end{cases}$$

will be k times continuously differentiable on $\mathbb{R}$ but will fail to be $k+1$ times differentiable at $x = 0$. More dramatically, an analysis similar to the one we used on the Weierstrass nowhere differentiable function shows that the function

$$g_k(x) = \sum_{j=1}^{\infty} \frac{3^j}{4^{j+jk}} \sin(4^j x)$$

is k times continuously differentiable on $\mathbb{R}$ but will not be $k+1$ times differentiable at any point (this function, with $k = 0$, was Weierstrass's original example). □

A more refined notion of smoothness/continuity of functions is that of Hölder continuity or Lipschitz continuity (see Section 5.3). If f is a function on an open interval I and if $0 < \alpha \leq 1$, then we say that f satisfies a *Lipschitz condition* of order α on I if there is a constant M such that for all $s, t \in I$ we have

$$|f(s) - f(t)| \le M \cdot |s - t|^\alpha.$$

Such a function is said to be of class $\text{Lip}_\alpha(I)$. Clearly a function of class Lip_α is uniformly continuous on I. For, if $\epsilon > 0$, then we may take $\delta = (\epsilon/M)^{1/\alpha}$: it follows that, for $|s - t| < \delta$, we have

$$|f(s) - f(t)| \le M \cdot |s - t|^\alpha < M \cdot \epsilon/M = \epsilon.$$

Interestingly, when $\alpha > 1$ the class Lip_α contains only constant functions. For in this instance the inequality

$$|f(s) - f(t)| \le M \cdot |s - t|^\alpha$$

leads to

$$\left| \frac{f(s) - f(t)}{s - t} \right| \le M \cdot |s - t|^{\alpha - 1}.$$

Because $\alpha - 1 > 0$, letting $s \to t$ yields that $f'(t)$ exists for every $t \in I$ and equals 0. It follows from Corollary 6.18 of the last section that f is constant on I.

Instead of trying to extend the definition of $\text{Lip}_\alpha(I)$ to $\alpha > 1$ it is customary to define classes of functions $C^{k,\alpha}$, for $k = 0, 1, \ldots$ and $0 < \alpha \le 1$, by the condition that f be of class C^k on I and that $f^{(k)}$ be an element of $\text{Lip}_\alpha(I)$. We leave it as an exercise for you to verify that $C^{k,\alpha} \subseteq C^{\ell,\beta}$ if either $k > \ell$ or both $k = \ell$ and $\alpha \ge \beta$.

In more advanced studies in analysis, it is appropriate to replace $\text{Lip}_1(I)$, and more generally $C^{k,1}$, with another space (invented by Antoni Zygmund, 1900–1992) defined in a more subtle fashion using second differences as in Example 6.23. These matters exceed the scope of this book, but we shall make a few remarks about them in the exercises.

Exercises

1. Suppose that f is a C^2 function on $\mathbb{R}$ and that $|f''(x)| \le C$ for all x. Prove that

$$\left| \frac{f(x + h) + f(x - h) - 2f(x)}{h^2} \right| \le C.$$

2. Fix a positive integer k. Give an example of two functions f and g neither of which is in C^k but such that $f \cdot g \in C^k$.

3. Fix a positive integer ℓ and define $f(x) = |x|^{\ell}$. In which class C^k does f lie? In which class $C^{k,\alpha}$ does it lie?

4. In the text we give sufficient conditions for the inclusion $C^{k,\alpha} \subseteq C^{\ell,\beta}$. Show that the inclusion is strict if either $k > \ell$ or $k = \ell$ and $\alpha > \beta$.

5. Suppose that f is a continuously differentiable function on an interval I and that $f'(x)$ is never zero. Prove that f is invertible. Then, prove that f^{-1} is differentiable. Finally, use the Chain Rule on the identity $f(f^{-1}) = x$ to derive a formula for $(f^{-1})'$.

6. Suppose that a function f on the interval $(0, 1)$ has left derivative equal to zero at every point. What conclusion can you draw?

7. We know that the first derivative can be characterized by the Newton quotient. Find an analogous characterization of second derivatives. What about third derivatives?

8. Use l'Hôpital's Rule to analyze the limit

$$\lim_{x \to +\infty} x^{1/x}.$$

*9. We know (see Section 8.4) that a continuous function on the interval $[0, 1]$ can be uniformly approximated by polynomials. But, if the function f is continuously differentiable on $[0, 1]$, then we can actually say something about the *rate* of approximation. That is, if $\epsilon > 0$, then f can be approximated uniformly within ϵ by a polynomial of degree not greater than $N = N(\epsilon)$. Calculate $N(\epsilon)$.

*10. In which class $C^{k,\alpha}$ is the function $x \cdot \ln|x|$ on the interval $[-1/2, 1/2]$? How about the function $x/\ln|x|$?

*11. Give an example of a function on $\mathbb{R}$ such that

$$\left| \frac{f(x + h) + f(x - h) - 2f(x)}{h} \right| \le C$$

for all x and all $h \ne 0$ but f is not in $\text{Lip}_1(\mathbb{R})$. (**Hint:** See Exercise **10**.)

12. Give an example of a function that is Lipschitz but not of class C^1.

13. Give an example of a function that is C^1 but its first derivative does not satisfy any Lipschitz condition.

*14. Suppose that $f(x, y)$ is Lipschitz in x for each fixed y and Lipschitz in y for each fixed x. Does it follow that f is Lipschitz as a function of (x, y)?

*15. A Lipschitz function can be written as the difference of two monotone increasing functions. Explain.

7

The Integral

7.1 Partitions and the Concept of Integral

We learn in calculus that it is often useful to think of an integral as representing area. However, this is but one of many important applications of integration theory. The integral is a generalization of the summation process. That is the point of view that we shall take in the present chapter.

Definition 7.1: Let $[a, b]$ be a closed interval in $\mathbb{R}$. A finite, ordered set of points $\mathcal{P} = \{x_0, x_1, x_2, \ldots, x_{k-1}, x_k\}$ such that

$$a = x_0 \leq x_1 \leq x_2 \leq \cdots \leq x_{k-1} \leq x_k = b$$

is called a *partition* of $[a, b]$. Refer to Figure 7.1.

If $\mathcal{P}$ is a partition of $[a, \ b]$, then we let I_j denote the interval $[x_{j-1}, x_j]$, $j = 1, 2, \ldots, k$. The symbol Δ_j denotes the *length* of I_j. The *mesh* of $\mathcal{P}$, denoted by $m(\mathcal{P})$, is defined to be $\max_j \Delta_j$.

The points of a partition need not be equally spaced, nor must they be distinct from each other.

Example 7.2: The set $\mathcal{P} = \{0, 1, 1, 9/8, 2, 5, 21/4, 23/4, 6\}$ is a partition of the interval $[0,6]$ with mesh 3 (because $I_5 = [2, 5]$, with length 3, is the longest interval in the partition). See Figure 7.2. □

Definition 7.3: Let $[a, b]$ be an interval and let f be a function with domain $[a, b]$. If $\mathcal{P} = \{x_0, x_1, x_2, \ldots, x_{k-1}, x_k\}$ is a partition of $[a, b]$ and if, for each j, s_j is an element of I_j, then the corresponding *Riemann sum* is defined to be

$$\mathcal{R}(f, \mathcal{P}) = \sum_{j=1}^{k} f(s_j) \Delta_j.$$

Example 7.4: Let $f(x) = x^2 - x$ and $[a, b] = [1, 4]$. Define the partition $\mathcal{P} = \{1, 3/2, 2, 7/3, 4\}$ of this interval. Then, a Riemann sum for this f and $\mathcal{P}$ is

DOI: 10.1201/9781003222682-8

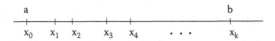

FIGURE 7.1
A partition.

FIGURE 7.2
The partition in Example 7.2.

$$\mathcal{R}(f, \mathcal{P}) = (1^2 - 1) \cdot \frac{1}{2} + ((7/4)^2 - (7/4)) \cdot \frac{1}{2}$$
$$+ ((7/3)^2 - (7/3)) \cdot \frac{1}{3} + (3^2 - 3) \cdot \frac{5}{3}$$
$$= \frac{10103}{864}. \qquad \qquad \Box$$

Notice that we have complete latitude in choosing each point s_j from the corresponding interval I_j. While at first confusing, we will find this freedom to be a powerful tool when proving results about the integral.

The first main step in the theory of the Riemann integral is to determine a method for "calculating the limit of the Riemann sums" of a function as the mesh of the partitions tends to zero. There are in fact several methods for doing so. We have chosen the simplest one.

Definition 7.5: Let $[a, b]$ be an interval and f a function with domain $[a, b]$. We say that *the Riemann sums of f tend to a limit ℓ as $m(\mathcal{P})$ tends to 0* if, for any $\epsilon > 0$, there is a $\delta > 0$ such that, if $\mathcal{P}$ is any partition of $[a, b]$ with $m(\mathcal{P}) < \delta$, then $|\mathcal{R}(f, \mathcal{P}) - \ell| < \epsilon$ for every choice of $s_j \in I_j$.

It will turn out to be critical for the success of this definition that we require that *every* partition of mesh smaller than δ satisfy the conclusion of the definition. The theory does not work effectively if for every $\epsilon > 0$ there is a $\delta > 0$ and *some* partition $\mathcal{P}$ of mesh less than δ which satisfies the conclusion of the definition.

Definition 7.6: A function f on a closed interval $[a, b]$ is said to be *Riemann integrable* on $[a, b]$ if the Riemann sums of $\mathcal{R}(f, \mathcal{P})$ tend to a finite limit ℓ as $m(\mathcal{P})$ tends to zero.

The value ℓ of the limit, when it exists, is called the *Riemann integral* of f over $[a, b]$ and is denoted by

$$\int_a^b f(x) dx.$$

Remark 7.7: We mention now a useful fact that will be formalized in later sections. Suppose that f is Riemann integrable on $[a, b]$ with the value of the integral being ℓ. Let $\epsilon > 0$. Then, as stated in the definition (with $\epsilon/2$ replacing ϵ), there is a $\delta > 0$ such that, if Q is a partition of $[a, b]$ of mesh smaller than δ, then $|\mathcal{R}(f, Q) - \ell| < \epsilon/2$. It follows that, if $\mathcal{P}$ and $\mathcal{P}'$ are partitions of $[a, b]$ of mesh smaller than δ, then

$$|\mathcal{R}(f, \mathcal{P}) - \mathcal{R}(f, \mathcal{P}')| \leq |\mathcal{R}(f, \mathcal{P}) - \ell| + |\ell - \mathcal{R}(f, \mathcal{P}')| < \frac{\epsilon}{2} + \frac{\epsilon}{2} = \epsilon.$$

This is like a Cauchy condition.

Note, however, that we may choose $\mathcal{P}'$ to equal the partition $\mathcal{P}$. Also we may for each j choose the point s_j, where f is evaluated for the Riemann sum over $\mathcal{P}$, to be a point where f very nearly assumes its supremum on I_j. Likewise we may for each j choose the point s'_j, where f is evaluated for the Riemann sum over $\mathcal{P}'$, to be a point where f very nearly assumes its infimum on I_j. It easily follows that, when the mesh of $\mathcal{P}$ is less than δ, then

$$\sum_j \left(\sup_{I_j} f - \inf_{I_j} f \right) \Delta_j \leq \epsilon. \tag{7.7.1}$$

This consequence of integrability will prove useful to us in some of the discussions in this and the next section. In the exercises we shall consider in detail the assertion that integrability implies (7.7.1) and the converse as well. □

Definition 7.8: If $\mathcal{P}$, $\mathcal{P}'$ are partitions of $[a, b]$, then their *common refinement* is the union of all the points of $\mathcal{P}$ and $\mathcal{P}'$. See Figure 7.3.

We record now a technical lemma that will be used in several of the proofs that follow:

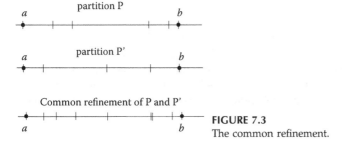

FIGURE 7.3
The common refinement.

Lemma 7.9: *Let f be a function with domain the closed interval [a, b]. The Riemann integral*

$$\int_a^b f(x)dx$$

exists if and only if, for every $\epsilon > 0$, there is a $\delta > 0$ such that, if P and P' are partitions of $[a, b]$ with $m(P) < \delta$ and $m(P') < \delta$, then their common refinement Q has the property that

$$|R(f, P) - R(f, Q)| < \epsilon$$

and   (7.9.1)

$$|R(f, P') - R(f, Q)| < \epsilon.$$

Proof: If f is Riemann integrable, then the assertion of the lemma follows immediately from the definition of the integral.

For the converse note that (7.9.1) certainly implies that, if $\epsilon > 0$, then there is a $\delta > 0$ such that, if P and P' are partitions of $[a, b]$ with $m(P) < \delta$ and $m(P') < \delta$, then

$$|R(f, P) - R(f, P')| < \epsilon \tag{7.9.2}$$

(just use the triangle inequality).

Now, for each $\epsilon_j = 2^{-j}, j = 1, 2, \ldots$, we can choose a $\delta_j > 0$ as in (7.9.2). Let S_j be the *closure* of the set

$$\{R(f, P): m(P) < \delta_j\}.$$

By the choice of δ_j, the set S_j is contained in a closed interval of length not greater than $2\epsilon_j$.

On the one hand,

$$\bigcap_j S_j$$

must be nonempty since it is the decreasing intersection of compact sets. On the other hand, the length estimate implies that the intersection must be contained in a closed interval of length 0—that is, the intersection is a point. That point is then the limit of the Riemann sums, that is, it is the value of the Riemann integral.   □

The most important, and perhaps the simplest, fact about the Riemann integral is that a large class of familiar functions is Riemann integrable.

Theorem 7.10: *Let f be a continuous function on a nontrivial closed, bounded interval $I = [a, b]$. Then, f is Riemann integrable on $[a, b]$.*

Proof: We use the lemma. Given $\epsilon > 0$, choose (by the uniform continuity of f on I—Theorem 5.27) a $\delta > 0$ such that, whenever $|s - t| < \delta$, then

$$|f(s) - f(t)| < \frac{\epsilon}{b - a}. \tag{7.10.1}$$

Let $\mathcal{P}$ and $\mathcal{P}'$ be any two partitions of $[a, b]$ of mesh smaller than δ. Let Q be the common refinement of $\mathcal{P}$ and $\mathcal{P}'$.

Now we let I_j denote the intervals arising in the partition $\mathcal{P}$ (and having length Δ_j) and $\tilde{I}_\ell$ the intervals arising in the partition Q (and having length $\tilde{\Delta}_\ell$). Since the partition Q contains every point of $\mathcal{P}$, plus some additional points as well, every $\tilde{I}_\ell$ is contained in some I_j. Fix j and consider the expression

$$\left| f(s_j)\Delta_j - \sum_{\tilde{I}_\ell \subseteq I_j} f(t_\ell)\tilde{\Delta}_\ell \right|. \tag{7.10.2}$$

We write

$$\Delta_j = \sum_{\tilde{I}_\ell \subseteq I_j} \tilde{\Delta}_\ell.$$

This equality enables us to rearrange (7.10.2) as

$$\left| f(s_j) \cdot \sum_{\tilde{I}_\ell \subseteq I_j} \tilde{\Delta}_\ell - \sum_{\tilde{I}_\ell \subseteq I_j} f(t_\ell)\tilde{\Delta}_\ell \right|$$

$$= \left| \sum_{\tilde{I}_\ell \subseteq I_j} \left[f(s_j) - f(t_\ell) \right] \tilde{\Delta}_\ell \right|$$

$$\leq \sum_{\tilde{I}_\ell \subseteq I_j} \left| f(s_j) - f(t_\ell) \right| \tilde{\Delta}_\ell.$$

But each of the points t_ℓ is in the interval I_j, as is s_j. So they differ by less than δ. Therefore, by (7.10.1), the last expression is less than

$$\sum_{\tilde{I}_\ell \subseteq I_j} \frac{\epsilon}{b - a}\tilde{\Delta}_\ell = \frac{\epsilon}{b - a}\sum_{\tilde{I}_\ell \subseteq I_j} \tilde{\Delta}_\ell$$

$$= \frac{\epsilon}{b - a} \cdot \Delta_j.$$

Now we conclude the argument by writing

$$|\mathcal{R}(f, \mathcal{P}) - \mathcal{R}(f, \mathcal{Q})| = \left| \sum_j f(s_j)\Delta_j - \sum_\ell f(t_\ell)\tilde{\Delta}_\ell \right|$$

$$\leq \sum_j \left| f(s_j)\Delta_j - \sum_{\tilde{I}_\ell \subseteq I_j} f(t_\ell)\tilde{\Delta}_\ell \right|$$

$$< \sum_j \frac{\epsilon}{b-a} \cdot \Delta_j$$

$$= \frac{\epsilon}{b-a} \cdot \sum_j \Delta_j$$

$$= \frac{\epsilon}{b-a} \cdot (b - a)$$

$$= \epsilon.$$

The estimate for $|\mathcal{R}(f, \mathcal{P}') - \mathcal{R}(f, \mathcal{Q})|$ is identical and we omit it. The result now follows from Lemma 7.9.   □

In the exercises we will ask you to extend the theorem to the case of functions f on $[a, b]$ that are bounded and have finitely many, or even countably many, discontinuities.

We conclude this section by noting an important fact about Riemann integrable functions. A Riemann integrable function on an interval $[a, b]$ *must be bounded*. If it were not, then one could choose the points s_j in the construction of $\mathcal{R}(f, \mathcal{P})$ so that $f(s_j)$ is arbitrarily large, and the Riemann sums would become arbitrarily large, hence cannot converge. You will be asked in the exercises to work out the details of this assertion.

****

Georg Friedrich Bernhard Riemann

Georg Friedrich Bernhard Riemann (1826–1866) was a German mathematician who made contributions to analysis, number theory, and differential geometry. In the field of real analysis, he is known for the first rigorous formulation of the integral, the Riemann integral, and his work on Fourier series. His contributions to complex analysis include most notably the introduction of Riemann surfaces, breaking new ground in a natural, geometric treatment of complex analysis. He is also the partial namesake of the Cauchy-Riemann equations. His famous 1859 paper on the prime-counting function, containing the original statement of the Riemann hypothesis, is regarded as one of the most influential papers in all of mathematics. Through his pioneering contributions to differential geometry, Riemann created the now-active field of differential geometry. This in turn laid the foundations of the mathematics of general relativity. He died young because of poverty and ill health. There is no telling what he might have accomplished if he had lived a normal life.

Riemann was born on 17 September 1826 in Breselenz, a village near Dannenberg in the Kingdom of Hanover. His father, Friedrich Bernhard Riemann, was a poor Lutheran pastor in Breselenz who fought in the Napoleonic Wars. His mother, Charlotte Ebell, died before her children had reached adulthood.

During 1840, Riemann went to Hanover to live with his grandmother and attend lyceum (middle school years). After the death of his grandmother in 1842, he attended high school at the Johanneum Lüneburg. In high school, Riemann studied the Bible intensively, but he was often distracted by mathematics.

During the spring of 1846, his father, after gathering enough money, sent Riemann to the University of Göttingen, where he planned to study towards a degree in Theology. However, once there, he began studying mathematics under Carl Friedrich Gauss. Riemann transferred to the University of Berlin in 1847. During his time of study, Carl Gustav Jacob Jacobi, Peter Gustav Lejeune Dirichlet, Jakob Steiner, and Gotthold Eisenstein were teaching. He stayed in Berlin for two years and returned to Göttingen in 1849.

Riemann held his first lectures in 1854, which founded the field of Riemannian geometry. In 1857, there was an attempt to promote Riemann to extraordinary professor status at the University of Göttingen. Although this attempt failed, it did result in Riemann finally being granted a regular salary. Riemann was the first to suggest using dimensions higher than merely three or four in order to describe physical reality.

In 1862 he married Elise Koch and they had a daughter Ida Schilling who was born on 22 December 1862.

Riemann fled Göttingen when the armies of Hanover and Prussia clashed there in 1866. He died of tuberculosis during his third journey to Italy in Selasca where he was buried in the cemetery in Biganzolo (Verbania).

Riemann was a dedicated Christian, the son of a Protestant minister, and saw his life as a mathematician as another way to serve God. At the time of his death, he was reciting the Lord's Prayer with his wife and died before they finished saying the prayer. Meanwhile, in Göttingen his housekeeper discarded some of the papers in his office, including much unpublished work. Riemann refused to publish incomplete work, and some deep insights may have been lost forever.

****

Exercises

1. If f is a Riemann integrable function on $[a, b]$, then show that f must be a bounded function.

2. Define the *Dirichlet function* to be

$$f(x) = \begin{cases} 1 & \text{if } x \text{ is rational} \\ 0 & \text{if } x \text{ is irrational} \end{cases}$$

Prove that the Dirichlet function is not Riemann integrable on the interval $[a, b]$.

3. Define

$$g(x) = \begin{cases} x \cdot \sin(1/x) & \text{if } x \neq 0 \\ 0 & \text{if } x = 0 \end{cases}$$

Is g Riemann integrable on the interval $[-1, 1]$?

4. To what extent is the following statement true? If f is Riemann integrable on $[a, b]$, then $1/f$ is Riemann integrable on $[a, b]$.

5. Show that any Riemann integrable function is the pointwise limit of continuous functions.

6. Write the Riemann sum for the function $f(x) = \sin(\pi x^2)$ on the interval $[1, 3]$ with a partition of five equally spaced points.

7. Prove that, if f is continuous on the interval $[a, b]$ except for finitely many discontinuities of the first kind, and if f is bounded, then f is Riemann integrable on $[a, b]$.

8. Do Exercise **7** with the phrase "finitely many" replaced by "countably many."

9. Provide the details of the assertion that, if f is Riemann integrable on the interval $[a, b]$ then, for any $\epsilon > 0$, there is a $\delta > 0$ such that, if $\mathcal{P}$ is a partition of mesh less than δ, then

$$\sum_j \left(\sup_{I_j} f - \inf_{I_j} f \right) \Delta_j < \epsilon.$$

[Hint: Follow the scheme presented in Remark 7.7. Given $\epsilon > 0$, choose $\delta > 0$ as in the definition of the integral. Fix a partition $\mathcal{P}$ with mesh smaller than δ. Let $K + 1$ be the number of points in $\mathcal{P}$. Choose points $t_j \in I_j$ so that $|f(t_j) - \sup_{I_j} f| < \epsilon/(2(K + 1))$; also choose points $t'_j \in I_j$ so that $|f(t'_j) - \inf_{I_j} f| < \epsilon/(2(K + 1))$. By applying the definition of the integral to this choice of t_j and t'_j we find that

$$\sum_j \left(\sup_{I_j} f - \inf_{I_j} f \right) \Delta_j < 2\epsilon.$$

The result follows.]

*10. Give an example of a function f such that f^2 is Riemann integrable but f is not.

11. The collection of Riemann integrable functions is closed under which arithmetic operations?

12. If f is Riemann integrable, then f^2 is Riemann integrable. Why is this true?

13. Refer to Exercise 12. If f and g are Riemann integrable, then $f \cdot g$ is Riemann integrable. Why is this true? [**Hint:** Look at $(f + g)^2$.]

7.2 Properties of the Riemann Integral

We begin this section with a few elementary properties of the integral that reflect its linear nature.

Theorem 7.11: *Let $[a, b]$ be a nonempty, bounded interval, let f and g be Riemann integrable functions on the interval, and let α be a real number. Then, $f \pm g$ and $\alpha \cdot f$ are integrable and we have*

a. $\int_a^b f(x) \pm g(x)dx = \int_a^b f(x)dx \pm \int_a^b g(x)dx;$

b. $\int_a^b \alpha \cdot f(x)dx = \alpha \cdot \int_a^b f(x)dx.$

Proof: For **(a)**, let

$$A = \int_a^b f(x)d\dot{x}$$

and

$$B = \int_a^b g(x)dx.$$

Let $\epsilon > 0$. Choose a $\delta_1 > 0$ such that if $\mathcal{P}$ is a partition of $[a, b]$ with mesh less than δ_1, then

$$|\mathcal{R}(f, \mathcal{P}) - A| < \frac{\epsilon}{2}.$$

Similarly choose a $\delta_2 > 0$ such that if $\mathcal{P}$ is a partition of $[a, b]$ with mesh less than δ_2 then

$$|\mathcal{R}(f, \mathcal{P}) - B| < \frac{\epsilon}{2}.$$

Let $\delta = \min\{\delta_1, \delta_2\}$. If $\mathcal{P}'$ is any partition of $[a, b]$ with $m(\mathcal{P}') < \delta$ then

$$
\begin{aligned}
|\mathcal{R}(f \pm g, \mathcal{P}') - (A \pm B)| &= |\mathcal{R}(f, \mathcal{P}') \pm \mathcal{R}(g, \mathcal{P}') - (A \pm B)| \\
&\leq |\mathcal{R}(f, \mathcal{P}') - A| + |\mathcal{R}(g, \mathcal{P}') - B| \\
&< \frac{\epsilon}{2} + \frac{\epsilon}{2} \\
&= \epsilon.
\end{aligned}
$$

This means that the integral of $f \pm g$ exists and equals $A \pm B$, as we were required to prove.

The proof of **(b)** follows similar lines but is much easier and we leave it as an exercise for you.

Theorem 7.12: *If c is a point of the interval $[a, b]$ and if f is Riemann integrable on both $[a, c]$ and $[c, b]$ then f is integrable on $[a, b]$ and*

$$\int_a^c f(x)dx + \int_c^b f(x)dx = \int_a^b f(x)dx.$$

Proof: Let us write

$$A = \int_a^c f(x)dx$$

and

$$B = \int_c^b f(x)dx.$$

Now pick $\epsilon > 0$. There is a $\delta_1 > 0$ such that if $\mathcal{P}$ is a partition of $[a, c]$ with mesh less than δ_1 then

$$|\mathcal{R}(f, \mathcal{P}) - A| < \frac{\epsilon}{3}.$$

Similarly, choose $\delta_2 > 0$ such that if $\mathcal{P}'$ is a partition of $[c, b]$ with mesh less than δ_2 then

$$|\mathcal{R}(f, \mathcal{P}') - B| < \frac{\epsilon}{3}.$$

Let M be an upper bound for $|f|$ (recall, from the remark at the end of

Section 7.1, that a Riemann integrable function must be bounded). Set $\delta = \min\{\delta_1, \delta_2, \epsilon/(6M)\}$. Now let $\mathcal{V} = \{v_1, \ldots, v_k\}$ be any partition of $[a, b]$ with mesh less than δ. There is a last point v_n which is in $[a, c]$ and a first point v_{n+1} in $[c, b]$. Observe that $\mathcal{P} = \{v_0, \ldots, v_n, c\}$ is a partition of $[a, c]$ with mesh smaller than δ_1 and $\mathcal{P}' = \{c, v_{n+1}, \ldots, v_k\}$ is a partition of $[c, b]$ with mesh smaller than δ_2. Let us rename the elements of $\mathcal{P}$ as $\{p_0, \ldots, p_{n+1}\}$ and the elements of $\mathcal{P}'$ as $\{p'_0, \cdots p'_{k-n+1}\}$. Notice that $p_{n+1} = p'_0 = c$. For each j let s_j be a point chosen in the interval $I_j = [v_{j-1}, v_j]$ from the partition $\mathcal{V}$.
Then we have

$$|\mathcal{R}(f, \mathcal{V}) - [A + B]|$$

$$= \left| \left(\sum_{j=1}^{n} f(s_j)\Delta_j - A \right) + f(s_{n+1})\Delta_{n+1} + \left(\sum_{j=n+2}^{k} f(s_j)\Delta_j - B \right) \right|$$

$$= \left| \left(\sum_{j=1}^{n} f(s_j)\Delta_j + f(c)\cdot(c - v_n) - A \right) \right.$$

$$+ \left(f(c)\cdot(v_{n+1} - c) + \sum_{j=n+2}^{k} f(s_j)\Delta_j - B \right)$$

$$+ (f(s_{n+1}) - f(c))\cdot(c - v_n) + (f(s_{n+1}) - f(c))\cdot(v_{n+1} - c) \Big|$$

$$\leq \left| \left(\sum_{j=1}^{n} f(s_j)\Delta_j + f(c)\cdot(c - v_n) - A \right) \right.$$

$$+ \left| \left(f(c)\cdot(v_{n+1} - c) + \sum_{j=n+2}^{k} f(s_j)\Delta_j - B \right) \right|$$

$$+ |(f(s_{n+1}) - f(c))\cdot(v_{n+1} - v_n)|$$

$$= |\mathcal{R}(f, \mathcal{P}) - A| + |\mathcal{R}(f, \mathcal{P}') - B|$$

$$+ |(f(s_{n+1}) - f(c))\cdot(v_{n+1} - v_n)|$$

$$< \frac{\epsilon}{3} + \frac{\epsilon}{3} + 2M\cdot\delta$$

$$\leq \epsilon$$

by the choice of δ.
  This shows that f is integrable on the entire interval $[a, b]$ and the value of the integral is

$$A + B = \int_a^c f(x)dx + \int_c^b f(x)dx. \qquad \square$$

Remark 7.13: The last proof illustrates why it is useful to be able to choose the $s_j \in I_j$ arbitrarily.

□

Example 7.14: If we adopt the convention that

$$\int_b^a f(x)dx = -\int_a^b f(x)dx$$

(which is consistent with the way that the integral was defined in the first place), then Theorem 7.12 is true even when c is not an element of $[a, b]$. For instance, suppose that $c < a < b$. Then, by Theorem 7.12,

$$\int_c^a f(x)dx + \int_a^b f(x)dx = \int_c^b f(x)dx.$$

But this may be rearranged to read

$$\int_a^b f(x)dx = -\int_c^a f(x)dx + \int_c^b f(x)dx = \int_a^c f(x)dx + \int_c^b f(x)dx. \qquad \square$$

One of the basic tools of analysis is to perform estimates. Thus we require certain fundamental inequalities about integrals. These are recorded in the next theorem.

Theorem 7.15: *Let f and g be integrable functions on a nonempty interval $[a, b]$. Then*

   i. $\left| \int_a^b f(x)dx \right| \leq \int_a^b |f(x)| dx;$

   ii. *If $f(x) \leq g(x)$ for all $x \in [a, b]$ then $\int_a^b f(x)dx \leq \int_a^b g(x)dx.$*

Proof: If $\mathcal{P}$ is any partition of $[a, b]$ then

$$|\mathcal{R}(f, \mathcal{P})| \leq \mathcal{R}(|f|, \mathcal{P}).$$

The first assertion follows.

    Next, for part **(ii)**,

$$\mathcal{R}(f, \mathcal{P}) \leq \mathcal{R}(g, \mathcal{P}).$$

This inequality implies the second assertion. $\qquad \square$

Example 7.16: We may estimate the integral

$$\int_0^1 \sin^3 x \; dx$$

as follows. We apply Theorem 7.15(i) to see that

$$\left| \int_0^1 \sin^3 x \ dx \right| \le \int_0^1 |\sin^3 x| \, dx.$$

Now we apply Theorem 7.15(ii) to determine that

$$\int_0^1 |\sin^3 x| \, dx \le \int_0^1 1 \ dx = 1. \qquad \square$$

Exercises

1. Suppose that f is a continuous, nonnegative function on the interval $[0, 1]$. Let M be the maximum of f on the interval. Prove that

$$\lim_{n \to \infty} \left[\int_0^1 f(t)^n dt \right]^{1/n} = M.$$

2. Let f be a bounded function on an unbounded interval of the form $[A, \infty)$. We say that f is integrable on $[A, \infty)$ if f is integrable on every compact subinterval of $[A, \infty)$ and

$$\lim_{B \to +\infty} \int_A^B f(x) dx$$

exists and is finite.

Assume that f is nonnegative and Riemann integrable on $[1, N]$ for every $N > 1$ and that f is decreasing. Show that f is Riemann integrable on $[1, \infty)$ if and only if $\sum_{j=1}^{\infty} f(j)$ is finite.

Suppose that g is nonnegative and integrable on $[1, \infty)$. If $0 \le |f(x)| \le g(x)$ for $x \in [1, \infty)$, and f is integrable on compact subintervals of $[1, \infty)$, then prove that f is integrable on $[1, \infty)$.

3. Let f be a function on an interval of the form $(a, b]$ such that f is integrable on compact subintervals of $(a, b]$. If

$$\lim_{\varepsilon \to 0^+} \int_{a+\varepsilon}^b f(x) dx$$

exists and is finite then we say that f is integrable on $(a, b]$. Prove that, if we restrict attention to bounded f, then in fact this definition gives rise to no new integrable functions. However, there are unbounded functions that can now be integrated. Give an example.

4. If $\int_3^6 f(x)dx = 2$ and $\int_3^8 f(x)dx = 5$, then calculate $\int_8^6 f(x)dx$.

5. Fix a continuous function g on the interval $[0, 1]$. Define

$$Tf = \int_0^1 f(x)g(x)dx$$

for f integrable on $[0, 1]$. Prove that

$$|Tf| \le C \int_0^1 |f(x)|\, dx.$$

What does the constant C depend on?

6. Let f and g be continuous functions on the interval $[a, b]$. Prove that

$$\int_a^b |f(x) \cdot g(x)|\, dx \le \int_a^b |f(x)|^2\, dx^{1/2} \cdot \int_a^b |g(x)|^2\, dx^{1/2}.$$

7. Prove part (b) of Theorem 7.11.

*8. Let $1 < p < \infty$ and $q = p/(p-1)$. Let f and g be continuous functions on the interval $[a, b]$. Prove that

$$\int_a^b |f(x) \cdot g(x)|\, dx \le \int_a^b |f(x)|^p\, dx^{1/p} \cdot \int_a^b |g(x)|^q\, dx^{1/q}.$$

*9. Prove that

$$\lim_{\eta \to 0^+} \int_\eta^{1/\eta} \frac{\cos(2r) - \cos r}{r}\, dr$$

exists.

*10. Suppose that f is a Riemann integrable function on the interval $[0, 1]$. Let $\epsilon > 0$. Show that there is a polynomial p so that

$$\int_0^1 |f(x) - p(x)|\, dx < \varepsilon.$$

11. Refer to Exercise 3. Calculate

$$\int_0^1 x^{-1/2}dx.$$

12. Refer to Exercise 3. Calculate

$$\int_1^2 (x - 1)^{-1/3} dx.$$

13. Refer to Exercise 2. Calculate

$$\int_1^\infty x^{-3/2} dx.$$

14. Refer to Exercise 2. Calculate

$$\int_1^\infty \frac{1}{x^2 \log x} dx.$$

7.3 Change of Variable and Related Ideas

Another fundamental operation in the theory of the integral is "change of variable" (sometimes called the "u-substitution" in calculus books). We next turn to a careful formulation and proof of this operation. First we need a lemma:

Lemma 7.17: *If f is a Riemann integrable function on $[a, b]$ and if ϕ is a continuous function on a compact interval that contains the range of f then $\phi \circ f$ is Riemann integrable.*

Proof: Let $\epsilon > 0$. Since ϕ is a continuous function on a compact set, it is uniformly continuous (Theorem 5.27). Let $\delta > 0$ be selected such that **(i)** $\delta < \epsilon$ and **(ii)** if $|x - y| < \delta$ then $|\phi(x) - \phi(y)| < \epsilon$.

Now the hypothesis that f is Riemann integrable implies that there exists a $\tilde{\delta} > 0$ such that if $\mathcal{P}$ and $\mathcal{P}'$ are partitions of $[a, b]$ and $m(\mathcal{P})$, $m(\mathcal{P}') < \tilde{\delta}$ then (by Lemma 7.9), for the common refinement Q of $\mathcal{P}$ and $\mathcal{P}'$, it holds that

$$|\mathcal{R}(f, \mathcal{P}) - \mathcal{R}(f, Q)| < \delta^2 \quad \text{and} \quad |\mathcal{R}(f, \tilde{\mathcal{P}}) - \mathcal{R}(f, Q)| < \delta^2.$$

Fix such a $\mathcal{P}$, $\mathcal{P}'$ and Q. Let J_ℓ be the intervals of Q and I_j the intervals of $\mathcal{P}$. Each J_ℓ is contained in some $I_{j(\ell)}$. We write

$$|\mathcal{R}(\phi \circ f, \mathcal{P}) - \mathcal{R}(\phi \circ f, \mathcal{Q})|$$

$$= \left| \sum_j \phi \circ f(t_j)\Delta_j - \sum_\ell \phi \circ f(s_\ell)\Delta_\ell \right|$$

$$= \left| \sum_j \sum_{J_\ell \subseteq I_j} \phi \circ f(t_j)\Delta_\ell - \sum_j \sum_{J_\ell \subseteq I_j} \phi \circ f(s_\ell)\Delta_\ell \right|$$

$$= \left| \sum_j \sum_{J_\ell \subseteq I_j} \left[\phi \circ f(t_j) - \phi \circ f(s_\ell) \right]\Delta_\ell \right|$$

$$\leq \left| \sum_j \sum_{J_\ell \subseteq I_j, \ell \in G} \left[\phi \circ f(t_j) - \phi \circ f(s_\ell) \right]\Delta_\ell \right|$$

$$+ \left| \sum_j \sum_{J_\ell \subseteq I_j, \ell \in B} \left[\phi \circ f(t_j) - \phi \circ f(s_\ell) \right]\Delta_\ell \right|,$$

where we put ℓ in G if $J_\ell \subseteq I_{j(\ell)}$ and $0 \leq \left(\sup_{I_{j(\ell)}} f - \inf_{I_{j(\ell)}} f \right) < \delta$; otherwise we put ℓ into B. Notice that

$$\sum_{\ell \in B} \delta \Delta_\ell \leq \sum_{\ell \in B} \left(\sup_{I_{j(\ell)}} f - \inf_{I_{j(\ell)}} f \right) \cdot \Delta_\ell$$

$$= \sum_{j=1}^{k} \sum_{J_\ell \subseteq I_j} \left(\sup_{I_j} f - \inf_{I_j} f \right) \cdot \Delta_\ell$$

$$= \sum_{j=1}^{k} \left(\sup_{I_j} f - \inf_{I_j} f \right) \Delta_j$$

$$< \delta^2$$

by the choice of $\tilde{\delta}$ (and Remark 7.7). Therefore

$$\sum_{\ell \in B} \Delta_\ell < \delta.$$

Let M be an upper bound for $|\phi|$ (Corollary 5.22). Then

$$\left| \sum_j \sum_{J_\ell \subseteq I_j, \ell \in B} \left(\phi \circ f(t_j) - \phi \circ f(s_\ell) \right)\Delta_\ell \right| \leq \left| \sum_j \sum_{J_\ell \subseteq I_j, \ell \in B} (2 \cdot M)\Delta_\ell \right|$$

$$\leq 2 \cdot \delta \cdot M$$

$$< 2M\varepsilon.$$

Also

$$\left| \sum_j \sum_{I_t \subseteq I_j, t \in G} \left(\phi \circ f(t_j) - \phi \circ f(s_t) \right) \Delta_t \right| \le \left| \sum_j \sum_{I_t \subseteq I_j, t \in G} \epsilon \Delta_t \right|$$

since, for $t \in G$, we know that $|f(\alpha) - f(\beta)| < \delta$ for any $\alpha, \beta \in I_{j(t)}$. However, the last line does not exceed $(b - a) \cdot \epsilon$. Putting together our estimates, we find that

$$|\mathcal{R}(\phi \circ f, \mathcal{P}) - \mathcal{R}(\phi \circ f, Q)| < \epsilon \cdot (2M + (b - a)).$$

By symmetry, an analogous inequality holds for $\mathcal{P}'$. By Lemma 7.9, this is what we needed to prove. □

An easier result is that, if f is Riemann integrable on an interval $[a, b]$ and if $\mu: [\alpha, \beta] \rightarrow [a, b]$ is continuously differentiable, then $f \circ \mu$ is Riemann integrable (see the exercises).

Corollary 7.18: *If f and g are Riemann integrable on $[a, b]$, then so is the function $f \cdot g$.*

Proof: By Theorem 7.11, $f + g$ is integrable. By the lemma, $(f + g)^2$ $= f^2 + 2f \cdot g + g^2$ is integrable. But the lemma also implies that f^2 and g^2 are integrable (here we use the function $\phi(x) = x^2$). It results, by subtraction, that $2 \cdot f \cdot g$ is integrable. Hence $f \cdot g$ is integrable. □

Theorem 7.19: *Let f be an integrable function on an interval $[a, b]$ of positive length. Let ψ be a continuously differentiable function from another interval $[\alpha, \beta]$ of positive length into $[a, b]$. Assume that ψ is increasing, one-to-one, and onto. Then*

$$\int_a^b f(x)dx = \int_\alpha^\beta f(\psi(x)) \cdot \psi'(x)dx.$$

Proof: Since f is integrable, its absolute value is bounded by some number M. Fix $\epsilon > 0$. Since ψ' is continuous on the compact interval $[\alpha, \beta]$, it is uniformly continuous (Theorem 5.27). Hence we may choose $\delta > 0$ so small that if $|s - t| < \delta$ then $|\psi'(s) - \psi'(t)| < \epsilon/(M \cdot (\beta - \alpha))$. If $\mathcal{P} = \{p_0, ..., p_k\}$ is any partition of $[a, b]$ then there is an associated partition $\tilde{\mathcal{P}} = \{\psi^{-1}(p_0), ..., \psi^{-1}(p_k)\}$ of $[\alpha, \beta]$. For simplicity denote the points of $\tilde{\mathcal{P}}$ by $\tilde{p}_j$. Let us choose the partition $\mathcal{P}$ so fine that the mesh of $\tilde{\mathcal{P}}$ is less than δ. If t_j

are points of $I_j = \left[p_{j-1}, p_j\right]$ then there are corresponding points $s_j = \psi^{-1}\left(t_j\right)$ of $\tilde{I}_j = [\tilde{p}_{j-1}, \tilde{p}_j]$. Then we have

$$\sum_{j=1}^{k} f\left(t_j\right)\Delta_j = \sum_{j=1}^{k} f\left(t_j\right)\left(p_j - p_{j-1}\right)$$

$$= \sum_{j=1}^{k} f\left(\psi\left(s_j\right)\right)\left(\psi\left(\tilde{p}_j\right) - \psi\left(\tilde{p}_{j-1}\right)\right)$$

$$= \sum_{j=1}^{k} f\left(\psi\left(s_j\right)\right)\psi'\left(u_j\right)\left(\tilde{p}_j - \tilde{p}_{j-1}\right),$$

where we have used the Mean Value Theorem in the last line to find each u_j. Our problem at this point is that $f \circ \psi$ and ψ' are evaluated at different points. So we must do some estimation to correct that problem.

The last displayed line equals

$$\sum_{j=1}^{k} f\left(\psi\left(s_j\right)\right)\psi'\left(s_j\right)\left(\tilde{p}_j - \tilde{p}_{j-1}\right) + \sum_{j=1}^{k} f\left(\psi\left(s_j\right)\right)\left(\psi'\left(u_j\right) - \psi'\left(s_j\right)\right)\left(\tilde{p}_j - \tilde{p}_{j-1}\right).$$

The first sum is a Riemann sum for $f\left(\psi(x)\right)\cdot\psi'(x)$ and the second sum is an error term. Since the points u_j and s_j are elements of the same interval $\tilde{I}_j$ of length less than δ, we conclude that $\left|\psi'\left(u_j\right) - \psi'\left(s_j\right)\right| < \epsilon/(M\cdot|\beta - \alpha|)$. Thus the error term in absolute value does not exceed

$$\sum_{j=1}^{k} M\cdot\frac{\epsilon}{M\cdot|\beta - \alpha|}\cdot\left(\tilde{p}_j - \tilde{p}_{j-1}\right) = \frac{\epsilon}{\beta - \alpha}\sum_{j=0}^{k}\left(\tilde{p}_j - \tilde{p}_{j-1}\right) = \epsilon.$$

This shows that every Riemann sum for f on $[a, b]$ with sufficiently small mesh corresponds to a Riemann sum for $f\left(\psi(x)\right)\cdot\psi'(x)$ on $[\alpha, \beta]$ plus an error term of size less than ϵ. A similar argument shows that every Riemann sum for $f\left(\psi(x)\right)\cdot\psi'(x)$ on $[\alpha, \beta]$ with sufficiently small mesh corresponds to a Riemann sum for f on $[a, b]$ plus an error term of magnitude less than ϵ. The conclusion is then that the integral of f on $[a, b]$ (which exists by hypothesis) and the integral of $f\left(\psi(x)\right)\cdot\psi'(x)$ on $[\alpha, \beta]$ (which exists by the corollary to the lemma) have the same value.   □

Example 7.20: Let us analyze the integral

$$\int_{0}^{1} \sin(x^3 + x) \cdot (3x^2 + 1)dx.$$

We let $f(t) = \sin t$ and $\psi(x) = x^3 + x$. Then we see that the integral has the form

$$\int_0^1 f \circ \psi(x) \cdot \psi'(x)dx.$$

Here $\psi: [0, 1] \to [0, 2]$.

By the theorem, this integral is equal to

$$\int_0^2 f(t)dt = \int_0^2 \sin t \; dt = -\cos 2 + 1. \qquad \square$$

We conclude this section with the very important

Theorem 7.21: (Fundamental Theorem of Calculus) *Let f be an integrable function on the interval $[a, b]$. For $x \in [a, b]$ we define*

$$F(x) = \int_a^x f(s)ds.$$

If f is continuous at $x \in (a, b)$ then

$$F'(x) = f(x).$$

Proof: Fix $x \in (a, b)$. Let $\epsilon > 0$. Choose, by the continuity of f at x, a $\delta > 0$ such that $|s - x| < \delta$ implies $|f(s) - f(x)| < \epsilon$. We may assume that $\delta < \min\{x - a, b - x\}$. If $|t - x| < \delta$ then

$$\left| \frac{F(t) - F(x)}{t - x} - f(x) \right| = \left| \frac{\int_a^t f(s)ds - \int_a^x f(s)ds}{t - x} - f(x) \right|$$

$$= \left| \frac{\int_x^t f(s)ds}{t - x} - \frac{\int_x^t f(x)ds}{t - x} \right|$$

$$= \left| \frac{\int_x^t (f(s) - f(x))ds}{t - x} \right|.$$

Notice that we rewrote $f(x)$ as the integral with respect to a dummy variable s over an interval of length $|t - x|$ divided by $(t - x)$. Assume for the moment that $t > x$. Then the last line is dominated by

$$\frac{\int_x^t |f(s) - f(x)| \; ds}{t - x} \leq \frac{\int_x^t \epsilon \; ds}{t - x}$$

$$= \epsilon.$$

A similar estimate holds when $t < x$ (simply reverse the limits of integration).

This shows that

$$\lim_{t \to x} \frac{F(t) - F(x)}{t - x}$$

exists and equals $f(x)$. Thus $F'(x)$ exists and equals $f(x)$.   □

In the exercises we shall consider how to use the theory of one-sided limits to make the conclusion of the Fundamental Theorem true on the entire interval $[a, b]$. We conclude with

Corollary 7.22: *If f is a continuous function on $[a, b]$ and if G is any continuously differentiable function on $[a, b]$ whose derivative equals f on (a, b) then*

$$\int_a^b f(x)dx = G(b) - G(a).$$

Proof: Define F as in the theorem. Since F and G have the same derivative on (a, b), they differ by a constant (Corollary 6.18). Then

$$\int_a^b f(x)dx = F(b) = F(b) - F(a) = G(b) - G(a)$$ □

as desired.

Example 7.23: Let

$$f(x) = \int_0^{x^2} \cos(e^t)dt.$$

What is the derivative of f?

It is not possible to actually evaluate the given integral, but we can still answer the question. Let $g(s) = \int_0^s \cos(e^t)dt$ and let $h(x) = x^2$. Then $f = g \circ h$. Therefore

$$f'(x) = g'(h(x)) \cdot h'(x).$$

Now the Fundamental Theorem of Calculus tells us that

$$g'(s) = \cos(e^s).$$

And obviously $h'(x) = 2x$.

In conclusion,

$$f'(x) = \cos\left(e^{x^2}\right) \cdot 2x.$$

□

Exercises

1. Imitate the proof of the Fundamental Theorem of Calculus in this section to show that, if f is continuous on $[a, b]$ and if we define

$$F(x) = \int_a^x f(t)\,dt,$$

then the one-sided derivative $F'(a)$ exists and equals $f(a)$ in the sense that

$$\lim_{t \to a^+} \frac{F(t) - F(a)}{t - a} = f(a).$$

Formulate and prove an analogous statement for the one-sided derivative of F at b.

2. Let f be a continuously differentiable function on the interval $[0, 2\pi]$. Further assume that $f(0) = f(2\pi)$ and $f'(0) = f'(2\pi)$. For $n \in \mathbb{N}$ define

$$\hat{f}(n) = \frac{1}{2\pi} \int_0^{2\pi} f(x)\sin\,nx\,\,dx.$$

Prove that

$$\sum_{n=1}^{\infty} |\hat{f}(n)|^2$$

converges. [**Hint:** Use integration by parts to obtain a favorable estimate on $|\hat{f}(n)|$.]

3. Let $f_1, f_2, \ldots$ be Riemann integrable functions on $[0,1]$. Suppose that $f_1(x) \geq f_2(x) \geq \cdots$ for every x and that $\lim_{j\to\infty} f_j(x) \equiv f(x)$ exists and is finite for every x. Is it the case that f is Riemann integrable?

4. Give an example of a function f that is not Riemann integrable but such that f^2 is Riemann integrable.

5. Prove that if f is Riemann integrable on the interval $[a, b]$, then f^2 is Riemann integrable on $[a, b]$.

6. Define

$$f(x) = \int_0^{\cos x^2} e^{\sin x} dx.$$

   Calculate $f''(x)$.

7. Give three intuitive reasons why differentiation and integration should be inverse operations.

8. Give an example of an integrable function f and a point x_0 so that

$$F(x) = \int_0^x f(t) dt$$

   is defined but $F'(x_0) \neq f(x_0)$.

9. Calculate the integral

$$\int_0^1 x^2 dx$$

   with the original definition of the integral using Riemann sums. Now calculate the integral using the Fundamental Theorem of Calculus. Confirm that both of your answers are the same.

*10. Integration by parts gives a way to think about the integral of a product of functions. But there is no formula for the integral of a quotient of functions. Explain why.

11. There are many reasons why

$$\int f(x) \cdot g(x) \, dx \neq \int f(x) dx \cdot \int g(x) dx.$$

   Give at least two such reasons.

*12. Let f be continuous on the interval $[0,1]$. Then it is the case that

$$\left| \int_0^1 f(x) dx \right|^2 \leq \int_0^1 |f(x)|^2 \, dx.$$

   Explain why.

13. If f and g are continuous on the interval $[0, 1]$, then it holds that

$$\left| \int_0^1 f(x) \cdot g(x) dx \right| \le \int_0^1 |f(x)|^2 dx^{1/2} \cdot \int_0^1 |g(x)|^2 dx^{1/2}.$$

Explain why.

7.4 Another Look at the Integral

For many purposes, such as integration by parts, it is natural to formulate the integral in a more general context than we have considered in the first two sections. Our new formulation is called the *Riemann–Stieltjes integral* and is described below.

Fix an interval $[a, b]$ and a monotonically increasing function α on $[a, b]$. If $\mathcal{P} = \{p_0, p_1, \ldots, p_k\}$ is a partition of $[a,b]$, then let $\Delta \alpha_j = \alpha(p_j) - \alpha(p_{j-1})$. Let f be a bounded function on $[a, b]$ and define the *upper Riemann sum* $\mathcal{U}$ of f with respect to α and the *lower Riemann sum* $\mathcal{L}$ of f with respect to α as follows:

$$\mathcal{U}(f, \mathcal{P}, \alpha) = \sum_{j=1}^{k} M_j \Delta \alpha_j$$

and

$$\mathcal{L}(f, \mathcal{P}, \alpha) = \sum_{j=1}^{k} m_j \Delta \alpha_j.$$

Here the notation M_j denotes the supremum of f on the interval $I_j = \left[p_{j-1}, p_j\right]$ and m_j denotes the infimum of f on I_j.

In the special case $\alpha(x) = x$ the Riemann sums discussed here have a form similar to the Riemann sums considered in the first two sections. Moreover,

$$\mathcal{L}(f, \mathcal{P}, \alpha) \le R(f, \mathcal{P}) \le \mathcal{U}(f, \mathcal{P}, \alpha).$$

We define

$$I^*(f) = \inf \mathcal{U}(f, \mathcal{P}, \alpha)$$

and

$$I_*(f) = \sup \mathcal{L}(f, \mathcal{P}, \alpha).$$

Here the supremum and infimum are taken with respect to all partitions of the interval $[a, b]$. These are, respectively, the *upper* and *lower integrals* of f with respect to α on $[a, b]$.

By definition it is always true that, for any partition $\mathcal{P}$,

$$\mathcal{L}(f, \mathcal{P}, \alpha) \leq I_*(f) \leq I^*(f) \leq \mathcal{U}(f, \mathcal{P}, \alpha).$$

It is natural to declare the integral to exist when the upper and lower integrals agree:

Definition 7.24: Let α be an increasing function on the interval $[a, b]$ and let f be a bounded function on $[a, b]$. We say that the *Riemann–Stieltjes integral of f with respect to α* exists if

$$I^*(f) = I_*(f).$$

When the integral exists we denote it by

$$\int_a^b f \, d\alpha.$$

Notice that the definition of Riemann–Stieltjes integral is different from the definition of Riemann integral that we used in the preceding sections. It turns out that, when $\alpha(x) = x$, the two definitions are equivalent (this assertion is explored in the exercises). In the present generality it is easier to deal with upper and lower integrals in order to determine the existence of integrals.

Definition 7.25: Let $\mathcal{P}$ and Q be partitions of the interval $[a, b]$. If each point of $\mathcal{P}$ is also an element of Q then we call Q a *refinement* of $\mathcal{P}$.

Notice that the refinement Q is obtained by adding points to $\mathcal{P}$. The mesh of Q will be less than or equal to that of $\mathcal{P}$. The following lemma enables us to deal effectively with our new language.

Lemma 7.26: *Let $\mathcal{P}$ be a partition of the interval $[a, b]$ and f a function on $[a, b]$. Fix an increasing function α on $[a, b]$. If Q is a refinement of $\mathcal{P}$ then*

$$\mathcal{U}(f, Q, \alpha) \leq \mathcal{U}(f, \mathcal{P}, \alpha)$$

and

$$\mathcal{L}(f, Q, \alpha) \geq \mathcal{L}(f, \mathcal{P}, \alpha).$$

Proof: Since Q is a refinement of $\mathcal{P}$ it holds that any interval I_ℓ arising from Q is contained in some interval $J_{j(\ell)}$ arising from $\mathcal{P}$. Let M_{I_ℓ} be the supremum of f on I_ℓ and $M_{J_{j(\ell)}}$ the supremum of f on the interval $J_{j(\ell)}$. Then $M_{I_\ell} \leq M_{J_{j(\ell)}}$. We conclude that

$$\mathcal{U}(f, Q, \alpha) = \sum_\ell M_{I_\ell} \Delta\alpha_\ell \leq \sum_\ell M_{J_{j(\ell)}} \Delta\alpha_\ell.$$

We rewrite the right-hand side as

$$\sum_j M_{J_j}\left(\sum_{I_\ell \subseteq J_j} \Delta\alpha_\ell \right).$$

However, because α is monotone, the inner sum simply equals $\alpha(p_j) - \alpha(p_{j-1}) = \Delta\alpha_j$. Thus the last expression is equal to $\mathcal{U}(f, \mathcal{P}, \alpha)$, as desired. In conclusion, $\mathcal{U}(f, Q, \alpha) \leq \mathcal{U}(f, \mathcal{P}, \alpha)$.

A similar argument applies to the lower sums. □

Example 7.27: Let $[a, b] = [0, 10]$ and let $\alpha(x)$ be the *greatest integer function*.[1] That is, $\alpha(x)$ is the greatest integer that does not exceed x. So, for example, $\alpha(0.5) = 0$, $\alpha(2) = 2$, and $\alpha(-3/2) = -2$. Then α is an increasing function on $[0, 10]$. Let f be any continuous function on $[0, 10]$. We shall determine whether

$$\int_0^{10} f \, d\alpha$$

exists and, if it does, calculate its value.

Let $\mathcal{P}$ be a partition of $[0,10]$. By the lemma, it is to our advantage to assume that the mesh of $\mathcal{P}$ is smaller than 1. Observe that $\Delta\alpha_j$ equals the number of integers that lie in the interval I_j—that is, either 0 or 1. Let $I_{j_0}, I_{j_2}, \ldots I_{j_{10}}$ be, in sequence, the intervals from the partition which do in fact contain each distinct integer (the first of these contains 0, the second contains 1, and so on up to 10). Then

$$\mathcal{U}(f, \mathcal{P}, \alpha) = \sum_{\ell=0}^{10} M_{j_\ell} \Delta\alpha_{j_\ell} = \sum_{\ell=1}^{10} M_{j_\ell}$$

and

$$\mathcal{L}(f, \mathcal{P}, \alpha) = \sum_{\ell=0}^{10} m_{j_\ell} \Delta\alpha_{j_\ell} = \sum_{\ell=1}^{10} m_{j_\ell}$$

because any term in these sums corresponding to an interval not containing an integer must have $\Delta\alpha_j = 0$. Notice that $\Delta\alpha_{j_0} = 0$ since $\alpha(0) = \alpha(p_1) = 0$.

Let $\epsilon > 0$. Since f is uniformly continuous on $[0, 10]$, we may choose a $\delta > 0$ such that $|s - t| < \delta$ implies that $|f(s) - f(t)| < \epsilon/20$. If $m(\mathcal{P}) < \delta$ then it follows that $\left|f(\ell) - M_{j_\ell}\right| < \epsilon/20$ and $\left|f(\ell) - m_{j_\ell}\right| < \epsilon/20$ for $\ell = 0, 1, \dots 10$. Therefore

$$\mathcal{U}(f, \mathcal{P}, \alpha) < \sum_{\ell=1}^{10} \left(f(\ell) + \frac{\epsilon}{20}\right)$$

and

$$\mathcal{L}(f, \mathcal{P}, \alpha) > \sum_{\ell=1}^{10} \left(f(\ell) - \frac{\epsilon}{20}\right).$$

Rearranging these inequalities leads to

$$\mathcal{U}(f, \mathcal{P}, \alpha) < \left(\sum_{\ell=1}^{10} f(\ell)\right) + \frac{\epsilon}{2}$$

and

$$\mathcal{L}(f, \mathcal{P}, \alpha) > \left(\sum_{\ell=1}^{10} f(\ell)\right) - \frac{\epsilon}{2}.$$

Thus, since $I_*(f)$ and $I^*(f)$ are trapped between $\mathcal{U}$ and $\mathcal{L}$, we conclude that

$$|I_*(f) - I^*(f)| < \epsilon.$$

We have seen that, if the partition is fine enough, then the upper and lower integrals of f with respect to α differ by at most ϵ. It follows that $\int_0^{10} f d\,\alpha$ exists. Moreover,

$$\left|I^*(f) - \sum_{\ell=1}^{10} f(\ell)\right| < \epsilon$$

and

$$\left| I_*(f) - \sum_{\ell=1}^{10} f(\ell) \right| < \epsilon.$$

We conclude that

$$\int_0^{10} f \, d\alpha = \sum_{\ell=1}^{10} f(\ell). \qquad \Box$$

The example demonstrates that the language of the Riemann–Stieltjes integral allows us to think of the integral as a generalization of the summation process. This is frequently useful, both philosophically and for practical reasons.

The next result, sometimes called Riemann's lemma, is crucial for proving the existence of Riemann–Stieltjes integrals.

Proposition 7.28: (Riemann's Lemma) *Let α be an increasing function on $[a, b]$ and f a bounded function on the interval. The Riemann–Stieltjes integral of f with respect to α exists if and only if, for every $\epsilon > 0$, there is a partition $\mathcal{P}$ such that*

$$|\mathcal{U}(f, \mathcal{P}, \alpha) - \mathcal{L}(f, \mathcal{P}, \alpha)| < \epsilon. \tag{7.28.1}$$

Proof: First assume that (7.28.1) holds. Fix $\epsilon > 0$. Since $\mathcal{L} \le I_* \le I^* \le \mathcal{U}$, inequality (7.28.1) implies that

$$|I^*(f) - I_*(f)| < \epsilon.$$

But this means that $\int_a^b f \, d\alpha$ exists.

Conversely, assume that the integral exists. Fix $\epsilon > 0$. Choose a partition Q_1 such that

$$|\mathcal{U}(f, Q_1, \alpha) - I^*(f)| < \epsilon/2.$$

Likewise choose a partition Q_2 such that

$$|\mathcal{L}(f, Q_2, \alpha) - I_*(f)| < \epsilon/2.$$

Since $I_*(f) = I^*(f)$ it follows that

$$|\mathcal{U}(f, Q_1, \alpha) - \mathcal{L}(f, Q_2, \alpha)| < \epsilon. \tag{7.28.2}$$

Let $\mathcal{P}$ be the common refinement of Q_1 and Q_2. Then we have, again by Lemma 7.26, that

$$\mathcal{L}(f, Q_2, \alpha) \le \mathcal{L}(f, \mathcal{P}, \alpha) \le \int_a^b f \, d\alpha \le \mathcal{U}(f, \mathcal{P}, \alpha) \le \mathcal{U}(f, Q_1, \alpha).$$

But, by (7.28.2), the expressions on the far left and on the far right of these inequalities differ by less than ϵ. Thus $\mathcal{P}$ satifies the condition (7.28.1). $\quad\square$

We note in passing that the basic properties of the Riemann integral noted in Section 7.2 (Theorems 7.11 and 7.12) hold without change for the Riemann–Stieltjes integral. The proofs are left as exercises for you (use Riemann's lemma!).

Exercises
1. Define $\beta(x)$ by the condition that $\beta(x) = x + k$ when $k \le x < k + 1$. Calculate

$$\int_2^6 t^2 \, d\beta(t).$$

2. Let $\alpha(x)$ be the greatest integer function as discussed in the text. Define the "fractional part" function by the formula $\gamma(x) = x - \alpha(x)$. Explain why this function has the name "fractional part." Note that γ is not monotone increasing, but it is at least *piecewise* monotone increasing. So the Riemann–Stieltjes integral with respect to γ still makes sense. Calculate

$$\int_0^5 x \, d\gamma(x).$$

3. If p is a polynomial and $\int_a^b p \, d\alpha = 0$ for every choice of α, then what can you conclude about p?

4. Suppose that α and β are monotonic polynomials on the interval $[a, b]$. If $\int f \, d\alpha = \int f \, d\beta$ for every choice of f, then what can you conclude about α and β?

5. Let $\alpha(x)$ be the greatest integer function and $f(x) = x^2$. Calculate $\int_0^3 f \, d\alpha(x)$.

6. Let $f(x) = \alpha(x)$ be the greatest integer function. Calculate $\int_0^4 f \, d\alpha$.

7. State and prove a version of Theorem 7.11 for Riemann–Stieltjes integrals.

8. State and prove a version of Theorem 7.12 for Riemann–Stieltjes integrals.

9. Let $f(x) = x^2$ and $\alpha(x) = x^3$. Calculate

$$\int_0^\pi f \; d\alpha.$$

10. State and prove a result to the effect that, when $\alpha(x) = x$, then the Riemann–Stieltjes integral is equivalent with the classical Riemann integral.

11. Any series can be represented as a Riemann–Stieltjes integral. But the converse is not true. Explain.

*12. The Riemann-Stieltjes integral puts summation by parts into a very natural and general context. Explain.

13. Calculate

$$\int_0^1 x \; dx^2.$$

14. Calculate

$$\int_0^1 \sin x \; dx^2.$$

7.5 Advanced Results on Integration Theory

We now turn to establishing the existence of certain Riemann–Stieltjes integrals.

Theorem 7.29: *Let f be continuous on* $[a, b]$ *and assume that* α *is monotonically increasing. Then*

$$\int_a^b f \; d\alpha$$

exists.

Proof: We may assume that α is nonconstant; otherwise there is nothing to prove.

Pick $\epsilon > 0$. By the uniform continuity of f we may choose a $\delta > 0$ such that if $|s - t| < \delta$ then $|f(s) - f(t)| < \epsilon / (\alpha(b) - \alpha(a))$. Let $\mathcal{P}$ be any partition of $[a, b]$ that has mesh smaller than δ. Then

$$|\mathcal{U}(f, \mathcal{P}, \alpha) - \mathcal{L}(f, \mathcal{P}, \alpha)| = \left| \sum_j M_j \Delta\alpha_j - \sum_j m_j \Delta\alpha_j \right|$$

$$= \sum_j |M_j - m_j| \Delta\alpha_j$$

$$< \sum_j \frac{\epsilon}{\alpha(b) - \alpha(a)} \Delta\alpha_j$$

$$= \frac{\epsilon}{\alpha(b) - \alpha(a)} \cdot \sum_j \Delta\alpha_j$$

$$= \epsilon.$$

Here, of course, we have used the monotonicity of α to observe that the last sum collapses to $\alpha(b) - \alpha(a)$. By Riemann's lemma, we see that the proof is complete.   □

Notice how simple Riemann's lemma is to use. You may find it instructive to compare the proofs of this section with the rather difficult proofs in Section 7.2. What we are learning is that a good definition (and accompanying lemma(s)) can, in the end, make everything much clearer and simpler. Now we establish a companion result to the first one.

Theorem 7.30: *If α is an increasing and continuous function on the interval $[a, b]$ and if f is monotonic on $[a, b]$ then $\int_a^b f \, d\alpha$ exists.*

Proof: We may assume that $\alpha(b) > \alpha(a)$ and that f is monotone *increasing*. Let $L = \alpha(b) - \alpha(a)$ and $M = f(b) - f(a)$. Pick $\epsilon > 0$. Choose a positive integer k so that

$$\frac{L \cdot M}{k} < \epsilon.$$

Let $p_0 = a$ and choose p_1 to be the first point to the right of p_0 such that $\alpha(p_1) - \alpha(p_0) = L/k$ (this is possible, by the Intermediate Value Theorem, since α is continuous). Continuing, choose p_j to be the first point to the right of p_{j-1} such that $\alpha(p_j) - \alpha(p_{j-1}) = L/k$. This process will terminate after k steps and we will have $p_k = b$. Then $\mathcal{P} = \{p_0, p_1, \ldots, p_k\}$ is a partition of $[a, b]$.

Next observe that, for each j, the value M_j of sup f on I_j is $f(p_j)$ since f is increasing. Similarly the value m_j of inf f on I_j is $f(p_{j-1})$. We find therefore that

$$\mathcal{U}(f, \mathcal{P}, \alpha) - \mathcal{L}(f, \mathcal{P}, \alpha) = \sum_{j=1}^{k} M_j \Delta \alpha_j - \sum_{j=1}^{k} m_j \Delta \alpha_j$$

$$= \sum_{j=1}^{k} \left((M_j - m_j) \frac{L}{k} \right)$$

$$= \frac{L}{k} \sum_{j=1}^{k} \left(f(x_j) - f(x_{j-1}) \right)$$

$$= \frac{L \cdot M}{k}$$

$$< \epsilon.$$

Therefore inequality (7.28.1) of Riemann's lemma is satisfied and the integral exists.   □

One of the useful features of Riemann–Stieltjes integration is that it puts integration by parts into a very natural setting. We begin with a lemma.

Lemma 7.31: *Let f be continuous on an interval $[a, b]$ and let g be monotone increasing and continuous on that interval. If G is an antiderivative for g then*

$$\int_a^b f(x) g(x) dx = \int_a^b f \ dG.$$

Proof: Apply the Mean Value Theorem to the Riemann sums for the integral on the right.   □

Theorem 7.32: (Integration by Parts) *Suppose that both f and g are continuous, increasing functions on the interval $[a, b]$. Let F be an antiderivative for f on $[a, b]$ and G an antiderivative for g on $[a, b]$. Then we have*

$$\int_a^b F \ dG = [F(b) \cdot G(b) - F(a) \cdot G(a)] - \int_a^b G \ dF.$$

Proof: Notice that, by the preceding lemma, both integrals exist. Set $P(x) = F(x) \cdot G(x)$. Then P has a continuous derivative on the interval $[a, b]$. Thus the Fundamental Theorem of Calculus applies and we may write

$$P(b) - P(a) = \int_a^b P'(x) \ dx = [F(b) \cdot G(b) - F(a) \cdot G(a)].$$

Now, writing out P' explicitly, using Leibnitz's Rule for the derivative of a product, we obtain

$$\int_a^b F(x) g(x) \ dx = [F(b)G(b) - F(a)G(a)] - \int_a^b G(x) f(x) dx.$$

But the lemma allows us to rewrite this equation as

$$\int_a^b F \; dG = [F(b)G(b) - F(a)G(a)] - \int_a^b G(x) \; dF.$$

□

Remark 7.33: The integration by parts formula can also be proved by applying *summation* by parts to the Riemann sums for the integral

$$\int_a^b F \; dG.$$

This method is explored in the exercises.

□

We have already observed that the Riemann–Stieltjes integral

$$\int_a^b f \; d\alpha$$

is linear in f; that is,

$$\int_a^b (f + g) \; d\alpha = \int_a^b f \; d\alpha + \int_a^b g \; d\alpha$$

and

$$\int_a^b c \cdot f \; d\alpha = c \cdot \int_a^b f \; d\alpha$$

when both f and g are Riemann–Stieltjes integrable with respect to α and for any constant c. We also would expect, from the very way that the integral is constructed, that it would be linear in the α entry. But we have not even defined the Riemann–Stieltjes integral for nonincreasing α. And what of a function α that is the difference of two increasing functions? Such a function need not be monotone. Is it possible to identify which functions α can be decomposed as sums or differences of monotonic functions? It turns out that there is a satisfactory answer to these questions, and we should like to discuss these matters briefly.

Definition 7.34: If α is a monotonically *decreasing* function on $[a, b]$ and f is a function on $[a, b]$ then we define

$$\int_a^b f \; d\alpha = - \int_a^b f \; d(-\alpha)$$

when the right side exists.

The definition exploits the simple observation that if α is decreasing then $-\alpha$ is increasing; hence the preceding theory applies to the function $-\alpha$.

Next we have

Definition 7.35: Let α be a function on $[a, b]$ that can be expressed as

$$\alpha(x) = \alpha_1(x) - \alpha_2(x),$$

where both α_1 and α_2 are increasing. Then, for any f on $[a, b]$, we define

$$\int_a^b f \, d\alpha = \int_a^b f \, d\alpha_1 - \int_a^b f \, d\alpha_2,$$

provided that both integrals on the right exist.

Now, by the very way that we have formulated our definitions, $\int_a^b f \, d\alpha$ is linear in both the f entry and the α entry. But the definitions are not satisfactory unless we can identify those α that can actually occur in the last definition. This leads us to a new class of functions.

Definition 7.36: Let f be a function on the interval $[a, b]$. For $x \in [a, b]$ we define

$$Vf(x) = \sup \sum_{j=1}^{k} \left| f(p_j) - f(p_{j-1}) \right|,$$

where the supremum is taken over all partitions $\mathcal{P}$ of the interval $[a, x]$.

If $Vf \equiv Vf(b) < \infty$, then the function f is said to be of *bounded variation* on the interval $[a, b]$. In this circumstance the quantity $Vf(b)$ is called the *total variation* of f on $[a, b]$.

A function of bounded variation has the property that its graph does not have unbounded total oscillation.

Example 7.37: Define $f(x) = \sin x$, with domain the interval $[0, 2\pi]$. Let us calculate Vf. Let $\mathcal{P}$ be a partition of $[0, 2\pi]$. Since adding points to the partition only makes the sum

$$\sum_{j=1}^{k} \left| f(p_j) - f(p_{j-1}) \right|$$

larger (by the triangle inequality), we may as well suppose that

$$\mathcal{P} = \{p_0, p_1, p_2, \ldots, p_k\}$$

contains the points $\pi/2, 3\pi/2$. Assume that $p_{\ell_1} = \pi/2$ and $p_{\ell_2} = 3\pi/2$. Then

$$\sum_{j=1}^{k} \left| f(p_j) - f(p_{j-1}) \right| = \sum_{j=1}^{\ell_1} \left| f(p_j) - f(p_{j-1}) \right|$$

$$+ \sum_{j=\ell_1+1}^{\ell_2} \left| f(p_j) - f(p_{j-1}) \right|$$

$$+ \sum_{j=\ell_2+1}^{k} \left| f(p_j) - f(p_{j-1}) \right|.$$

However, f is increasing on the interval $[0, \pi/2] = [0, p_{\ell_1}]$. Therefore the first sum is just

$$\sum_{j=1}^{\ell_1} f(p_j) - f(p_{j-1}) = f(p_{\ell_1}) - f(p_0) = f(\pi/2) - f(0) = 1.$$

Similarly, f is monotone on the intervals $[\pi/2, 3\pi/2] = [p_{\ell_1}, p_{\ell_2}]$ and $[3\pi/2, 2\pi]$ $= [p_{\ell_2}, p_k]$. Thus the second and third sums equal $f(p_{\ell_1}) - f(p_{\ell_2}) = 2$ and $f(p_k) - f(p_{\ell_2}) = 1$ respectively. It follows that

$$Vf = Vf(2\pi) = 1 + 2 + 1 = 4.$$

Of course $Vf(x)$ for any $x \in [0, 2\pi]$ can be computed by similar means (see the exercises). □

Example 7.38: In general, if f is a continuously differentiable function on an interval $[a, b]$ then

$$Vf(x) = \int_a^x |f'(t)| \, dt.$$

This assertion will be explored in the exercises. □

Example 7.39: The function $f(x) = \cos x$ on the interval $[0, 2\pi]$ is of bounded variation. And in fact

$$Vf = 4$$

because the function goes from 1 down to 0 and then from 0 down to −1 and finally from −1 up to 0 and finally from 0 up to 1.

Alternatively, one can obtain the same answer by calculating the integral

$$Vf = \int_0^{2\pi} |f'(x)| \, dx = \int_0^{2\pi} |\sin x| \, dx. \qquad \Box$$

Lemma 7.40: *Let f be a function of bounded variation on the interval [a, b]. Then the function Vf is increasing on [a, b].*

Proof: Let $s < t$ be elements of $[a, b]$. Let $\mathcal{P} = \{p_0, p_1, \ldots, p_k\}$ be a partition of $[a, s]$. Then $\tilde{\mathcal{P}} = \{p_0, p_1, \ldots, p_k, t\}$ is a partition of $[a, t]$ and

$$\sum_{j=1}^{k} |f(p_j) - f(p_{j-1})|$$

$$\leq \sum_{j=1}^{k} |f(p_j) - f(p_{j-1})| + |f(t) - f(p_k)|$$

$$\leq Vf(t).$$

Taking the supremum on the left over all partitions $\mathcal{P}$ of $[a, s]$ yields that

$$Vf(s) \leq Vf(t). \qquad \Box$$

Lemma 7.41: *Let f be a function of bounded variation on the interval [a, b] . Then the function Vf − f is increasing on the interval [a, b].*

Proof: Let $s < t$ be elements of $[a, b]$. Pick $\epsilon > 0$. By the definition of Vf we may choose a partition $\mathcal{P} = \{p_0, p_1, \ldots, p_k\}$ of the interval $[a, s]$ such that

$$Vf(s) - \epsilon < \sum_{j=1}^{k} \left| f(p_j) - f(p_{j-1}) \right|. \qquad (7.41.1)$$

But then $\tilde{\mathcal{P}} = \{p_0, p_1, \ldots, p_k, t\}$ is a partition of $[a, t]$ and we have that

$$\sum_{j=1}^{k} \left| f(p_j) - f(p_{j-1}) \right| + |f(t) - f(s)| \leq Vf(t).$$

Using (7.41.1), we may conclude that

$$Vf(s) - \epsilon + f(t) - f(s) < \sum_{j=1}^{k} \left| f(p_j) - f(p_{j-1}) \right| + |f(t) - f(s)| \le Vf(t).$$

We conclude that

$$Vf(s) - f(s) < Vf(t) - f(t) + \epsilon.$$

Since the inequality holds for every $\epsilon > 0$, we see that the function $Vf - f$ is increasing.   □

Now we may combine the last two lemmas to obtain our main result:

Proposition 7.42: *If a function f is of bounded variation on $[a, b]$, then f may be written as the difference of two increasing functions. Conversely, the difference of two increasing functions is a function of bounded variation.*

Proof: If f is of bounded variation write $f = Vf - (Vf - f) \equiv f_1 - f_2$. By the lemmas, both f_1 and f_2 are increasing.

For the converse, assume that $f = f_1 - f_2$ with f_1, f_2 increasing. Then it is easy to see that

$$Vf(b) \le \left| f_1(b) - f_1(a) \right| + \left| f_2(b) - f_2(a) \right|.$$

Thus f is of bounded variation.   □

Now the main point of this discussion is the following theorem:

Theorem 7.43: *If f is a continuous function on $[a, b]$ and if α is of bounded variation on $[a, b]$ then the integral*

$$\int_a^b f \, d\alpha$$

exists and is finite.

If g is of bounded variation on $[a, b]$ and if β is a continuous function of bounded variation on $[a, b]$ then the integral

$$\int_a^b g \, d\beta$$

exists and is finite.

Proof: Write the function(s) of bounded variation as the difference of increasing functions. Then apply Theorems 7.29 and 7.30. □

Exercises

1. Prove that the integral

$$\int_0^\infty \frac{\sin x}{x} dx$$

   exists.

2. Prove that, if f is a continuously differentiable function on the interval $[a, b]$, then

$$Vf = \int_a^b |f'(x)| dx.$$

   [**Hint:** You will prove two inequalities. For one, use the Fundamental Theorem of Calculus. For the other, use the Mean Value Theorem.]

3. Give an example of a continuous function on the interval $[0, 1]$ that is not of bounded variation.

4. Let β be a nonnegative, increasing function on the interval $[a, b]$. Set $m = \beta(a)$ and $M = \beta(b)$. For any number λ lying between m and M, set $S_\lambda = \{x \in [a, b]: \beta(x) > \lambda\}$. Prove that S_λ must be an interval. Let $\ell(\lambda)$ be the length of S_λ. Then prove that

$$\int_a^b \beta(t)^p dt = -\int_m^M s^p \, d\ell(s)$$
$$= \int_0^M \ell(s) \cdot p \cdot s^{p-1} ds.$$

5. If φ is a convex function on the real line, then prove that, for f integrable on $[0, 1]$,

$$\varphi\left(\int_0^1 f(x) dx\right) \le \int_0^1 \varphi(f(x)) dx.$$

6. Give an example of a continuously differentiable function on an open interval that is not of bounded variation.

7. Prove that a continuously differentiable function on a compact interval is of bounded variation.

8. Let $f(x) = \sin x$ on the interval $[0, 2\pi]$. Calculate $Vf(x)$ for any $x \in [0, 2\pi]$.

9. Let $f(x) = x^2$ and $\alpha(x) = \sin x$. Calculate

$$\int_0^\pi f \, d\alpha.$$

10. Calculate the total variation of the function

$$f(x) = \sin(jx)$$

on the interval $(0, \pi)$.

11. Provide a detailed proof of Lemma 7.31.

12. Show that the function $f(x) = \sin(1/x)$ is of infinite total variation on the interval $(0, 1)$.

13. What is the total variation of the function $f(x) = \sin x$ on the interval $[0, 2\pi]$?

14. If f, g are both of bounded variation on the interval $[0, 1]$ then what can you say about the total variation of $f \cdot g$ on $[0, 1]$?

Note

1 In many texts the greatest integer in x is denoted by $[x]$. We do not use that notation here because it could get confused with our notation for a closed interval.

8

Sequences and Series of Functions

8.1 Partial Sums and Pointwise Convergence

A *sequence of functions* is usually written

$$f_1, f_2, \ldots \quad \text{or} \quad \{f_j\}_{j=1}^{\infty}.$$

We will generally assume that the functions f_j all have the same domain S.

Definition 8.1: A sequence of functions $\{f_j\}_{j=1}^{\infty}$ with domain $S \subseteq \mathbb{R}$ is said to *converge pointwise* to a limit function f on S if, for each $x \in S$, the sequence of numbers $\{f_j(x)\}$ converges to $f(x)$.

Example 8.2: Define $f_j(x) = x^j$ with domain $S = \{x : 0 \leq x \leq 1\}$. If $0 \leq x < 1$, then $f_j(x) \to 0$. However, $f_j(1) \to 1$. Therefore the sequence f_j converges pointwise to the function

$$f(x) = \begin{cases} 0 & \text{if } 0 \leq x < 1 \\ 1 & \text{if } x = 1 \end{cases}$$

See Figure 8.1. We see that, even though the f_j are each continuous, the limit function f is not. □

Here are some of the basic questions that we must ask about a sequence of functions f_j that converges to a function f on a domain S:

1. If the functions f_j are continuous, then is f continuous?
2. If the functions f_j are integrable on an interval I, then is f integrable on I?
3. If f is integrable on I, then does the sequence $\int_I f_j(x)dx$ converge to $\int_I f(x)dx$?

DOI: 10.1201/9781003222682-9

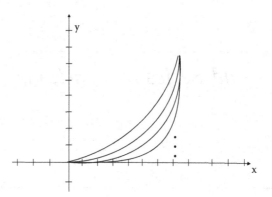

FIGURE 8.1
The sequence {x^j}.

4. If the functions f_j are differentiable, then is f differentiable?
5. If f is differentiable, then does the sequence f'_j converge to f'?

We see from Example 8.2 that the answer to the first question is "no": Each of the f_j is continuous but f is not. It turns out that, in order to obtain a favorable answer to our questions, we must consider a stricter notion of convergence of functions. This motivates the next definition.

Definition 8.3: Let f_j be a sequence of functions on a domain S. We say that the functions f_j *converge uniformly* to f on S if, given $\epsilon > 0$, there is an $N > 0$ such that, for any $j > N$ and any $x \in S$, it holds that $|f_j(x) - f(x)| < \epsilon$.

Notice that the special feature of uniform convergence is that the rate at which $f_j(x)$ converges is independent of $x \in S$. In Example 8.1, $f_j(x)$ is converging very rapidly to zero for x near zero but arbitrarily slowly to zero for x near 1—see Figure 8.1. In the next example we shall prove this assertion rigorously:

Example 8.4: The sequence $f_j(x) = x^j$ does *not* converge uniformly to the limit function

$$f(x) = \begin{cases} 0 & \text{if } 0 \le x < 1 \\ 1 & \text{if } x = 1 \end{cases}$$

on the domain $S = [0, 1]$. In fact it does not even do so on the smaller domain $[0, 1]$. To see this notice that, no matter how large j is, we have by the Mean Value Theorem that

$$f_j(1) - f_j(1 - 1/(2j)) = \frac{1}{2j} \cdot f'_j(\xi)$$

for some ξ between $1 - 1/(2j)$ and 1. But $f'_j(x) = j \cdot x^{j-1}$ hence $|f'_j(\xi)| < j$ and we conclude that

$$\left| f_j(1) - f_j(1 - 1/(2j)) \right| < \frac{1}{2}$$

or

$$f_j(1 - 1/(2j)) > f_j(1) - \frac{1}{2} = \frac{1}{2} .$$

In conclusion, no matter how large j, there will be values of x (namely, $x = 1 - 1/(2j)$) at which $f_j(x)$ is at least distance $1/2$ from the limit 0. We conclude that the convergence is not uniform. □

Theorem 8.5: *If f_j are continuous functions on a set S that converge uniformly on S to a function f then f is also continuous.*

Proof: Let $\epsilon > 0$. Choose an integer N so large that, if $j > N$, then $|f_j(x) - f(x)| < \epsilon/3$ for all $x \in S$. Fix $P \in S$. Choose $\delta > 0$ so small that if $|x - P| < \delta$ then $|f_N(x) - f_N(P)| < \epsilon/3$. For such x we have

$$|f(x) - f(P)| \leq \left| f(x) - f_N(x) \right| + \left| f_N(x) - f_N(P) \right| + \left| f_N(P) - f(P) \right|$$
$$< \frac{\epsilon}{3} + \frac{\epsilon}{3} + \frac{\epsilon}{3}$$

by the way that we chose N and δ. But the last line sums to ϵ, proving that f is continuous at P. Since $P \in S$ was chosen arbitrarily, we are done. □

Example 8.6: Define functions

$$f_j(x) = \begin{cases} 0 & \text{if } x = 0 \\ j & \text{if } 0 < x \leq 1/j \\ 0 & \text{if } 1/j < x \leq 1 \end{cases}$$

for $j = 2, 3, \ldots$. Then, $\lim_{j \to \infty} f_j(x) = 0 \equiv f(x)$ for all x in the interval $I = [0, 1]$. However

$$\int_0^1 f_j(x)\,dx = \int_0^{1/j} j\,dx = 1$$

for every j. Thus the f_j converge to the integrable limit function $f(x) \equiv 0$ (with integral 0), but their integrals do not converge to the integral of f. Of course. the f_j do not converge uniformly.   □

Example 8.7: Let $q_1, q_2, \dots$ be an enumeration of the rationals in the interval $I = [0, 1]$. Define functions

$$f_j(x) = \begin{cases} 1 & \text{if } x \in \{q_1, q_2, \dots, q_j\} \\ 0 & \text{if } x \notin \{q_1, q_2, \dots, q_j\} \end{cases}$$

Then, the functions f_j converge pointwise to the Dirichlet function f which is equal to 1 on the rationals and 0 on the irrationals. Each of the functions f_j has integral 0 on I. But the function f is not Riemann integrable on I.   □

The last two examples show that something more than pointwise convergence is needed in order for the integral to respect the limit process.

Theorem 8.8: *Let f_j be integrable functions on a nontrivial bounded interval $[a, b]$ and suppose that the functions f_j converge uniformly to the limit function ... Then, f is integrable on $[a, b]$ and*

$$\lim_{j \to \infty} \int_a^b f_j(x)\,dx = \int_a^b f(x)\,dx .$$

Proof: Pick $\epsilon > 0$. Choose N so large that, if $j > N$, then $|f_j(x) - f(x)| < \epsilon/[2(b - a)]$ for all $x \in [a, b]$. Notice that, if $j, k > N$, then

$$\left| \int_a^b f_j(x)\,dx - \int_a^b f_k(x)\,dx \right| \le \int_a^b |f_j(x) - f_k(x)|\,dx. \tag{8.8.1}$$

But $|f_j(x) - f_k(x)| \le |f_j(x) - f(x)| + |f(x) - f_k(x)| < \epsilon/(b - a)$. Therefore, line (8.8.1) does not exceed

$$\int_a^b \frac{\epsilon}{b - a}\,dx = \epsilon.$$

Thus the numbers $\int_a^b f_j(x)\,dx$ form a Cauchy sequence. Let the limit of this sequence be called A. Notice that, if we let $k \to \infty$ in the inequality

$$\left| \int_a^b f_j(x)\,dx - \int_a^b f_k(x)\,dx \right| \le \epsilon,$$

then we obtain

$$\left| \int_a^b f_j(x)\,dx - A \right| \le \epsilon$$

for all $j \ge N$. This estimate will be used below.

By hypothesis there is a $\delta > 0$ such that, if $\mathcal{P} = \{p_1, \ldots, p_k\}$ is a partition of $[a, b]$ with $m(\mathcal{P}) < \delta$, then

$$\left| \mathcal{R}(f_N, \mathcal{P}) - \int_a^b f_N(x)\,dx \right| < \epsilon.$$

But then, for such a partition, we have

$$|\mathcal{R}(f, \mathcal{P}) - A| \le \left| \mathcal{R}(f, \mathcal{P}) - \mathcal{R}(f_N, \mathcal{P}) \right| + \left| \mathcal{R}(f_N, \mathcal{P}) - \int_a^b f_N(x)\,dx \right|$$
$$+ \left| \int_a^b f_N(x)\,dx - A \right|.$$

We have already noted that, by the choice of N, the third term on the right is smaller than ϵ. The second term is smaller than ϵ by the way that we chose the partition $\mathcal{P}$. It remains to examine the first term. Now

$$\left| \mathcal{R}(f, \mathcal{P}) - \mathcal{R}(f_N, \mathcal{P}) \right| = \left| \sum_{j=1}^k f(s_j)\Delta_j - \sum_{j=1}^k f_N(s_j)\Delta_j \right|$$
$$\le \sum_{j=1}^k \left| f(s_j) - f_N(s_j) \right| \Delta_j$$
$$< \sum_{j=1}^k \frac{\epsilon}{2(b-a)} \Delta_j$$
$$= \frac{\epsilon}{2(b-a)} \sum_{j=1}^k \Delta_j$$
$$= \frac{\epsilon}{2}.$$

Therefore $|\mathcal{R}(f, \mathcal{P}) - A| < 3\epsilon$ when $m(\mathcal{P}) < \delta$. This shows that the function f is integrable on $[a, b]$ and has integral with value A. $\qquad \square$

We have succeeded in answering questions **(1)** and **(2)** that were raised at the beginning of the section. In the next section we will answer questions **(3)**, **(4)**, and **(5)**.

Example 8.9: Define

$$f_j(x) = \begin{cases} 0 & \text{if } x \le j \\ x - j & \text{if } j < x \le j+1 \\ (j+2) - x & \text{if } j+1 < x \le j+2 \\ 0 & \text{if } j+2 < x. \end{cases}$$

Then,

$$\int f_j(x)\,dx = 1$$

for each j. But

$$\lim_{j \to \infty} f_j(x) = 0$$

for every x. So we see that

$$1 = \lim_{j \to \infty} \int f_j(x)\,dx \ne \int \lim_{j \to \infty} f_j(x)\,dx = \int 0 \; dx = 0. \qquad \square$$

Exercises

1. If $f_j \to f$ uniformly on a domain S and if f_j, f never vanish on S, then does it follow that the functions $1/f_j$ converge uniformly to $1/f$ on S?

2. Write out the first five partial sums for the series

$$\sum_{j=1}^{\infty} \frac{\sin^3 j}{j^2}.$$

3. Write a series of polynomials that converges to $f(x) = \sin x^2$. Can you prove that it converges?

4. Write a series of trigonometric functions that converges to $f(x) = x$. Can you prove that it converges?

5. Write a series of piecewise linear functions that converges to $f(x) = x^2$ on the interval $[0, 1]$. Can you prove that it converges?

6. Write a series of functions that converges pointwise on $[0, 1]$ but does not converge uniformly on any proper subinterval. [**Hint:** First consider a sequence.]

7. Give an example of a Taylor series that converges uniformly on compact sets to its limit function.

8. Prove that the series

$$\sum_{j=1}^{\infty} \frac{\sin jx}{j^2}$$

converges uniformly to a continuous function on the interval $[0, 1]$.

9. A Taylor series will never converge only pointwise. Explain.

10. Define

$$f_j(x) = \begin{cases} 1 + 1/j & \text{if } x < j \\ 1/j & \text{if } x \geq j \end{cases}$$

Show that f_j converges to the identically 1 function pointwise but not uniformly.

11. Define

$$f_j(x) = \begin{cases} 0 & \text{if } x \leq 0 \\ x^2/j & \text{if } x > 0 \end{cases}$$

Prove that each f_j is continuous, and the sequence $\{f_j\}$ converges pointwise to the identically 0 function. But the sequence does not converge uniformly.

12. Show that, if $\sum_j f'_j$ converges uniformly on $[0, 1]$ (where the prime stands for the derivative), and if $f_j(0) = 0$ for all j, then $\sum_j f_j$ converges uniformly on compact sets.

13. TRUE or FALSE: If $\sum_j f_j$ converges absolutely and uniformly and $\sum_j g_j$ converges absolutely and uniformly on a compact interval $[a, b]$, then so does $\sum_j f_j g_j$.

14. Write the function $f(x) = \sin x$ on the interval $[0, 2\pi]$ as the uniform limit of functions that are discontinuous.

15. Suppose that f_j are piecewise linear, continuous functions that converge uniformly on the interval $[0, 1]$ to a limit function f. What can you say about f? Is it piecewise linear? Is it continuous?

16. Suppose that the functions f_j are continuously differentiable on $[0, 1]$ and that they converge uniformly to a limit function g. It does *not* follow that the f'_j converge uniformly. Explain.

8.2 More on Uniform Convergence

In general, limits do not commute. Since the integral is defined with a limit, and since we saw in the last section that integrals do not always respect limits of functions, we know some concrete instances of non-commutation of limits. The fact that continuity is defined with a limit, and that the limit of continuous functions need not be continuous, gives even more examples of situations in which limits do not commute. Let us now turn to a situation in which limits *do* commute:

Theorem 8.10: *Fix a set S and a point* $s \in S$. *Assume that the functions* f_j *converge uniformly on the domain* $S \setminus \{s\}$ *to a limit function* f. *Suppose that each function* $f_j(x)$ *has a limit* α_j *as* $x \to s$. *Then* f *itself has a limit* α_0 *as* $x \to s$ *and*

$$\alpha_0 = \lim_{x \to s} f(x) = \lim_{j \to \infty} \lim_{x \to s} f_j(x) = \lim_{j \to \infty} \alpha_j.$$

Because of the way that f *is defined, we may rewrite this conclusion as*

$$\lim_{x \to s} \lim_{j \to \infty} f_j(x) = \lim_{j \to \infty} \lim_{x \to s} f_j(x).$$

In other words, the limits $\lim_{x \to s}$ *and* $\lim_{j \to \infty}$ *commute.*

Proof: Let $\alpha_j = \lim_{x \to s} f_j(x)$. Let $\epsilon > 0$. There is a number $N > 0$ (independent of $x \in S \setminus \{s\}$) such that $j > N$ implies that $|f_j(x) - f(x)| < \epsilon/4$. Fix $j, k > N$. Choose $\delta > 0$ such that $0 < |x - s| < \delta$ implies both that $|f_j(x) - \alpha_j| < \epsilon/4$ and $|f_k(x) - \alpha_k| < \epsilon/4$. Then

$$\left| \alpha_j - \alpha_k \right| \le \left| \alpha_j - f_j(x) \right| + \left| f_j(x) - f(x) \right| + \left| f(x) - f_k(x) \right| + \left| f_k(x) - \alpha_k \right|.$$

The first and last expressions are less than $\epsilon/4$ by the choice of x. The middle two expressions are less than $\epsilon/4$ by the choice of N (and therefore of j and k). We conclude that the sequence α_j is Cauchy. Let α be the limit of that sequence.

Letting $k \to \infty$ in the inequality

$$\left| \alpha_j - \alpha_k \right| < \epsilon$$

that we obtained above yields

$$\left|\alpha_j - \alpha\right| \le \epsilon$$

for $j > N$. Now, with δ as above and $0 < |x - s| < \delta$, we have

$$\left|f(x) - \alpha\right| \le \left|f(x) - f_j(x)\right| + \left|f_j(x) - \alpha_j\right| + \left|\alpha_j - \alpha\right|.$$

By the choices we have made, the first term is less than $\epsilon/4$, the second is less than $\epsilon/4$, and the third is less than or equal to ϵ. Altogether, if $0 < |x - s| < \delta$ then $|f(x) - \alpha| < 2\epsilon$. This is the desired conclusion. □

Example 8.11: Consider the example

$$f_j(x) = x^j$$

on the interval $[0, 1]$. We see that

$$\lim_{j \to \infty} f_j(x) = 0 \equiv f(x)$$

for $0 \le x < 1$. Thus

$$\lim_{x \to 1^-} f(x) = 0 .$$

But

$$\lim_{j \to \infty} \lim_{x \to 1^-} f_j(x) = \lim_{j \to \infty} 1 = 1.$$

Thus the two dual limits in the theorem are unequal in this example. But of course the functions f_j do not converge uniformly. □

Parallel with our notion of Cauchy sequence of numbers, we have a concept of Cauchy sequence of functions in the uniform sense:

Definition 8.12: A sequence of functions f_j on a domain S is called *a uniformly Cauchy sequence* if, for each $\epsilon > 0$, there is an $N > 0$ such that, if $j, k > N$, then

$$\left|f_j(x) - f_k(x)\right| < \epsilon \qquad \forall\, x \in S.$$

The key point for "uniformly Cauchy" sequence of functions is that the choice of N does not depend on x.

Proposition 8.13: *A sequence of functions f_j is uniformly Cauchy on a domain S if and only if the sequence converges uniformly to a limit function f on the domain S.*

Proof: The proof is straightforward and is assigned as an exercise.   □

We will use the last two results in our study of the limits of differentiable functions. First we consider an example.

Example 8.14: Define the function

$$f_j(x) = \begin{cases} 0 & \text{if } x \le 0 \\ jx^2 & \text{if } 0 < x \le 1/(2j) \\ x - 1/(4j) & \text{if } 1/(2j) < x < \infty \end{cases}$$

We leave it as an exercise for you to check that the functions f_j converge uniformly on the entire real line to the function

$$f(x) = \begin{cases} 0 & \text{if } x \le 0 \\ x & \text{if } x > 0 \end{cases}$$

(draw a sketch to help you see this). Notice that each of the functions f_j is continuously differentiable on the entire real line, but f is not differentiable at 0.   □

It turns out that we must strengthen our convergence hypotheses if we want the limit process to respect differentiation. The basic result is this:

Theorem 8.15: *Suppose that a sequence f_j of differentiable functions on an open interval I converges pointwise to a limit function f. Suppose further that the sequence f'_j converges uniformly on I to a limit function g. Then the limit function f is differentiable on I and $f'(x) = g(x)$ for all $x \in I$.*

Proof: Let $\epsilon > 0$. The sequence $\{f'_j\}$ is uniformly Cauchy. Therefore we may choose N so large that $j, k > N$ implies that

$$\left| f'_j(x) - f'_k(x) \right| < \frac{\epsilon}{2} \quad \forall x \in I. \tag{8.15.1}$$

Fix a point $P \in I$. Define

$$\mu_j(x) = \frac{f_j(x) - f_j(P)}{x - P}$$

for $x \in I$, $x \neq P$. It is our intention to apply Theorem 8.10 above to the functions μ_j.

First notice that, for each j, we have

$$\lim_{x \to P} \mu_j(x) = f'_j(P).$$

Thus

$$\lim_{j \to \infty} \lim_{x \to P} \mu_j(x) = \lim_{j \to \infty} f'_j(P) = g(P).$$

That calculates the limits in one order.

On the other hand,

$$\lim_{j \to \infty} \mu_j(x) = \frac{f(x) - f(P)}{x - P} \equiv \mu(x)$$

for $x \in I \setminus \{P\}$. If we can show that this convergence is uniform then Theorem 8.10 applies and we may conclude that

$$\lim_{x \to P} \mu(x) = \lim_{j \to \infty} \lim_{x \to P} \mu_j(x) = \lim_{j \to \infty} f'_j(P) = g(P).$$

But this just says that f is differentiable at P and the derivative equals g. That is the desired result.

To verify the uniform convergence of the μ_j, we apply the Mean Value Theorem to the function $f_j - f_k$. For $x \neq P$ we have

$$\left| \mu_j(x) - \mu_k(x) \right| = \frac{1}{|x - P|} \cdot \left| \left(f_j(x) - f_k(x) \right) - \left(f_j(P) - f_k(P) \right) \right|$$

$$= \frac{1}{|x - P|} \cdot |x - P| \cdot \left| \left(f_j - f_k \right)'(\xi) \right|$$

$$= \left| \left(f_j - f_k \right)'(\xi) \right|$$

for some ξ between x and P. But line (8.15.1) guarantees that the last line does not exceed $\epsilon/2$. That shows that the μ_j converge uniformly and concludes the proof. □

Remark 8.16: A little additional effort shows that we need only assume in the theorem that the functions f_j converge at a single point x_0 in the domain. One of the exercises asks you to prove this assertion.

Notice further that, if we make the additional assumption that each of the functions f'_j is continuous, then the proof of the theorem becomes much easier. For then

$$f_j(x) = f_j(x_0) + \int_{x_0}^x f'_j(t)\,dt$$

by the Fundamental Theorem of Calculus. The hypothesis that the f'_j converge uniformly then implies, by Theorem 8.8, that the integrals converge to

$$\int_{x_0}^x g(t)\,dt.$$

The hypothesis that the functions f_j converge at x_0 then allows us to conclude that the sequence $f_j(x)$ converges for every x to $f(x)$ and

$$f(x) = f(x_0) + \int_{x_0}^x g(t)\,dt.$$

The Fundamental Theorem of Calculus then yields that $f' = g$ as desired. □

Example 8.17: Consider the sequence of functions $f_j(x) = j^{-1/2}\sin(jx)$. This sequence converges uniformly to the identically zero function $f(x) \equiv 0$. But $f'_j = j^{1/2}\cos(jx)$ does not converge at any point.

We can sum up this result by saying that

$$\lim_{j\to\infty}\frac{d}{dx}f_j(x) \ne \frac{d}{dx}\lim_{j\to\infty}f_j(x).$$

□

Exercises

1. Prove that, if a series of continuous functions converges uniformly, then the sum function is also continuous.

2. If a sequence of functions f_j on a domain $S \subseteq \mathbb{R}$ has the property that $f_j \to f$ uniformly on S, then does it follow that $(f_j)^2 \to f^2$ uniformly on S? What simple additional hypothesis will make your answer affirmative?

3. Let f_j be a uniformly convergent sequence of functions on a common domain S. What would be suitable conditions on a function ϕ to guarantee that $\phi \circ f_j$ converges uniformly on S?

4. Prove that a sequence $\{f_j\}$ of functions converges pointwise if and only if the series

$$ f_1 + \sum_{j=2}^{\infty} \left(f_j - f_{j-1} \right) $$

converges pointwise. Prove the same result for uniform convergence.

5. Assume that f_j are continuous functions on the interval $[0, 1]$. Suppose that $\lim_{j \to \infty} f_j(x)$ exists for each $x \in [0, 1]$ and defines a function f on $[0, 1]$. Further suppose that $f_1 \leq f_2 \leq \cdots$. Can you conclude that f is continuous?

6. Let $f: \mathbb{R} \to \mathbb{R}$ be a function. We say that f is *piecewise constant* if the real line can be written as the infinite pairwise disjoint union of intervals and f is constant on each of those intervals. Now let φ be a continuous function on $[a, b]$. Show that φ can be uniformly approximated by piecewise constant functions.

7. Refer to Exercise **6** for terminology. Let f be a piecewise constant function. Show that f is the pointwise limit of polynomials.

8. Prove Proposition 8.13. Refer to the parallel result in Chapter 3 for some hints.

9. Prove the assertion made in Remark 8.16 that Theorem 8.15 is still true if the functions f_j are assumed to converge at just one point (and also that the derivatives f'_j converge uniformly).

*10. A function is called "piecewise linear" if it is (i) continuous and (ii) its graph consists of finitely many linear segments. Prove that a continuous function on an interval $[a, b]$ is the uniform limit of a sequence of piecewise linear functions.

*11. Construct a sequence of continuous functions $f_j(x)$ that has the property that $f_j(q)$ increases monotonically to $+\infty$ for each rational q but such that, at uncountably many irrationals x, $|f_j(x)| \leq 1$ for infinitely many j.

*12. Show that the collection of continuous functions on the interval $[0, 1]$ is a normed linear space that is complete when equipped with the uniform norm. That is, $\|f\| \equiv \max_{[0,1]} |f(x)|$. Here "complete" means that any Cauchy sequence has a limit in the space.

13. Let f be a continous function on $[0, 1]$. Prove that f is the uniform limit of functions f_j that are discontinuous at every point.

14. Let $f(x) = \log |x|$ on the set $(-1, 1) \setminus \{0\}$. Show that f is the pointwise limit of polynomials on this set.

8.3 Series of Functions

Definition 8.18: The formal expression

$$\sum_{j=1}^{\infty} f_j(x),$$

where the f_j are functions on a common domain S, is called a *series of functions*. For $N = 1, 2, 3, \ldots$ the expression

$$S_N(x) = \sum_{j=1}^{N} f_j(x) = f_1(x) + f_2(x) + \ldots + f_N(x)$$

is called the Nth *partial sum* for the series. In case

$$\lim_{N \to \infty} S_N(x)$$

exists and is finite then we say that the series *converges* at x. Otherwise we say that the series *diverges* at x.

Notice that the question of convergence of a series of functions, which should be thought of as an *addition process*, reduces to a question about the *sequence* of partial sums. Sometimes, as in the next example, it is convenient to begin the series at some index other than $j = 1$.

Example 8.19: Consider the series

$$\sum_{j=0}^{\infty} x^j.$$

This is the geometric series from Proposition 3.15. It converges absolutely for $|x| < 1$ and diverges otherwise.

By the formula for the partial sums of a geometric series,

$$S_N(x) = \frac{1 - x^{N+1}}{1 - x}.$$

For $|x| < 1$ we see that

$$S_N(x) \to \frac{1}{1 - x}.$$ □

Definition 8.20: Let

$$\sum_{j=1}^{\infty} f_j(x)$$

be a series of functions on a domain S. If the partial sums $S_N(x)$ converge uniformly on S to a limit function $g(x)$ then we say that the series *converges uniformly* on S.

Of course all of our results about uniform convergence of *sequences* of functions translate, via the sequence of partial sums of a series, to results about uniformly convergent series of functions. For example,

a. If f_j are continuous functions on a domain S and if the series

$$\sum_{j=1}^{\infty} f_j(x)$$

converges uniformly on S to a limit function f then f is also continuous on S.

b. If f_j are integrable functions on $[a, b]$ and if

$$\sum_{j=1}^{\infty} f_j(x)$$

converges uniformly on $[a, b]$ to a limit function f then f is also integrable on $[a, b]$ and

$$\int_a^b f(x)dx = \sum_{j=1}^{\infty} \int_a^b f_j(x)dx.$$

You will be asked to provide details of these assertions, as well as a statement and proof of a result about derivatives of series, in the exercises.

Meanwhile we turn to an elegant test for uniform convergence that is due to Weierstrass.

Theorem 8.21: (The Weierstrass M-Test) *Let* $\{f_j\}_{j=1}^{\infty}$ *be functions on a common domain S. Assume that each* $|f_j|$ *is bounded on S by a constant* M_j *and that*

$$\sum_{j=1}^{\infty} M_j < \infty.$$

Then the series

$$\sum_{j=1}^{\infty} f_j \tag{8.21.1}$$

converges uniformly on the set S.

Proof: By hypothesis, the sequence T_N of partial sums of the series $\sum_{j=1}^{\infty} M_j$ is Cauchy. Given $\epsilon > 0$ there is therefore a number K so large that $q > p > K$ implies that

$$\sum_{j=p+1}^{q} M_j = \left| T_q - T_p \right| < \epsilon.$$

We may conclude that the partial sums S_N of the original series $\sum f_j$ satisfy, for $q > p > K$,

$$\left| S_q(x) - S_p(x) \right| = \left| \sum_{j=p+1}^{q} f_j(x) \right|$$

$$\leq \sum_{j=p+1}^{q} \left| f_j(x) \right| \leq \sum_{j=p+1}^{q} M_j < \epsilon$$

Thus the partial sums $S_N(x)$ of the series (8.21.1) are uniformly Cauchy. The series (8.21.1) therefore converges uniformly. □

Example 8.22: Let us consider the series

$$f(x) = \sum_{j=1}^{\infty} 2^{-j} \sin(2^j x).$$

The sine terms oscillate so erratically that it would be difficult to calculate partial sums for this series. However, noting that the jth summand

$f_j(x) = 2^{-j} \sin(2^j x)$ is dominated in absolute value by 2^{-j}, we see that the Weierstrass M-Test applies to this series. We conclude that the series converges uniformly on the entire real line.

By property **(a)** of uniformly convergent series of continuous functions that was noted above, we may conclude that the function f defined by our series is continuous. It is also 2π-periodic: $f(x + 2\pi) = f(x)$ for every x since this assertion is true for each summand. Since the continuous function f restricted to the compact interval $[0, 2\pi]$ is uniformly continuous (Theorem 5.27), we may conclude that f is uniformly continuous on the entire real line.

However, it turns out that f is nowhere differentiable. The proof of this assertion follows lines similar to the treatment of nowhere differentiable functions in Theorem 6.6. The details will be covered in an exercise.   □

Exercises

1. Prove Dini's theorem: If f_j are continuous functions on a compact set $K, f_1(x) \le f_2(x) \le \ldots$ for all $x \in K$, and the f_j converge to a continuous function f on K then in fact the f_j converge *uniformly* to f on K.

2. Use the concept of boundedness of a function to show that the functions $\sin x$ and $\cos x$ cannot be polynomials.

3. Prove that, if p is any polynomial, then there is an N large enough that $e^x > |p(x)|$ for $x > N$. Conclude that the function e^x is not a polynomial.

4. Find a way to prove that $\tan x$ and $\ln x$ are not polynomials.

5. Prove that the series

$$\sum_{j=1}^{\infty} \frac{\sin jx}{j}$$

converges uniformly on compact intervals that do not contain odd multiples of $\pi/2$. (**Hint:** Sum by parts and the result will follow.)

6. Suppose that the sequence $f_j(x)$ on the interval $[0, 1]$ satisfies $|f_j(s) - f_j(t)| \le |s - t|$ for all $s, t \in [0, 1]$. Further assume that the f_j converge pointwise to a limit function f on the interval $[0, 1]$. Does the series converge uniformly?

7. Prove a comparison test for uniform convergence of series: if f_j, g_j are functions and $0 \le f_j \le g_j$ and the series Σg_j converges uniformly then so also does the series Σf_j.

8. Show by giving an example that the converse of the Weierstrass M-Test is false.

9. Show that if f_j are continuous functions on a domain S and if the series

$$\sum_{j=1}^{\infty} f_j(x)$$

converges uniformly on S to a limit function f then f is also continuous on S.

10. Prove that if a series $\sum_{j=1}^{\infty} f_j$ of integrable functions on an interval $[a, b]$ is uniformly convergent on $[a, b]$ then the sum function f is integrable and

$$\int_a^b f(x)dx = \sum_{j=1}^{\infty} \int_a^b f_j(x)dx.$$

*11. Give an example of a series of functions on the interval $[0, 1]$ that converges pointwise but does not converge uniformly on any subinterval.

12. Formulate and prove a result about the derivative of the sum of a convergent series of differentiable functions.

*13. Let $0 < \alpha \leq 1$. Prove that the series

$$\sum_{j=1}^{\infty} 2^{-j\alpha} \sin(2^j x)$$

defines a function f that is nowhere differentiable. To achieve this end, follow the scheme that was used to prove Theorem 6.6: **a)** Fix x; **b)** for h small, choose M such that 2^{-M} is approximately equal to $|h|$; **c)** break the series up into the sum from 1 to $M - 1$, the single summand $j = M$, and the sum from $j = M + 1$ to ∞. The middle term has very large Newton quotient and the first and last terms are relatively small.

*14. Prove that the sequence of functions $f_j(x) = \sin(jx)$ has no subsequence that converges at every x.

15. The sequence $\log jx$ does not converge, either pointwise or uniformly, on the interval $[1, 2]$. Prove this statement.

16. Write the function $\sin x$ as the uniform limit of polynomials on the interval $[0, \pi]$.

17. If $f_j \to f$ uniformly on $[0, 1]$ then does it follow that $e^{f_j(x)}$ converges to $e^{f(x)}$ uniformly?

8.4 The Weierstrass Approximation Theorem

The name Weierstrass has occurred frequently in this chapter. In fact Karl Weierstrass (1815–1897) revolutionized analysis with his examples and theorems. This section is devoted to one of his most striking results. We introduce it with a motivating discussion.

It is natural to wonder whether the usual functions of calculus—sin x, cos x, and e^x, for instance—are actually polynomials of some very high degree. Since polynomials are so much easier to understand than these transcendental functions, an affirmative answer to this question would certainly simplify mathematics. Of course a moment's thought shows that this wish is impossible: a polynomial of degree k has at most k real roots. Since sine and cosine have infinitely many real roots they cannot be polynomials. A polynomial of degree k has the property that if it is differentiated enough times (namely, $k + 1$ times) then the derivative is zero. Since this is not the case for e^x, we conclude that e^x cannot be a polynomial. The exercises of the last section discuss other means for distinguishing the familiar transcendental functions of calculus from polynomial functions.

In calculus we learned of a formal procedure, called Taylor series, for associating polynomials with a given function f. In some instances these polynomials form a sequence that converges back to the original function. Of course the method of the Taylor expansion has no hope of working unless f is infinitely differentiable. Even then, it turns out that the Taylor series rarely converges back to the original function—see the discussion at the end of Section 9.2. Nevertheless, Taylor's theorem with remainder might cause us to speculate that any reasonable function can be approximated in some fashion by polynomials. In fact the theorem of Weierstrass gives a spectacular affirmation of this speculation:

Theorem 8.23: (Weierstrass Approximation Theorem) *Let f be a continuous function on an interval* $[a, b]$*. Then there is a sequence of polynomials* $p_j(x)$ *with the property that the sequence* p_j *converges uniformly on* $[a, b]$ *to* f*.*

In a few moments we shall prove this theorem in detail. Let us first consider some of its consequences. A restatement of the theorem would be that, given a continuous function f on $[a, b]$ and an $\epsilon > 0$, there is a polynomial p such that

$$|f(x) - p(x)| < \epsilon$$

for every $x \in [a, b]$. If one were programming a computer to calculate values of a fairly wild function f, the theorem guarantees that, up to a given degree of accuracy, one could use a polynomial instead (which would in fact be much easier for the computer to handle). Advanced techniques can even tell what degree of polynomial is needed to achieve a given degree of accuracy. The proof that we shall present also suggests how this might be done.

Let f be the Weierstrass nowhere differentiable function. The theorem guarantees that, on any compact interval, f is the uniform limit of polynomials. Thus even the uniform limit of infinitely differentiable functions need not be differentiable—even at one point. This explains why the hypotheses of Theorem 8.15 needed to be so stringent.

We shall break up the proof of the Weierstrass Approximation Theorem into a sequence of lemmas.

Lemma 8.24: *Let ψ_j be a sequence of continuous functions on the interval $[-1, 1]$ with the following properties:*

  i. $\psi_j(x) \geq 0$ *for all x;*
  ii. $\int_{-1}^{1} \psi_j(x)dx = 1$ *for each j;*
  iii. *For any $\delta > 0$ we have*

$$\lim_{j \to \infty} \int_{\delta \leq |x| \leq 1} \psi_j(x)dx = 0.$$

If f is a continuous function on the real line which is identically zero off the interval $[0, 1]$ then the functions

$$f_j(x) = \int_{-1}^{1} \psi_j(t)f(x - t)dt$$

converge uniformly on the interval $[0, 1]$ to $f(x)$.

Proof: By multiplying f by a constant we may assume that $\sup |f| = 1$. Let $\epsilon > 0$. Since f is uniformly continuous on the interval $[0, 1]$ we may choose a $\delta > 0$ such that if $x, t \in [0, 1]$ and if $|x - t| < \delta$ then $|f(x) - f(t)| < \epsilon/2$. By property **(iii)** above, we may choose an N so large that $j > N$ implies that $|\int_{\delta \leq |t| \leq 1} \psi_j(t)dt| < \epsilon/4$. Then, for any $x \in [0, 1]$, we have

$$\left|f_j(x) - f(x)\right| = \left|\int_{-1}^{1} \psi_j(t)f(x - t)dt - f(x)\right|$$

$$= \left|\int_{-1}^{1} \psi_j(t)f(x - t)dt - \int_{-1}^{1} \psi_j(t)f(x)dt\right|.$$

Notice that, in the last line, we have used fact **(ii)** about the functions ψ_j to multiply the term $f(x)$ by 1 in a clever way. Now we may combine the two integrals to find that the last line

$$= \left| \int_{-1}^{1} (f(x-t) - f(x)) \psi_j(t) dt \right|$$

$$\leq \int_{-\delta}^{\delta} |f(x-t) - f(x)| \psi_j(t) dt$$

$$+ \int_{\delta \leq |t| \leq 1} |f(x-t) - f(x)| \psi_j(t) dt$$

$$= A + B.$$

To estimate term A, we recall that, for $|t| < \delta$, we have $|f(x-t) - f(x)| < \epsilon/2$; hence

$$A \leq \int_{-\delta}^{\delta} \frac{\epsilon}{2} \psi_j(t) dt \leq \frac{\epsilon}{2} \cdot \int_{-1}^{1} \psi_j(t) dt = \frac{\epsilon}{2}.$$

For B we write

$$B \leq \int_{\delta \leq |t| \leq 1} 2 \cdot \sup |f| \cdot \psi_j(t) dt$$

$$\leq 2 \cdot \int_{\delta \leq |t| \leq 1} \psi_j(t) dt$$

$$< 2 \cdot \frac{\epsilon}{4} = \frac{\epsilon}{2},$$

where in the penultimate line we have used the choice of j. Adding together our estimates for A and B, and noting that these estimates are independent of the choice of x, yields the result. □

Lemma 8.25: *Define $\psi_j(t) = k_j \cdot (1 - t^2)^j$, where the positive constants k_j are chosen so that $\int_{-1}^{1} \psi_j(t) dt = 1$. Then the functions ψ_j satisfy the properties* **(i)–(iii)** *of the last lemma.*

Proof: Of course property **(ii)** is true by design. Property **(i)** is obvious. In order to verify property **(iii)**, we need to estimate the size of k_j.
    Notice that

$$\int_{-1}^{1} (1 - t^2)^j dt = 2 \cdot \int_{0}^{1} (1 - t^2)^j dt$$

$$\geq 2 \cdot \int_{0}^{1/\sqrt{j}} (1 - t^2)^j dt$$

$$\geq 2 \cdot \int_{0}^{1/\sqrt{j}} (1 - jt^2) dt,$$

where we have used the binomial theorem. But this last integral is easily evaluated and equals $4/(3\sqrt{j})$. We conclude that

$$\int_{-1}^{1} (1 - t^2)^j dt > \frac{1}{\sqrt{j}}.$$

As a result, $k_j < \sqrt{j}$.

Now, to verify property **(iii)** of the lemma, we notice that, for $\delta > 0$ fixed and $\delta \leq |t| \leq 1$, it holds that

$$\left|\psi_j(t)\right| \leq k_j \cdot (1 - \delta^2)^j \leq \sqrt{j} \cdot (1 - \delta^2)^j$$

and this expression tends to 0 as $j \to \infty$. Thus $\psi_j \to 0$ uniformly on $\{t: \delta \leq |t| \leq 1\}$. It follows that the ψ_j satisfy property **(iii)** of the lemma. □

Proof of the Weierstrass Approximation Theorem: We may assume without loss of generality (just by changing coordinates) that f is a continuous function on the interval $[0, 1]$. After adding a linear function (which is a polynomial) to f, we may assume that $f(0) = f(1) = 0$. Thus f may be continued/extended to be a continuous function which is identically zero on $\mathbb{R} \setminus [0, 1]$.

Let ψ_j be as in Lemma 8.25 and form f_j as in Lemma 8.24. Then we know that the f_j converge uniformly on $[0, 1]$ to f. Finally,

$$f_j(x) = \int_{-1}^{1} \psi_j(t) f(x - t) dt$$
$$= \int_{0}^{1} \psi_j(x - t) f(t) dt$$
$$= k_j \int_{0}^{1} (1 + (x - t)^2)^j f(t) dt.$$

But multiplying out the expression $(1 + (x - t)^2)^j$ in the integrand then shows that f_j is a polynomial of degree at most $2j$ in x. Thus we have constructed a sequence of polynomials f_j that converges uniformly to the function f on the interval $[0, 1]$.   □

Example 8.26: The Weierstrass nowhere differentiable function is a continuous function on $[0, 1]$ that is not differentiable at any point. Nevertheless, it is (by the Weierstrass Approximation Theorem) uniformly approximable by polynomials.

Of course the uniform limit of polynomials will be continuous, so we can only consider continuous functions in this context.

Exercises

1. If f is a continuous function on the interval $[a, b]$ and if

$$\int_a^b f(x)p(x)dx = 0$$

for every polynomial p, then prove that f must be the zero function. (**Hint:** Use Weierstrass's Approximation Theorem.)

2. Let $\{f_j\}$ be a sequence of continuous functions on the real line. Suppose that the f_j converge uniformly to a function f. Prove that

$$\lim_{j \to \infty} f_j(x + 1/j) = f(x)$$

uniformly on any bounded interval.
Can any of these hypotheses be weakened?

3. Prove that the Weierstrass Approximation Theorem fails if we restrict attention to polynomials of degree less than or equal to 1000.

4. Is the Weierstrass Approximation Theorem true if we restrict ourselves to only using polynomials of even degree?

5. Is the Weierstrass Approximation Theorem true if we restrict ourselves to only using polynomials with coefficients of size not exceeding 1?

6. Use the polar form of complex numbers (that is, $z = re^{i\theta}$) to show that, on the unit circle, trigonometric polynomials and ordinary polynomials are really the same thing.

7. The Weierstrass approximation theorem says that, if f is a continuous function on $[0, 1]$, then there is a sequence of polynomials p_j that converges uniformly on $[0, 1]$ to f. Now take f to be continuously differentiable. The Weierstrass theorem applies to give a sequence p_j that converges to f. What can you say about p'_j converging to f'?

*8. Use the Weierstrass Approximation Theorem and Mathematical Induction to prove that, if f is k times continuously differentiable on an interval $[a, b]$, then there is a sequence of polynomials p_j with the property that

$$p_j \to f$$

uniformly on $[a, b]$,

$$p'_j \rightarrow f'$$

uniformly on $[a, b]$,

$$\cdots$$

$$p_j^{(k)} \rightarrow f^{(k)}$$

uniformly on $[a, b]$.

*9. Let $a < b$ be real numbers. Call a function of the form

$$f(x) = \begin{cases} 1 & \text{if } a \le x \le b \\ 0 & \text{if } x < a \quad \text{or} \quad x > b \end{cases}$$

a *characteristic function* for the interval $[a, b]$. Then a function of the form

$$g(x) = \sum_{j=1}^{k} a_j \cdot f_j(x),$$

with the f_j characteristic functions of intervals $[a_j, b_j]$, is called *simple*. Prove that any continuous function on an interval $[c, d]$ is the uniform limit of a sequence of simple functions. (**Hint:** The proof of this assertion is conceptually simple; do *not* imitate the proof of the Weierstrass Approximation Theorem.)

*10. Define a *trigonometric polynomial* to be a function of the form

$$\sum_{j=1}^{k} a_j \cdot \cos jx + \sum_{j=1}^{\ell} b_j \cdot \sin jx.$$

Prove a version of the Weierstrass Approximation Theorem on the interval $[0, 2\pi]$ for 2π-periodic continuous functions and with the phrase "trigonometric polynomial" replacing "polynomial." (**Hint:** Prove that

$$\sum_{\ell=-j}^{j} \left(1 - \frac{|\ell|}{j+1}\right)(\cos \ell t) = \frac{1}{j+1}\left(\frac{\sin \frac{j+1}{2}t}{\sin \frac{1}{2}t}\right)^2.$$

Use these functions as the ψ_js in the proof of Weierstrass's theorem.)

*11. There is a version of the Weierstrass Approximation Theorem on the unit square $[0, 1] \times [0, 1] \subseteq \mathbb{R}^2$. What should it say?

*12. Formulate a version of the Weierstrass approximation theorem for C^k functions. Indicate how the proof would work.

13. Let f be a C^1 function on the interval $[0, 1]$. If p_j are polynomials converging uniformly to f on $[0, 1]$, then what can you say about p'_j converging to f'?

*14. Let f be a continuous function on $[0, 2\pi]$. Can we realize f as a uniform limit of finite trigonometic sums? Can you reduce this question to the Weierstrass approximation theorem?

9

Elementary Transcendental Functions

9.1 Power Series

A series of the formpower series

$$\sum_{j=0}^{\infty} a_j (x - c)^j$$

is called a *power series* expanded about the point c. Our first task is to determine the nature of the set on which a power series converges.

Proposition 9.1: *Assume that the power series*

$$\sum_{j=0}^{\infty} a_j (x - c)^j$$

converges at the value $x = d$ with $d \neq c$. Let $r = |d - c|$. Then the series converges uniformly and absolutely on compact subsets of $I = \{x : |x - c| < r\}$.

Proof: We may take the compact subset of I to be $K = [c - s, c + s]$ for some number $0 < s < r$. For $x \in K$ it then holds that

$$\sum_{j=0}^{\infty} |a_j (x - c)^j| = \sum_{j=0}^{\infty} |a_j (d - c)^j| \cdot \left| \frac{x - c}{d - c} \right|^j.$$

In the sum on the right, the first expression in absolute values is bounded by some constant C (by the convergence hypothesis). The quotient in absolute values is majorized by $L = s/r < 1$. The series on the right is thus dominated by

$$\sum_{j=0}^{\infty} C \cdot L^j.$$

DOI: 10.1201/9781003222682-10

This geometric series converges. By the Weierstrass M-Test, the original series converges absolutely and uniformly on K.   □

An immediate consequence of the proposition is that the set on which the power series

$$\sum_{j=0}^{\infty} a_j(x - c)^j$$

converges is interval of convergence an interval centered about c. We call this set the *interval of convergence*. The series will converge absolutely and uniformly on compact subsets of the interval of convergence. The *radius* of the interval of convergence (called the *radius of convergence*) is defined to be half its length. Whether convergence holds at the endpoints of the interval will depend on the particular series being studied. Ad hoc methods must be used to check the endpoints. Let us use the notation C to denote the *open interval of convergence*.

It happens that, if a power series converges at either of the endpoints of its interval of convergence, then the convergence is uniform up to that endpoint. This is a consequence of Abel's partial summation test; details will be explored in the exercises.

Example 9.2: Consider the power series

$$\sum_{j=1}^{\infty} 2^j x^j.$$

We may apply the Root Test to this series to see that

$$|a_j|^{1/j} = |2^j x^j|^{1/j} = 2|x|.$$

This expression is less than 1 precisely when $|x| < 1/2$. Thus the open interval C of convergence for this power series is $(-1/2, 1/2)$. We can easily check by hand that the series does *not* converge at the endpoints.   □

On the interval of convergence C, the power series defines a function f. Such a function is said to be *real analytic*. More precisely, we have

Definition 9.3: A function f, with domain an open set $U \subseteq \mathbb{R}$ and range either the real or the complex numbers, is called *real analytic* if, for each $c \in U$, the function f may be represented by a convergent power series on an interval of positive radius centered at c:

$$f(x) = \sum_{j=0}^{\infty} a_j (x - c)^j.$$

Example 9.4: The function $f(x) = 1/(1 + x^2)$ is real analytic on the interval $(-1, 1)$. This is true because

$$\frac{1}{1 + x^2} = \sum_{j=0}^{\infty} (-x^2)^j.$$

In actuality, f is real analytic on the entire real line. But it requires power series centered at points other than the origin to see this. The entire matter is best explained in the context of complex variables, and this point of view is explained below. □

We need to know both the algebraic and the calculus properties of a real analytic function: is it continuous? differentiable? How does one add/subtract/multipy/divide two such functions?

Proposition 9.5: *Let*

$$\sum_{j=0}^{\infty} a_j (x - c)^j \quad and \quad \sum_{j=0}^{\infty} b_j (x - c)^j$$

be two power series with intervals of convergence C_1 and C_2 centered at c. Let $f_1(x)$ be the function defined by the first series on C_1 and $f_2(x)$ the function defined by the second series on C_2. Then, on their common domain $C = C_1 \cap C_1$, it holds that

1. $f(x) \pm g(x) = \sum_{j=0}^{\infty} (a_j \pm b_j)(x - c)^j$;
2. $f(x) \cdot g(x) = \sum_{m=0}^{\infty} \sum_{j+k=m} (a_j \cdot b_k)(x - c)^m$.

Proof: Let

$$A_N = \sum_{j=0}^{N} a_j (x - c)^j \quad and \quad B_N = \sum_{j=0}^{N} b_j (x - c)^j$$

be, respectively, the Nth partial sums of the power series that define f and g. If C_N is the Nth partial sum of the series

$$\sum_{j=0}^{\infty} (a_j \pm b_j)(x - c)^j$$

then

$$f(x) \pm g(x) = \lim_{N \to \infty} A_N \pm \lim_{N \to \infty} B_N = \lim_{N \to \infty} [A_N \pm B_N]$$

$$= \lim_{N \to \infty} C_N = \sum_{j=0}^{\infty} (a_j \pm b_j)(x - c)^j.$$

This proves (1).
   For (2), let

$$D_N = \sum_{m=0}^{N} \sum_{j+k=m} (a_j \cdot b_k)(x - c)^m \quad \text{and} \quad R_N = \sum_{j=N+1}^{\infty} b_j(x - c)^j.$$

We have

$$D_N = a_0 B_N + a_1(x - c)B_{N-1} + \cdots + a_N(x - c)^N B_0$$
$$= a_0(g(x) - R_N) + a_1(x - c)(g(x) - R_{N-1})$$
$$\quad + \cdots + a_N(x - c)^N (g(x) - R_0)$$
$$= g(x) \sum_{j=0}^{N} a_j(x - c)^j$$
$$\quad - [a_0 R_N + a_1(x - c)R_{N-1} + \cdots + a_N(x - c)^N R_0].$$

Clearly,

$$g(x) \sum_{j=0}^{N} a_j(x - c)^j$$

converges to $g(x)f(x)$ as N approaches ∞. In order to show that $D_N \to g \cdot f$, it will thus suffice to show that

$$|a_0 R_N + a_1(x - c)R_{N-1} + \cdots + a_N(x - c)^N R_0|$$

converges to 0 as N approaches ∞. Fix x. Now we know that

$$\sum_{j=0}^{\infty} a_j(x - c)^j$$

is absolutely convergent so we may set

$$A = \sum_{j=0}^{\infty} |a_j| |x - c|^j.$$

Also $\sum_{j=0}^{\infty} b_j (x - c)^j$ is convergent. Therefore, given $\epsilon > 0$, we can find N_0 so that $N > N_0$ implies $|R_N| < \epsilon$. Thus we have

$$
\begin{aligned}
& |a_0 R_N + a_1 (x - c) R_{N-1} + \cdots + a_N (x - c)^N R_0| \\
& \quad \le |a_0 R_N + \cdots + a_{N-N_0} (x - c)^{N-N_0} R_{N_0}| \\
& \quad + |a_{N-N_0+1} (x - c)^{N-N_0+1} R_{N_0-1} + \cdots + a_N (x - c)^N R_0| \\
& \quad \le \sup_{M \ge N_0} R_M \cdot \left(\sum_{j=0}^{\infty} |a_j| \, |x - c|^j \right) \\
& \quad + |a_{N-N_0+1} (x - c)^{N-N_0+1} R_{N_0-1} \cdots + a_N (x - c)^N R_0| \\
& \quad \le \epsilon \cdot A + |a_{N-N_0+1} (x - c)^{N-N_0+1} R_{N_0-1} \cdots + a_N (x - c)^N R_0| .
\end{aligned}
$$

Thus

$$
\begin{aligned}
& |a_0 R_N + a_1 (x - c) R_{N-1} + \cdots + a_N (x - c)^N R_0| \\
& \quad \le \epsilon \cdot A + M \cdot \sum_{j=N-N_0+1}^{N} |a_j| \, |x - c|^j ,
\end{aligned}
$$

where M is an upper bound for $|R_j(x)|$. Since the series defining A converges, we find on letting $N \to \infty$ that

$$
\limsup_{N \to \infty} |a_0 R_N + a_1 (x - c) R_{N-1} + \cdots + a_N (x - c)^N R_0| \le \epsilon \cdot A.
$$

Since $\epsilon > 0$ was arbitrary, we may conclude that

$$
\lim_{N \to \infty} |a_0 R_N + a_1 (x - c) R_{N-1} + \cdots + a_N (x - c)^N R_0| = 0. \qquad \square
$$

Remark 9.6: Observe that the form of the product of two power series provides some motivation for the form that the product of numerical series took in Theorem 3.49. $\qquad \square$

Next we turn to division real analytic functions! elementary operations on of real analytic functions. If f and g are real analytic functions, both defined on an open interval I, and if g does not vanish on I, then we would like f/g to be a well-defined real analytic function (it surely is a well-defined *function*) and we would like to be able to calculate its power series expansion by formal long division. This is what the next result tells us.

Proposition 9.7: *Let f and g be real analytic functions, both of which are defined on an open interval I. Assume that g does not vanish on I. Then the function*

$$h(x) = \frac{f(x)}{g(x)}$$

is real analytic on I. Moreover, if I is centered at the point c and if

$$f(x) = \sum_{j=0}^{\infty} a_j(x - c)^j \quad \text{and} \quad g(x) = \sum_{j=0}^{\infty} b_j(x - c)^j,$$

then the power series expansion of h about c may be obtained by formal long division of the latter series into the former. That is, the zeroeth coefficient c_0 of h is

$$c_0 = a_0/b_0,$$

the order one coefficient c_1 is

$$c_1 = \frac{1}{b_0}\left(a_1 - \frac{a_0 b_1}{b_0}\right),$$

etc.

Proof: If we can show that the power series

$$\sum_{j=0}^{\infty} c_j(x - c)^j$$

converges on I then the result on multiplication of series in Proposition 9.5 yields this new result. There is no loss of generality in assuming that $c = 0$. Assume for the moment that $b_1 \neq 0$.

Notice that one may check inductively that, for $j \geq 1$,

$$c_j = \frac{1}{b_0}(a_j - b_1 \cdot c_{j-1}). \tag{9.7.1}$$

Without loss of generality, we may scale the a_js and the b_js and assume that the radius of I is $1 + \epsilon$, some $\epsilon > 0$. Then we see from (9.7.1) that

$$|c_j| \leq C \cdot (|a_j| + |c_{j-1}|),$$

where $C = \max\{|1/b_0|, |b_1/b_0|\}$. It follows that

$$|c_j| \leq C' \cdot (1 + |a_j| + |a_{j-1}| + \cdots + |a_0|),$$

Since the radius of I exceeds 1, $\sum |a_j| < \infty$ and we see that the $|c_j|$ are bounded. Hence the power series with coefficients c_j has radius of convergence 1.

In case $b_1 = 0$ then the role of b_1 is played by the first nonvanishing b_m, $m > 1$. Then a new version of formula (9.7.1) is obtained and the argument proceeds as before. □

Example 9.8: In practice it is often useful to calculate f/g by expanding g in a "geometric series." To illustrate this idea, we assume for simplicity that f and g are real analytic in a neighborhood of 0. Then

$$\frac{f(x)}{g(x)} = f(x) \cdot \frac{1}{g(x)}$$

$$= f(x) \cdot \frac{1}{b_0 + b_1 x + \cdots}$$

$$= f(x) \cdot \frac{1}{b_0} \cdot \frac{1}{1 + (b_1/b_0)x + \cdots}.$$

Now we use the fact that, for β small,

$$\frac{1}{1 - \beta} = 1 + \beta + \beta^2 + \cdots .$$

Setting $\beta = -(b_1/b_0)x - (b_2/b_0)x^2 - \cdots$, we thus find that

$$\frac{f(x)}{g(x)} = \frac{f(x)}{b_0} \cdot (1 + [-(b_1/b_0)x - (b_2/b_0)x^2 - \cdots]$$

$$+ [-(b_1/b_0)x - (b_2/b_0)x^2 - \cdots]^2 + \cdots).$$

We explore this technique further in the exercises. □

Exercises
1. Prove that the composition of two real analytic functions, when the composition makes sense, is also real analytic.
2. Prove that

$$\sin^2 x + \cos^2 x = 1$$

directly from the power series expansions.

3. Verify the formula

$$\frac{1}{1-\beta} = 1 + \beta + \beta^2 + \cdots$$

for $|\beta| < 1$.

4. Use the technique described at the end of this section to calculate the first five terms of the power series expansion of $\sin x / e^x$ about the origin.

5. Show that the solution of the differential equation $y' + y = x$ will be real analytic.

6. Provide the details of the method for dividing real analytic functions that is described in Example 9.8.

*7. Let $f(x) = \sum_{j=0}^{\infty} a_j x^j$ be defined by a power series convergent on the interval $(-r, r)$ and let Z denote those points in the interval where f vanishes. Prove that if Z has an accumulation point in the interval then $f \equiv 0$. (**Hint:** If a is the accumulation point, expand f in a power series about a. What is the first nonvanishing term in that expansion?)

*8. Verify that the function

$$f(x) = \begin{cases} 0 & \text{if } x = 0 \\ e^{-1/x^2} & \text{if } x \neq 0 \end{cases}$$

is infinitely differentiable on all of $\mathbb{R}$ and that $f^{(k)}(0) = 0$ for every k. However, f is not real analytic.

*9. Prove the assertion from the text that, if a power series converges at an endpoint of the interval of convergence, then the convergence is uniform up to that endpoint.

*10. Prove Borel's theorem: If $\{a_j\}_{j=0}^{\infty}$ is any sequence of real numbers then there is an infinitely differentiable function f in a neighborhood of the origin whose power series coefficients at 0 are the a_j.

*11. Let f be a C^∞ function on the interval $(0, 1)$ whose derivatives of all orders at all points of the interval are positive. Prove that f is in fact real analytic.

12. Let f be real analytic on the interval $(-1, 1)$. Then it is *not* the case that if $U \subseteq \mathbb{R}$ is open then $f(U)$ is open. Give an example to explain why not.

9.2 More on Power Series: Convergence Issues

We now introduce the *Hadamard formula* for the radius of convergence of a power series.

Lemma 9.9: (Hadamard) *For the power series*

$$\sum_{j=0}^{\infty} a_j (x - c)^j,$$

define A and ρ by

$$A = \limsup_{n \to \infty} |a_n|^{1/n},$$

$$\rho = \begin{cases} 0 & \text{if } A = \infty, \\ 1/A & \text{if } 0 < A < \infty, \\ \infty & \text{if } A = 0, \end{cases}$$

then ρ is the radius of convergence of the power series about c.

Proof: Observing that

$$\limsup_{n \to \infty} |a_n (x - c)^n|^{1/n} = A |x - c|,$$

we see that the lemma is an immediate consequence of the Root Test. □

Example 9.10: Consider the power series

$$\sum_{n=1}^{\infty} n x^n.$$

We calculate that

$$A = \limsup_{n \to \infty} |a_n|^{1/n} = \limsup_{n \to \infty} n^{1/n} = 1.$$

It follows that the radius of convergence of the power series is $1/1 = 1$. So the open interval of convergence is $C = (-1, 1)$. The series does *not* converge at the endpoints. □

Corollary 9.11: *The power series*

$$\sum_{j=0}^{\infty} a_j (x - c)^j$$

has radius of convergence ρ if and only if, when $0 < R < \rho$, there exists a constant $0 < C = C_R$ such that

$$|a_j| \leq \frac{C}{R^j}.$$

Example 9.12: The series

$$\sum_{j=0}^{\infty} \frac{3^j}{j^2 + 1} x^j$$

satisfies

$$|a_j| \leq 3^j.$$

It follows from the corollary then that the radius of convergence of the series is $1/3$. □

From the power series

$$\sum_{j=0}^{\infty} a_j (x - c)^j$$

it is natural to create the *derived series*

$$\sum_{j=1}^{\infty} j a_j (x - c)^{j-1}$$

using term-by-term differentiation.

Proposition 9.13: *The radius of convergence of the derived series is the same as the radius of convergence of the original power series.*

Proof: We observe that

$$\lim_{j \to \infty} \sup |j a_j|^{1/j} = \lim_{j \to \infty} j^{-1/j} \lim_{j \to \infty} \sup |j a_j|^{1/j}$$

$$= \limsup_{j \to \infty} |a_j|^{1/j}.$$

So the result follows from the Hadamard formula. □

Proposition 9.14: *Let f be a real analytic function defined on an open interval I. Then f is continuous and has continuous, real analytic derivatives of all orders. In fact the derivatives of f are obtained by differentiating its series representation term by term.*

Proof: Since, for each $c \in I$, the function f may be represented by a convergent power series about c with positive radius of convergence, we see that, in a sufficiently small open interval about each $c \in I$, the function f is the uniform limit of a sequence of continuous functions: the partial sums of the power series representing f. It follows that f is continuous at c. Since the radius of convergence of the derived series is the same as that of the original series, it also follows that the derivatives of the partial sums converge uniformly on an open interval about c to a continuous function. It then follows from Theorem 8.15 that f is differentiable and its derivative is the function defined by the derived series. By mathematical induction, f has continuous derivatives of all orders at c. □

Example 9.15: The series

$$\sum_{j=0}^{\infty} x^j$$

has derived series

$$\sum_{j=0}^{\infty} j x^{j-1}.$$

Of course the original series converges to $1/(1 - x)$ and the derived series converges to $1/(1 - x)^2$. □

We can now show that a real analytic function has a unique power series representation at any point.

Corollary 9.16: *If the function f is represented by a convergent power series on an interval of positive radius centered at c,*

$$f(x) = \sum_{j=0}^{\infty} a_j (x - c)^j,$$

then the coefficients of the power series are related to the derivatives of the function by

$$a_n = \frac{f^{(n)}(c)}{n!}.$$

Proof: This follows readily by differentiating both sides of the above equation n times, as we may by the proposition, and evaluating at $x = c$. □

Example 9.17: The function

$$f(x) = x \sin x$$

has power series expansion about 0 with coefficients

$$a_0 = \frac{1}{0!}\frac{d^0}{dx^0}f(x)\bigg|_{x=0} = 0,$$

$$a_1 = \frac{1}{1!}\frac{d}{dx}f(x)\bigg|_{x=0} = 0,$$

$$a_2 = \frac{1}{2!}\frac{d^2}{dx^2}f(x)\bigg|_{x=0} = 1,$$

$$a_3 = \frac{1}{3!}\frac{d^3}{dx^3}f(x)\bigg|_{x=0} = 0,$$

$$a_4 = \frac{1}{4!}\frac{d^4}{dx^4}f(x)\bigg|_{x=0} = -\frac{1}{3!},$$

etc ...   □

Finally, we note that integration of power series is as well-behaved as differentiation.

Proposition 9.18: *The power series*

$$\sum_{j=0}^{\infty} a_j (x - c)^j$$

and the series

$$\sum_{j=0}^{\infty} \frac{a_j}{j+1}(x - c)^{j+1}$$

obtained from term-by-term integration have the same radius of convergence, and the function F defined by

$$F(x) = \sum_{j=0}^{\infty} \frac{a_j}{j+1}(x - c)^{j+1}$$

on the common interval of convergence satisfies

$$F'(x) = \sum_{j=0}^{\infty} a_j(x - c)^j = f(x).$$

Proof: The proof is left to the exercises. □

It is sometimes convenient to allow the variable in a power series to be a complex number. In this case we write

$$\sum_{j=0}^{\infty} a_j(z - c)^j,$$

where z is the complex argument. We now allow c and the a_js to be complex numbers as well. Noting that the elementary facts about series hold for complex series as well as real series (you should check this for yourself), we see that the arguments of this section show that the domain of convergence of a complex power series is a *disc* in the complex plane with radius ρ given as follows:

$$A = \limsup_{n \to \infty} |a_n|^{1/n}$$

$$\rho = \begin{cases} 0 & \text{if } A = \infty \\ 1/A & \text{if } 0 < A < \infty \\ \infty & \text{if } A = 0. \end{cases}$$

The proofs in this section apply to show that convergent complex power series may be added, subtracted, multiplied, and divided (provided that we do not divide by zero) on their common domains of convergence. They may also be differentiated and integrated term by term.

These observations about complex power series will be useful in the next section.

Example 9.19: The function $f(x) = 1/(1 + x^2)$ has power series expansion about the origin given by

$$\sum_{j=0}^{\infty} (-x^2)^j.$$

The radius of convergence of the power series is 1, and one might not have anticipated this fact by examining the formula for f.

But if instead one replaces x by z and examines the complex version of the function then one has

$$\tilde{f}(z) = \frac{1}{1 + z^2},$$

and one sees that this function has a singularity at $z = i$. That explains why the radius of convergence is 1. The power series about 0 cannot make sense at the singular point. $\square$

We conclude this section with a consideration of Taylor series:

Theorem 9.20: (Taylor's Expansion) *For k a nonnegative integer, let f be a $k + 1$ times continuously differentiable function on an open interval $I = (a - \varepsilon, a + \varepsilon)$. Then, for $x \in I$,*

$$f(x) = \sum_{j=0}^{k} f^{(j)}(a)\frac{(x - a)^j}{j!} + R_{k,a}(x),$$

where

$$R_{k,a}(x) = \int_{a}^{x} f^{(k+1)}(t)\frac{(x - t)^k}{k!}dt.$$

Proof: We apply integration by parts to the Fundamental Theorem of Calculus to obtain

$$f(x) = f(a) + \int_{a}^{x} f'(t)\, dt$$

$$= f(a) + \left(f'(t)\frac{(t - x)}{1!}\right)\Big|_{a}^{x} - \int_{a}^{x} f''(t)\frac{(t - x)}{1!}dt$$

$$= f(a) + f'(a)\frac{(x - a)}{1!} + \int_{a}^{x} f''(t)\frac{x - t}{1!}dt.$$

Notice that, when we performed the integration by parts, we used $t - x$ as an antiderivative for dt. This is of course legitimate, as a glance at the integration by parts theorem reveals. We have proved the theorem for the case $k = 1$. The result for higher values of k is obtained inductively by repeated applications of integration by parts. □

Taylor's theorem allows us to associate with any infinitely differentiable function a formal expansion of the form

$$\sum_{j=0}^{\infty} a_j (x - a)^j.$$

However, there is no guarantee that this series will converge; even if it does converge, it may not converge back to $f(x)$.

Example 9.21: An important example to keep in mind is the function

$$h(x) = \begin{cases} 0 & \text{if } x = 0 \\ e^{-1/x^2} & \text{if } x \neq 0. \end{cases}$$

This function is infinitely differentiable at every point of the real line (including the point 0—use l'Hôpital's Rule). However, all of its derivatives at $x = 0$ are equal to zero (this matter will be treated in the exercises). Therefore the formal Taylor series expansion of h about $a = 0$ is

$$\sum_{j=0}^{\infty} 0 \cdot (x - 0)^j = 0.$$

We see that the formal Taylor series expansion for h converges to the zero function at every x, but not to the original function h itself. □

In fact the theorem tells us that the Taylor expansion of a function f converges to f at a point x if and only if $R_{k,a}(x) \to 0$. In the exercises we shall explore the following more quantitative assertion:

An infinitely differentiable function f on an interval I has Taylor series expansion about $a \in I$ that converges back to f on a neighborhood J of a if and only if there are positive constants C, R such that, for every $x \in J$ and every k, it holds that

$$|f^{(k)}(x)| \leq C \cdot \frac{k!}{R^k}.$$

The function h in Example 9.21 should not be thought of as an isolated exception. For instance, we know from calculus that the function

$f(x) = \sin x$ has Taylor expansion that converges to f at every x. But then, for ε small, the function $g_\varepsilon(x) = f(x) + \varepsilon \cdot h(x)$ has Taylor series that does *not* converge back to $g_\varepsilon(x)$ for $x \neq 0$. Similar examples may be generated by using other real analytic functions in place of sine.

Remark 9.22: Real analytic functions have only been well understood for the past 100 years. As you can see from the results in this section, there is a very large difference between C^∞ functions and real analytic functions. It would be useful to have a scale of function spaces spanning the difference between these two types of functions, but nobody knows how to do this.

Exercises

1. Let f be an infinitely differentiable function on an interval I. If $a \in I$ and there are positive constants C, R such that, for every x in a neighborhood of a and every k, it holds that

$$|f^{(k)}(x)| \leq C \cdot \frac{k!}{R^k},$$

then prove that the Taylor series of f about a converges to $f(x)$. (**Hint:** estimate the error term.) What is the radius of convergence?

2. Let f be an infinitely differentiable function on an open interval I centered at a. Assume that the Taylor expansion of f about a converges to f at every point of I. Prove that there are constants C, R and a (possibly smaller) interval J centered at a such that, for each $x \in J$, it holds that

$$|f^{(k)}(x)| \leq C \cdot \frac{k!}{R^k}.$$

3. Give examples of power series, centered at 0, on the interval $(-1, 1)$, which (a) converge only on $(-1, 1)$, (b) converge only on $[-1, 1)$, (c) converge only on $(-1, 1]$, (d) converge only on $[-1, 1]$.

4. We know from the text that the real analytic function $1/(1 + x^2)$ is well defined on the entire real line. Yet its power series about 0 only converges on an interval of radius 1.
   How do matters differ for the function $1/(1 - x^2)$?

5. Prove Proposition 9.18.

*6. The function defined by a power series may extend continuously to an endpoint of the interval of convergence without the series converging at that endpoint. Give an example.

*7. Prove that, if a function on an interval I has derivatives of all orders which are positive at every point of I, then f is real analytic on I.

*8. What can you say about the set of convergence of a power series of two real variables?

*9. Show that the function

$$h(x) = \begin{cases} 0 & \text{if } x = 0 \\ e^{-1/x^2} & \text{if } x \neq 0. \end{cases}$$

is infinitely differentiable on the entire real line, but it is not real analytic.

*10. For which x, y does the two-variable power series

$$\sum_j 2^j x^j y^j$$

converge?

11. Verify that the function $f(x) = \sin x^2$ is real analytic on the entire real line.

12. Any closed set can be the zero set of a C^∞ function. But there are very specific restrictions on which sets can be zero sets of real analytic functions.
Explain.

13. Construct a real analytic function on $\mathbb{R}$ that vanishes at each point which is an integer multiple of $1/3$.

9.3 The Exponential and Trigonometric Functions

We begin by defining the exponential function:

Definition 9.23: The power series

$$\sum_{j=0}^{\infty} \frac{z^j}{j!}$$

converges, by the Ratio Test, for every complex value of z. The function defined thereby is called the *exponential function* and is written $\exp(z)$.

Proposition 9.24: *The function* $\exp(z)$ *satisfies*

$$\exp(a + b) = \exp(a) \cdot \exp(b)$$

for any complex numbers a *and* b.

Proof: We write the right-hand side as

$$\left(\sum_{j=0}^{\infty} \frac{a^j}{j!} \right) \cdot \left(\sum_{j=0}^{\infty} \frac{b^j}{j!} \right).$$

Now convergent power series may be multiplied term by term. We find that the last line equals

$$\sum_{j=0}^{\infty} \left(\sum_{\ell=0}^{j} \frac{a^{j-\ell}}{(j - \ell)!} \cdot \frac{b^\ell}{\ell!} \right). \tag{9.24.1}$$

However, the inner sum on the right side of this equation may be written as

$$\frac{1}{j!} \sum_{\ell=0}^{j} \frac{j!}{\ell!(j - \ell)!} a^{j-\ell} b^\ell = \frac{1}{j!}(a + b)^j.$$

It follows that line (9.24.1) equals $\exp(a + b)$.   □

Example 9.25: We set $e = \exp(1)$. This is consistent with our earlier treatment of the number e in Section 3.4. The proposition tells us that, for any positive integer k, we have

$$e^k = e \cdot e \cdots e = \exp(1) \cdot \exp(1) \cdots \exp(1) = \exp(k).$$

If m is another positive integer then

$$(\exp(k/m))^m = \exp(k) = e^k,$$

whence

$$\exp(k/m) = e^{k/m}.$$

We may extend this formula to *negative* rational exponents by using the fact that $\exp(a) \cdot \exp(-a) = 1$. Thus, for any rational number q,

$$\exp(q) = e^q. \qquad \square$$

Now note that the function exp is increasing and continuous. It follows (this fact is treated in the exercises) that if we set, for any $r \in \mathbb{R}$,

$$e^r = \sup\{e^q : q \in \mathbb{Q}, q < r\}$$

(this is a *definition* of the expression e^r) then $e^x = \exp(x)$ for every real x. [You may find it useful to review the discussion of exponentiation in Sections 2.4, 3.4; the presentation here parallels those treatments.] We will adhere to custom and write e^x instead of $\exp(x)$ when the argument of the function is real.

Proposition 9.26: *The exponential function e^x, for $x \in \mathbb{R}$, satisfies*

  a. $e^x > 0$ *for all x;*
  b. $e^0 = 1$;
  c. $(e^x)' = e^x$;
  d. e^x *is strictly increasing;*
  e. *the graph of e^x is asymptotic to the negative x-axis;*
  f. *for each integer $N > 0$ there is a number c_N such that $e^x > c_N \cdot x^N$ when $x > 0$.*

Proof: The first three statements are obvious from the power series expansion for the exponential function.

If $s < t$ then the Mean Value Theorem tells us that there is a number ξ between s and t such that

$$e^t - e^s = (t - s) \cdot e^\xi > 0;$$

hence the exponential function is strictly increasing.

By inspecting the power series we see that $e^x > 1 + x$ hence e^x increases to $+\infty$. Since $e^x \cdot e^{-x} = 1$ we conclude that e^{-x} tends to 0 as $x \to +\infty$. Thus the graph of the exponential function is asymptotic to the negative x-axis.

Finally, by inspecting the power series for e^x, we see that the last assertion is true with $c_N = 1/N!$. $\qquad \square$

Example 9.27: Let us think about 9.26(c). Which functions satisfy $y' = y$? We may rewrite this equation as

$$\frac{y'}{y} = 1.$$

Now integrate both sides to obtain

$$\ln |y| = x + C.$$

Exponentiation now yields

$$|y| = e^C \cdot e^x.$$

If we assume that y is a positive function then we can erase the absolute value signs on the left-hand side. And we can rename the constant e^C with the simpler name K. So our equation is

$$y = Ke^x.$$

We have discovered that, up to a constant factor, the exponential function is the only function that satisfies 9.26(c).   □

Now we turn to the trigonometric functions. The definition of the trigonometric functions that is found in calculus texts is unsatisfactory because it relies too heavily on a picture and because the continual need to subtract off superfluous multiples of 2π is clumsy. We have nevertheless used the trigonometric functions in earlier chapters to illustrate various concepts. It is time now to give a rigorous definition of the trigonometric functions that is independent of these earlier considerations.

Definition 9.28: The power series

$$\sum_{j=0}^{\infty} (-1)^j \frac{x^{2j+1}}{(2j + 1)!}$$

converges at every point of the real line (by the Ratio Test). The function that it defines is called the *sine* function and is usually written sin x.

The power series

$$\sum_{j=0}^{\infty} (-1)^j \frac{x^{2j}}{(2j)!}$$

converges at every point of the real line (by the Ratio Test). The function that it defines is called the *cosine* function and is usually written cos x.

Example 9.29: One advantage of having sine and cosine defined with power series is that we can actually use the series to obtain approximate numerical values for these functions. For instance, if we want to know the value of sin 1, we may write

$$\sin 1 \approx 1 - \frac{1^3}{3!} + \frac{1^5}{5!} = 1 - \frac{1}{6} + \frac{1}{120} = \frac{101}{120} \approx 0.84167 \,.$$

The true value of sin 1, determined with a calculator, is 0.84147. So this is a fairly good result. More accuracy can of course be obtained by using more terms of the series. □

You may recall that the power series that we use to define the sine and cosine functions are precisely the Taylor series expansions for the functions sine and cosine that were derived in your calculus text. But now we *begin* with the power series and must derive the properties of sine and cosine that we need *from these series*.

In fact the most convenient way to achieve this goal is to proceed by way of the exponential function. [The point here is mainly one of convenience. It can be verified by direct manipulation of the power series that $\sin^2 x + \cos^2 x = 1$ and so forth but the algebra is extremely unpleasant.] The formula in the next proposition is usually credited to Euler.

Proposition 9.30: *The exponential function and the functions sine and cosine are related by the formula (for x and y real and $i^2 = -1$)*

$$\exp(x + iy) = e^x \cdot (\cos y + i \sin y).$$

Proof: We shall verify the case $x = 0$ and leave the general case for the reader.

Thus we are to prove that

$$e^{iy} = \cos y + i \sin y. \tag{9.30.1}$$

Writing out the power series for the exponential, we find that the left-hand side of (9.30.1) is

$$\sum_{j=0}^{\infty} \frac{(iy)^j}{j!}$$

and this equals

$$\left[1 - \frac{y^2}{2!} + \frac{y^4}{4!} - + \cdots \right] + i \left[\frac{y}{1!} - \frac{y^3}{3!} + \frac{y^5}{5!} - + \cdots \right].$$

Of course the two series on the right are the familiar power series for cosine and sine as specified in Definition 9.28. Thus

$$e^{iy} = \cos y + i \sin y,$$

as desired. □

Example 9.31: We may calculate that

$$e^{i\pi/3} = \cos \frac{\pi}{3} + i \sin \frac{\pi}{3} = \frac{1}{2} + i\frac{\sqrt{3}}{2}. \qquad \square$$

In what follows, we think of the formula (9.30.1) as *defining* what we mean by e^{iy}. As a result,

$$e^{x+iy} = e^x \cdot e^{iy} = e^x \cdot (\cos y + i \sin y).$$

Notice that $e^{-iy} = \cos(-y) + i \sin(-y) = \cos y - i \sin y$ (we know that the sine function is odd and the cosine function even from their power series expansions).

Then formula (9.30.1) tells us that

$$\cos y = \frac{e^{iy} + e^{-iy}}{2}$$

and

$$\sin y = \frac{e^{iy} - e^{-iy}}{2i}.$$

Now we may prove:

Proposition 9.32: *For every real x it holds that*

$$\sin^2 x + \cos^2 x = 1 .$$

Proof: We see that

$$\sin^2 x + \cos^2 x = \left(\frac{e^{ix} - e^{-ix}}{2i}\right)^2 + \left(\frac{e^{ix} + e^{-ix}}{2}\right)^2$$

$$= \frac{e^{2ix} - 2 + e^{-2ix}}{-4} + \frac{e^{2ix} + 2 + e^{-2ix}}{4}$$

$$= 1.$$

That completes the proof. □

We list several other properties of the sine and cosine functions that may be proved by similar methods. The proofs are requested of you in the exercises.

Proposition 9.33: *The functions sine and cosine have the following properties:*

a. $\sin(s + t) = \sin s \cos t + \cos s \sin t$;
b. $\cos(s + t) = \cos s \cos t - \sin s \sin t$;
c. $\cos(2s) = \cos^2 s - \sin^2 s$;
d. $\sin(2s) = 2 \sin s \cos s$;
e. $\sin(-s) = -\sin s$;
f. $\cos(-s) = \cos s$;
g. $\sin'(s) = \cos s$;
h. $\cos'(s) = -\sin s$.

One important task to be performed in a course on the foundations of analysis is to define the number π and establish its basic properties. In a course on Euclidean geometry, the constant π is defined to be the ratio of the circumference of a circle to its diameter. Such a definition is not useful for our purposes (however, it *is* consistent with the definition about to be given here).

Observe that $\cos 0$ is the real part of e^{i0} which is 1. Thus if we set

$$\alpha = \inf\{x > 0 : \cos x = 0\}$$

then $\alpha > 0$ and, by the continuity of the cosine function, $\cos \alpha = 0$. We define $\pi = 2\alpha$.

Applying Proposition 9.32 to the number α yields that $\sin \alpha = \pm 1$. Since α is the *first* zero of cosine on the right half line, the cosine function must be positive on $(0, \alpha)$. But cosine is the derivative of sine. Thus the sine function is *increasing* on $(0, \alpha)$. Since $\sin 0$ is the imaginary part of e^{i0} which is 0, we conclude that $\sin \alpha > 0$ hence that $\sin \alpha = +1$.

Now we may apply parts (c) and (d) of Proposition 9.33 with $s = \alpha$ to conclude that $\sin \pi = 0$ and $\cos \pi = -1$. A similar calculation with $s = \pi$ shows that $\sin 2\pi = 0$ and $\cos 2\pi = 1$. Next we may use parts (a) and (b) of Proposition 9.33 to calculate that $\sin(x + 2\pi) = \sin x$ and $\cos(x + 2\pi) = \cos x$ for all x. In other words, the sine and cosine functions are 2π–periodic.

Example 9.34: The business of calculating a decimal expansion for π would take us far afield. One approach would be to utilize the already-noted fact that the sine function is strictly increasing on the interval $[0, \pi/2]$ hence its inverse function

$$\text{Sin}^{-1} : [0, 1] \rightarrow [0, \pi/2]$$

is well defined. Then one can determine (see Chapter 6) that

$$(\text{Sin}^{-1})'(x) = \frac{1}{\sqrt{1 - x^2}}.$$

By the Fundamental Theorem of Calculus,

$$\frac{\pi}{2} = \text{Sin}^{-1}(1) = \int_0^1 \frac{1}{\sqrt{1 - x^2}} dx.$$

By approximating the integral by its Riemann sums, one obtains an approximation to $\pi/2$ and hence to π itself. This approach will be explored in more detail in the exercises.

Let us for now observe that

$$\cos 2 = 1 - \frac{2^2}{2!} + \frac{2^4}{4!} - \frac{2^6}{6!} + - \cdots$$

$$= 1 - 2 + \frac{16}{24} - \frac{64}{720} + \cdots.$$

Since the series defining $\cos 2$ is an alternating series with terms that strictly decrease to zero in magnitude, we may conclude (following reasoning from Chapter 4) that the last line is less than the sum of the first three terms:

$$\cos 2 < -1 + \frac{2}{3} < 0.$$

It follows that $\alpha = \pi/2 < 2$ hence $\pi < 4$. A similar calculation of $\cos(3/2)$ would allow us to conclude that $\pi > 3$.   □

Exercises
1. Prove the equality $(\text{Sin}^{-1})'(x) = 1/\sqrt{1 - x^2}$.
2. Prove that

$$\cos 2x = \cos^2 x - \sin^2 x$$

   directly from the power series expansions.
3. Prove that

$$\sin 2x = 2 \sin x \cos x$$

   directly from the power series expansions.
4. Use one of the methods described at the end of Section 3 to calculate π to two decimal places.
*5. Prove that the trigonometric polynomials, that is to say, the functions of the form

$$p(x) = \sum_{j=-N}^{N} a_j e^{ijx},$$

   are dense in the continuous functions on $[0, 2\pi]$ in the uniform topology.
6. Find a formula for $\tan^4 x$ in terms of $\sin 2x$, $\sin 4x$, $\cos 2x$, and $\cos 4x$.
7. Prove Proposition 9.26(a), 9.26(b), 9.26(c).
8. Prove the general case of Proposition 9.30.
9. Derive a formula for $\cos 4x$ in terms of $\cos x$ and $\sin x$.
10. Provide the details of the assertion preceding Proposition 9.26 to the effect that if we define, for any real $\mathbb{R}$,

$$e^r = \sup\{e^q : q \in \mathbb{Q}, q < r\},$$

   then $e^x = \exp(x)$ for every real x.
11. Prove Proposition 9.33.

*12. Complete the following outline of a proof of Ivan Niven (see [NIV]) that π irrational:

   a. Define

$$f(x) = \frac{x^n (1 - x)^n}{n!},$$

where n is a positive integer to be selected later. For each $0 < x < 1$ we have

$$0 < f(x) < 1/n!. \qquad (*)$$

b. For every positive integer j we have $f^{(j)}(0)$ is an integer.
c. $f(1 - x) = f(x)$ hence $f^{(j)}(1)$ is an integer for every positive integer j.
d. Seeking a contradiction, assume that π is rational. Then π^2 is rational. Thus we may write $\pi^2 = a/b$, where a, b are positive integers and the fraction is in lowest terms.
e. Define

$$\begin{aligned} F(x) = b^n (\pi^{2n} f(x) \\ - \pi^{2n-2} f^{(2)}(x) + \pi^{2n-4} f^{(4)}(x) \\ - \cdots + (-1)^n f^{(2n)}(x)). \end{aligned}$$

Then $F(0)$ and $F(1)$ are integers.

f. We have

$$\begin{aligned} \frac{d}{dx} [F'(x)\sin(\pi x) \\ - \pi F(x)\cos(\pi x)] \\ = \pi^2 a^n f(x)\sin(\pi x). \end{aligned}$$

g. We have

$$\begin{aligned} \pi a^n \int_0^1 f(x)\sin(\pi x)dx \\ = \left[\frac{F'(x)\sin x}{\pi} - F(x)\cos \pi x \right]_0^1 \\ = F(1) + F(0). \end{aligned}$$

h. From this and $(*)$ we conclude that

$$\begin{aligned} 0 < \pi a^n \int_0^1 f(x)\sin(\pi x)\ dx \\ < \frac{\pi a^n}{n!} < 1. \end{aligned}$$

When n is sufficiently large this contradicts the fact that $F(0) + F(1)$ is an integer.

13. Verify from the definitions that $e^{a+b} = e^a \cdot e^b$.

14. Verify from the definitions that $(e^a)^b = e^{ab}$.

15. Verify Euler's formula:

$$e^{it} = \cos t + i \sin t$$

for $t \in \mathbb{R}$.

9.4 Logarithms and Powers of Real Numbers

Definition 9.35: Since the exponential function $\exp(x) = e^x$ is positive and strictly increasing it is a one-to-one function from $\mathbb{R}$ to $(0, \infty)$. Thus it has a well-defined inverse function that we call the *natural logarithm*. We write this function as $\ln x$.

Proposition 9.36: *The natural logarithm function has the following properties:*

a. $(\ln x)' = 1/x$;

b. $\ln x$ *is strictly increasing*;

c. $\ln(1) = 0$;

d. $\ln e = 1$;

e. *the graph of the natural logarithm function is asymptotic to the negative y axis*;

f. $\ln(s \cdot t) = \ln s + \ln t$;

g. $\ln(s/t) = \ln s - \ln t$.

Proof: These follow immediately from corresponding properties of the exponential function. For example, to verify part **(f)**, set $s = e^\sigma$ and $t = e^\tau$. Then

$$\begin{aligned}
\ln(s \cdot t) &= \ln(e^\sigma \cdot e^\tau) \\
&= \ln(e^{\sigma+\tau}) \\
&= \sigma + \tau \\
&= \ln s + \ln t.
\end{aligned}$$

The other parts of the proposition are proved similarly. □

Proposition 9.37: *If a and b are positive real numbers then*

$$a^b = e^{b \cdot \ln a}.$$

Proof: When b is an integer then the formula may be verified directly using Proposition 9.36, part **(f)**. For $b = m/n$ a rational number the formula follows by our usual trick of passing to nth roots. For arbitrary b we use a limiting argument as in our discussions of exponentials in Sections 2.3 and 9.3. □

Example 9.38: We have discussed several different approaches to the exponentiation process. We proved the existence of nth roots, $n \in \mathbb{N}$, as an illustration of the completeness of the real numbers (by taking the supremum of a certain set). We treated rational exponents by composing the usual arithmetic process of taking mth powers with the process of taking nth roots. Then, in Sections 2.3 and 9.3, we passed to arbitrary powers by way of a limiting process.

Proposition 9.37 gives us a unified and direct way to treat all exponentials at once. This unified approach will prove (see the next proposition) to be particularly advantageous when we wish to perform calculus operations on exponential functions. □

Proposition 9.39: *Fix $a > 0$. The function $f(x) = a^x$ has the following properties:*

  a. $(a^x)' = a^x \cdot \ln a$;
  b. $f(0) = 1$;
  c. *if $0 < a < 1$ then f is decreasing and the graph of f is asymptotic to the positive x-axis;*
  d. *if $1 < a$ then f is increasing and the graph of f is asymptotic to the negative x-axis.*

Proof: These properties follow immediately from corresponding properties of the function exp. As an instance, to prove part (a), we calculate that

$$(a^x)' = (e^{x \ln a})' = e^{x \ln a} \cdot \ln a = a^x \cdot \ln a.$$

The other parts of the proposition are proved in a similar fashion. Details are left to the exercises. □

The logarithm function arises, among other places, in the context of probability and in the study of entropy. The reason is that the logarithm function is uniquely determined by the way that it interacts with the operation of multiplication:

Theorem 9.40: *Let $\phi(x)$ be a continuously differentiable function with domain the positive reals and which satisfies the identity*

$$\phi(s \cdot t) = \phi(s) + \phi(t) \tag{9.40.1}$$

for all positive s and t. Then there is a constant $C > 0$ such that

$$\phi(x) = C \cdot \ln x$$

for all x.

Proof: Differentiate the equation (9.40.1) with respect to s to obtain

$$t \cdot \phi'(s \cdot t) = \phi'(s).$$

Now fix s and set $t = 1/s$ to conclude that

$$\phi'(1) \cdot \frac{1}{s} = \phi'(s).$$

We take the constant C to be $\phi'(1)$ and apply Proposition 9.36(a) to conclude that $\phi(s) = C \cdot \ln s + D$ for some constant D. But ϕ cannot satisfy (9.40.1) unless $D = 0$, so the theorem is proved. □

Observe that the *natural logarithm function* is then the unique continuously differentiable function that satisfies the condition (9.40.1) and whose derivative at 1 equals 1. That is the reason that the natural logarithm function (rather than the common logarithm, or logarithm to the base ten) is singled out as the focus of our considerations in this section.

Exercises
1. Calculate

$$\lim_{j \to \infty} \frac{j^{j/2}}{j!}.$$

2. At infinity, any nontrivial polynomial function dominates the natural logarithm function. Explain what this means, and prove it.
3. Give three distinct reasons why the natural logarithm function is not a polynomial.
4. Prove Proposition 9.39, parts (b), (c), (d), by following the hint provided.
5. Prove Proposition 9.36, except for part (f).

6. Prove that condition (9.40.1) implies that $\phi(1) = 0$. Assume that ϕ is differentiable at $x = 1$ but make no other hypothesis about the smoothness of ϕ. Prove that condition (9.40.1) then implies that ϕ is differentiable at every $x > 0$.

7. Show that the hypothesis of Theorem 9.40 may be replaced with $f \in \mathrm{Lip}_\alpha([0, 2\pi])$, some $\alpha > 0$.

8. Which function grows more quickly at infinity: $f(x) = x^k$ or $g(x) = |\ln x|^x$?

*9. The *Lambert W function* is defined implicitly by the equation

$$z = W(z) \cdot e^{W(z)}.$$

It is a fact that any elementary transcendental function may be expressed (with an elementary formula) in terms of the W function. Prove that this is so for the exponential function and the sine function.

*10. Prove Euler's formula relating the exponential to sine and cosine *not* by using power series, but rather by using differential equations.

11. For which positive exponents α does the series

$$\sum_{j=2}^{\infty} \frac{1}{j \cdot |\log j|^\alpha}$$

converge?

12. Define $a_1 = \log 100$, $a_2 = \log a_1$ and, in general, $a_j = \log a_{j-1}$. What is $\lim_{j \to \infty} a_j$?

13. Let a and b be positive real numbers. Define

$$\log_a b = \frac{\ln b}{\ln a}.$$

Prove that

$$a^{\log_a b} = b.$$

14. Refer to Exercise 13. Verify the properties of logarithm stated in Proposition 9.36 for the function $\log_a b$.

10

Functions of Several Variables

10.1 A New Look at the Basic Concepts of Analysis

A point of $\mathbb{R}^k$ is denoted $(x_1, x_2, \ldots, x_k)$. In the analysis of functions of one real variable, the domain of a function is typically an open interval. Since any open set in $\mathbb{R}^1$ is the disjoint union of open intervals, it is natural to work in the context of intervals. Such a simple situation is not obtained in the analysis of several variables. We will need some new notations and concepts in order to study functions in $\mathbb{R}^k$.

If $\mathbf{x} = (x_1, x_2, \ldots, x_k)$ is an element of $\mathbb{R}^k$, then we set

$$\|\mathbf{x}\| = \sqrt{(x_1)^2 + (x_2)^2 + \cdots + (x_k)^2}.$$

The expression $\|\mathbf{x}\|$ is commonly called the *norm* of $\mathbf{x}$. The norm of $\mathbf{x}$ measures the distance of $\mathbf{x}$ to the origin.

In general, we measure distance between two points $\mathbf{s} = (s_1, s_2, \ldots, s_k)$ and $\mathbf{t} = (t_1, t_2, \ldots, t_k)$ in $\mathbb{R}^k$ by the formula

$$\|\mathbf{s} - \mathbf{t}\| = \sqrt{(s_1 - t_1)^2 + (s_2 - t_2)^2 + \cdots + (s_k - t_k)^2}.$$

See Figure 10.1. Of course, this notion of distance can be justified by considerations using the Pythagorean theorem (see the exercises), but we treat this as a definition. The distance between the two points is nonnegative, and equals zero if and only if the two points are identical. Moreover, there is a triangle inequality shown as follows:

$$\|\mathbf{s} - \mathbf{t}\| \leq \|\mathbf{s} - \mathbf{u}\| + \|\mathbf{u} - \mathbf{t}\|.$$

We sketch a proof of this inequality in the exercises (by reducing it to the one-dimensional triangle inequality).

DOI: 10.1201/9781003222682-11

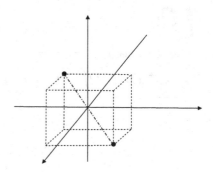

FIGURE 10.1
Distance in space.

Definition 10.1: If $x \in \mathbb{R}^k$ and $r > 0$, then the *open ball* with center x and radius r is the set

$$B(x, r) = \{t \in \mathbb{R}^k : \|x - t\| < r\}.$$

The *closed ball* with center x and radius r is the set

$$\bar{B}(x, r) = \{t \in \mathbb{R}^k : \|t - x\| \leq r\}.$$

Definition 10.2: A set $U \subseteq \mathbb{R}^k$ is said to be *open* if, for each $x \in U$, there is an $r > 0$ such that the ball $B(x, r)$ is contained in U.

Example 10.3: Let

$$S = \{x = (x_1, x_2, x_3) \in \mathbb{R}^3 : 1 < \|x\| < 2\}.$$

This set is open. See Figure 10.2. For, if $x \in S$, let $r = \min\{\|x\| - 1, 2 - \|x\|\}$. Then, $B(x, r)$ is contained in S for the following reason: if $t \in B(x, r)$, then

$$\|x\| \leq \|t - x\| + \|t\|$$

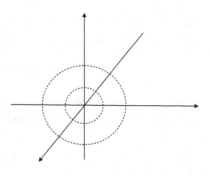

FIGURE 10.2
An open set.

hence,

$$\|t\| \geq \|x\| - \|t - x\| > \|x\| - r \geq \|x\| - (\|x\| - 1) = 1.$$

Likewise,

$$\|t\| \leq \|x\| + \|t - x\| < \|x\| + r \leq \|x\| + (2 - \|x\|) = 2.$$

It follows that $t \in S$; hence, $B(x, r) \subseteq S$. We conclude that S is open.

However, a moment's thought shows that S could not be written as a disjoint union of open balls, or open cubes, or any other regular type of open set. □

In this chapter, we consider functions with domain a set (usually open) in $\mathbb{R}^k$. See Figure 10.3. This means that the function f may be written in the form $f(x_1, x_2, \ldots, x_k)$. An example of such a function is $f(x_1, x_2, x_3, x_4) = x_1 \cdot (x_2)^4 - x_3 / x_4$ or $g(x_1, x_2, x_3) = (x_3)^2 \cdot \sin(x_1 \cdot x_2 \cdot x_3)$.

Definition 10.4: Let $E \subseteq \mathbb{R}^k$ be a set and let f be a real-valued function with domain E. Fix a point **P**, which is either in E or is an accumulation point of E (in the sense discussed in Chapter 4). We say that

$$\lim_{x \to P} f(x) = \ell,$$

with ℓ a real number if, for each $\epsilon > 0$ there is a $\delta > 0$ such that, when $x \in E$ and $0 < \|x - P\| < \delta$, then

$$|f(x) - \ell| < \epsilon.$$

Refer to Figure 10.4.

Compare this definition with the definition in Section 5.1: the only difference is that we now measure the distance between points of the domain of f using $\| \|$ instead of $| |$.

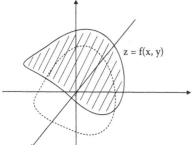

FIGURE 10.3
A function in space.

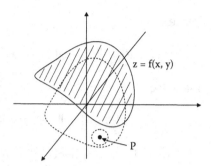

FIGURE 10.4
The limit of a function in space.

Example 10.5: The function

$$f(x_1, x_2, x_3) = \begin{cases} \frac{x_1 x_2}{x_1^2 + x_2^2 + x_3^2} & \text{if } (x_1, x_2, x_3) \neq 0 \\ 0 & \text{if } (x_1, x_2, x_3) = 0 \end{cases}$$

has no limit as $x \to 0$. For, if we take $x = (t, 0, 0)$, then we obtain the limit

$$\lim_{t \to 0} f(t, 0, 0) = 0$$

while if we take $x = (t, t, t)$, then we obtain the limit

$$\lim_{t \to 0} f(t, t, t) = \frac{1}{3}.$$

Thus, for $\epsilon < \frac{1}{6} = \frac{1}{2} \cdot \frac{1}{3}$, there will exist no δ satisfying the definition of limit. See Figure 10.5.

However, the function

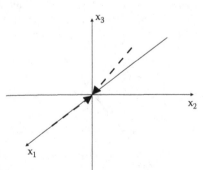

FIGURE 10.5
Approaching the origin from two different directions.

$$g(x_1, x_2, x_3, x_4) = x_1^2 + x_2^2 + x_3^2 + x_4^2$$

satisfies

$$\lim_{x \to 0} g(x) = 0$$

because given $\epsilon > 0$, we take $\delta = \sqrt{\epsilon/4}$. Then, $\|x - 0\| < \delta$ implies that $|x_j - 0| < \sqrt{\epsilon/4}$ for $j = 1, 2, 3, 4$; hence,

$$|g(x_1, x_2, x_3, x_4) - 0| < \left| \left(\frac{\sqrt{\epsilon}}{\sqrt{4}} \right)^2 + \left(\frac{\sqrt{\epsilon}}{\sqrt{4}} \right)^2 + \left(\frac{\sqrt{\epsilon}}{\sqrt{4}} \right)^2 + \left(\frac{\sqrt{\epsilon}}{\sqrt{4}} \right)^2 \right| = \epsilon.$$

Remark 10.6: Notice that, just as in the theory of one variable, the limit properties of f at a point P are independent of the *actual value* of f at P. □

Definition 10.7: Let f be a function with domain $E \subseteq \mathbb{R}^k$ and let $\mathbf{P} \in E$. We say that f is *continuous* at $\mathbf{P}$ if

$$\lim_{x \to \mathbf{P}} f(x) = f(\mathbf{P}).$$

See Figure 10.6.

The limiting process respects the elementary arithmetic operations, just as in the one-variable situation explored in Chapter 5. We will treat these matters in the exercises. Similarly, continuous functions are closed under the arithmetic operations (provided that we do not divide by zero). Next, we turn to the fundamental properties of the derivative. (We refer the reader to the Appendix for a review of linear algebra.) In what follows, we use the notation $^t M$ to denote the transpose of the matrix M. We need the transpose so that the indicated matrix multiplications make sense.

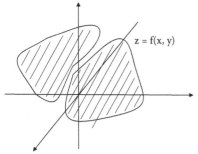

z = f(x, y)

FIGURE 10.6
A discontinuous function.

Definition 10.8: Let $f(x)$ be a scalar-valued function whose domain contains a ball $B(\mathbf{P}, r)$. We say that f is *differentiable* at $\mathbf{P}$ if there is a $1 \times k$ matrix $M_{\mathbf{P}} = M_{\mathbf{P}}(f)$ such that, for all $\mathbf{h} \in \mathbb{R}^k$ satisfying $\|\mathbf{h}\| < r$, it holds that

$$f(\mathbf{P} + \mathbf{h}) = f(\mathbf{P}) + M_{\mathbf{P}} \cdot {}^t\mathbf{h} + \mathcal{R}_{\mathbf{P}}(f, \mathbf{h}), \qquad (10.8.1)$$

where

$$\lim_{\mathbf{h} \to 0} \frac{\mathcal{R}_{\mathbf{P}}(f, \mathbf{h})}{\|\mathbf{h}\|} = 0.$$

The matrix $M_{\mathbf{P}} = M_{\mathbf{P}}(f)$ is called the *derivative* of f at $\mathbf{P}$.

Example 10.9: Consider the scalar-valued function $f(x, y) = x^2 - 2xy$ at the point $\mathbf{P} = (1, 2)$. Let $\mathbf{h} = (h_1, h_2)$. The correct 1×2 matrix $M_{\mathbf{P}}$ is $(-2, -2)$ as we are about to see. This is because

$$
\begin{aligned}
f(\mathbf{P} + \mathbf{h}) &= f(P_1 + h_1, P_2 + h_2) \\
&= f(1 + h_1, 2 + h_2) \\
&= (1 + h_1)^2 - 2(1 + h_1)(2 + h_2) \\
&= [-3] + [-2h_1 - 2h_2] + [h_1^2 - 2h_1 h_2] \\
&= f(P) + M_{\mathbf{P}} \cdot {}^t\mathbf{h} + \mathcal{R}_{\mathbf{P}}(f, \mathbf{h}).
\end{aligned}
$$

So, we have verified that $M_{\mathbf{P}} = (-2, -2)$ is the derivative of f at $\mathbf{P}$. □

Example 10.10: Consider the function $f(x, y) = 4 - \sqrt{x^2 + y^2}$. The graph of this function is the lower nappe of a cone. See Figure 10.7. It is easy to calculate, using $\mathbf{h} = (t, 0, 0)$ for $t < 0$ and $t > 0$, that this f is not differentiable at the origin. □

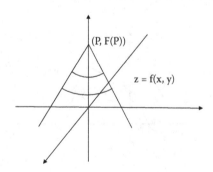

FIGURE 10.7
A function that is not differentiable at P.

The best way to begin to understand any new idea is to reduce it to a situation that we have already digested. If f is a function of one variable that is differentiable at $\mathbf{P} \in \mathbb{R}$, then there is a number M such that

$$\lim_{h \to 0} \frac{f(\mathbf{P} + h) - f(\mathbf{P})}{h} = M.$$

We may rearrange this equality as

$$\frac{f(\mathbf{P} + h) - f(\mathbf{P})}{h} - M = \mathcal{S}_\mathbf{P},$$

where $\mathcal{S}_\mathbf{P} \to 0$ as $h \to 0$. But, this may be rewritten as

$$f(\mathbf{P} + h) = f(\mathbf{P}) + M \cdot h + \mathcal{R}_\mathbf{P}(f, h), \tag{10.11}$$

where $\mathcal{R}_\mathbf{P} = h \cdot \mathcal{S}_\mathbf{P}$ and

$$\lim_{h \to 0} \frac{\mathcal{R}_\mathbf{P}(f, h)}{h} = 0.$$

Equation (10.11) is parallel to (10.8.1) that defines the concept of derivative. The role of the $1 \times k$ matrix $M_\mathbf{P}$ is played here by the numerical constant M. But, *a numerical constant is a 1×1 matrix*. Thus, our equation in one variable is a special case of the equation in k variables. In one variable, the matrix representing the derivative is just the singleton consisting of the numerical derivative.

Note in passing that (just as in the one-variable case) the way that we now define the derivative of a function of several variables is closely related to the Taylor expansion. The number M in the one-variable case is the coefficient of the first-order term in that expansion, which we know from Chapter 9 to be the first derivative.

What is the significance of the matrix $M_\mathbf{P}$ in our definition of derivative for a function of k real variables? Suppose that f is differentiable according to Definition 10.8. Let us attempt to calculate the "partial derivative" (as in calculus) with respect to x_1 of f. Let $\mathbf{h} = (h, 0, \ldots, 0)$. Then,

$$f(P_1 + h, P_2, \ldots, P_k) = f(\mathbf{P}) + M_\mathbf{P} \cdot \begin{pmatrix} h \\ 0 \\ \vdots \\ 0 \end{pmatrix} + \mathcal{R}_\mathbf{P}(f, \mathbf{h}).$$

Rearranging this equation, we have

$$\frac{f(P_1 + h, P_2, \ldots, P_k) - f(P)}{h} = (M_P)_1 + S_P,$$

where $S_P \to 0$ as $h \to 0$ and $(M_P)_1$ is the first entry of the $1 \times k$ matrix M_P.

But, letting $h \to 0$ in this last equation, we see that the partial derivative with respect to x_1 of the function f exists at P and equals $(M_P)_1$. A similar calculation shows that the partial derivative with respect to x_2 of the function f exists at P and equals $(M_P)_2$; likewise, the partial derivative with respect to x_j of the function f exists at P and equals $(M_P)_j$ for $j = 1, \ldots, k$.

We summarize with the following theorem:

Theorem 10.11: *Let f be a function defined on an open ball $B(P, r) \subseteq \mathbb{R}^k$ and suppose that f is differentiable at P with derivative the $1 \times k$ matrix M_P. Then, the first partial derivatives of f at P exist and they are, respectively, the entries of M_P. That is,*

$$(M_P)_1 = \frac{\partial}{\partial x_1} f(P), \ (M_P)_2 = \frac{\partial}{\partial x_2} f(P), \ \ldots, (M_P)_k = \frac{\partial}{\partial x_k} f(P).$$

Example 10.12: Let $f(x, y) = \sin x - 3y$ and let $P = (0, 0)$. Then

$$\frac{\partial f}{\partial x}(P) = \cos x \big|_{x=0, y=0} = 1$$

and

$$\frac{\partial f}{\partial y}(P) = -3.$$

So, we see that $M_P = (1, -3)$ and we are guaranteed that

$$f(P + h) = f(P) + M_P \cdot{}^t h + \mathcal{R}_P(f, h).$$

$\square$

Unfortunately, the converse of this theorem is not true: it is possible for the first partial derivatives of f to exist at a single point P without f being differentiable at P in the sense of Definition 10.8. Counterexamples will be explored in the exercises. In contrast, as the last example suggests, the two different notions of *continuous differentiablity* are the same. We formalize this statement with the following proposition:

Proposition 10.13: *Let f be a function defined on an open ball $B(\mathbf{P}, r)$. Assume that f is differentiable at each point of $B(\mathbf{P}, r)$ in the sense of Definition 10.8 and that the function*

$$x \mapsto M_x$$

is continuous in the sense that each of the functions

$$x \mapsto (M_x)_j$$

is continuous, $j = 1, 2, \ldots, k$. Then, each of the partial derivatives

$$\frac{\partial}{\partial x_1} f(x) \quad \frac{\partial}{\partial x_2} f(x) \ldots, \quad \frac{\partial}{\partial x_k} f(x)$$

exists for $x \in B(\mathbf{P}, r)$ and is continuous.

Conversely, if each of the partial derivatives exists on $B(\mathbf{P}, r)$ and is continuous at each point, then M_x exists at each point $x \in B(\mathbf{P}, r)$ and is continuous. The entries of M_x are given by the partial derivatives of f.

Proof: This is essentially a routine check of definitions. The only place where the continuity is used is in proving the converse: that the existence and continuity of the partial derivatives implies the existence of M_x. In proving the converse, you should apply the one-variable Taylor expansion to the function $t \mapsto f(x + t\mathbf{h})$. □

Exercises

1. Fix elements $\mathbf{s}, \mathbf{t}, \mathbf{u} \in \mathbb{R}^k$. First assume that these three points are colinear. By reduction to the one-dimensional case, prove the *triangle inequality*

   $$\|\mathbf{s} - \mathbf{t}\| \leq \|\mathbf{s} - \mathbf{u}\| + \|\mathbf{u} - \mathbf{t}\|.$$

   Now, establish the general case of the triangle inequality by comparison with the colinear case.

2. Give another proof of the triangle inequality by squaring both sides and invoking the Schwarz inequality.

3. Formulate and prove the elementary properties of limits for functions of k variables (refer to Chapter 5 for the one-variable analogues).

4. Give an example of an infinitely differentiable function with domain $\mathbb{R}^2$ such that $\{(x_1, x_2) : f(x_1, x_2) = 0\} = \{(x_1, x_2) : |x_1|^2 + |x_2|^2 \leq 1\}$.

5. Formulate a notion of uniform convergence for functions of k real variables. Prove that the uniform limit of a sequence of continuous functions is continuous.

6. Formulate a notion of "compact set" for subsets of $\mathbb{R}^k$. Prove that the continuous image, under a vector-valued function, of a compact set is compact.

7. Refer to Exercise **6**. Prove that if f is a continuous, scalar-valued function on a compact set, then f assumes both a maximum and a minimum value.

8. Give an example of a connected set in $\mathbb{R}^2$ with disconnected boundary.

9. Give an example of a disconnected set in $\mathbb{R}^2$ with infinitely many-connected components.

10. Justify our notion of distance in $\mathbb{R}^k$ using Pythagorean Theorem considerations.

11. If $\mathbf{s}, \mathbf{t} \in \mathbb{R}^k$, then prove that

$$\|\mathbf{s} + \mathbf{t}\| \geq \|\mathbf{s}\| - \|\mathbf{t}\|.$$

12. Prove Proposition 10.13.

*13. Give an example of a function f of two variables such that f has both first partial derivatives at a point P, yet f is not differentiable at P according to Definition 10.8.

14. Give an example of a function f of two variables such that f has both first partial derivatives at a point P, yet f is not continuous at P.

15. Give an example of a function $f(x, y)$ defined on $\mathbb{R}^2 \setminus \{0\}$ so that $\lim_{h \to 0} f(h, 0)$ exists, $\lim_{k \to 0} f(0, k)$ exists, but $\lim_{r \to 0} f(r)$ does not exist.

16. Let $f(x, y) = x^2 + y^2$. What is the linear function of best approximation to the graph of $z = f(x, y)$ at the point $(1, 1, 2)$?

17. Give an example of a function $f(x, y)$ of two real variables that is infinitely differentiable but not real analytic.

10.2 Properties of the Derivative

The arithmetic properties of the derivative—that is the sum and difference, scalar multiplication, product, and quotient rules—are straightforward and are left to the exercises for you to consider. However, the Chain Rule takes on a different form and requires careful consideration.

To treat meaningful instances of the Chain Rule, we must first discuss *vector-valued* functions. That is, we consider functions with domain a subset of $\mathbb{R}^k$ and range *either* $\mathbb{R}^1$ *or* $\mathbb{R}^2$ *or* $\mathbb{R}^m$ for some integer $m > 0$. When we consider vector-valued functions, it simplifies notation if we consider all vectors to be column vectors. This convention will be in effect for the rest of the chapter. (Thus, we will no longer use the "transpose" notation.) Note in passing that the expression $\|x\|$ means the same thing for a column vector as it does for a row vector—the square root of the sum of the squares of the components. Also, $f(x)$ means the same thing whether x is written as a row vector or a column vector.

Example 10.14: Define the function

$$f(x_1, x_2, x_3) = \begin{pmatrix} (x_1)^2 - x_2 \cdot x_3 \\ x_1 \cdot (x_2)^3 \end{pmatrix}.$$

This is a function with domain consisting of all triples of real numbers, or $\mathbb{R}^3$, and range consisting of all pairs of real numbers, or $\mathbb{R}^2$. For example,

$$f(-1, 2, 4) = \begin{pmatrix} -7 \\ -8 \end{pmatrix}.$$

$\square$

We say that a vector-valued function of k variables

$$f(x) = (f_1(x), f_2(x), \dots, f_m(x))$$

(where m is a positive integer) is differentiable at a point $\mathbf{P}$ if each of its component functions is differentiable in the sense of Section 1. For example, the function

$$f(x_1, x_2, x_3) = \begin{pmatrix} x_1 \cdot x_2 \\ (x_3)^2 \end{pmatrix}$$

is differentiable at all points (because $f_1(x_1, x_2, x_3) = x_1 \cdot x_2$ and $f_2(x_1, x_2, x_3) = x_3^2$ are differentiable) while the function

$$g(x_1, x_2, x_3) = \begin{pmatrix} x_2 \\ |x_3| - x_1 \end{pmatrix}$$

is not differentiable at points of the form $(x_1, x_2, 0)$.

It is a good exercise in matrix algebra (which you will be asked to do at the end of the section) to verify that a vector-valued function f is differentiable at a point $\mathbf{P}$ if and only if there is an $m \times k$ matrix (where k is the dimension of the domain and m the dimension of the range) $M_\mathbf{P}(f)$ such that

$$f(\mathbf{P} + \mathbf{h}) = f(\mathbf{P}) + M_\mathbf{P}(f)\mathbf{h} + \mathcal{R}_\mathbf{P}(f, \mathbf{h});$$

here, $\mathbf{h}$ is a $\mathbf{k}$-column vector and the remainder term $\mathcal{R}_\mathbf{P}$ is a column vector of length m satisfying

$$\frac{\|\mathcal{R}_\mathbf{P}(f, \mathbf{h})\|}{\|\mathbf{h}\|} \to 0$$

as $\mathbf{h} \to 0$. One nice consequence of this formula is that, by what we learned in the last section about partial derivatives, the entry in the ith row and jth column of the matrix $M_\mathbf{P}(f)$ is $\partial f_i / \partial x_j$. Here f_i is the ith component of the mapping f.

Of course, the Chain Rule provides a method for differentiating compositions of functions. What we will discover in this section is that the device of thinking of the derivative as a matrix occurring in an expansion of f about a point $\mathbf{P}$ makes the Chain Rule a very natural and easy result to derive. It will also prove to be a useful way of keeping track of information.

Theorem 10.15: *Let g be a function of k real variables taking values in $\mathbb{R}^m$ and let f be a function of m real variables taking values in $\mathbb{R}^n$. Suppose that the range of g is contained in the domain of f, so that $f \circ g$ makes sense. If g is differentiable at a point $\mathbf{P}$ in its domain and f is differentiable at $g(\mathbf{P})$ then $f \circ g$ is differentiable at $\mathbf{P}$ and its derivative is $M_{g(\mathbf{P})}(f) \cdot M_\mathbf{P}(g)$. We use the symbol $\cdot$ here to denote matrix multiplication.*

Proof: By the hypothesis about the differentiability of g,

$$
\begin{aligned}
(f \circ g)(\mathbf{P} + \mathbf{h}) &= f(g(\mathbf{P} + \mathbf{h})) \\
&= f(g(\mathbf{P}) + M_\mathbf{P}(g)\mathbf{h} + \mathcal{R}_\mathbf{P}(g, \mathbf{h})) \qquad (10.15.1) \\
&= f(g(\mathbf{P}) + \mathbf{k}),
\end{aligned}
$$

where

$$\mathbf{k} = M_\mathbf{P}(g)\mathbf{h} + \mathcal{R}_\mathbf{P}(g, \mathbf{h}).$$

But then, the differentiability of f at $g(\mathbf{P})$ implies that (10.15.1) equals

$$f(g(\mathbf{P})) + M_{g(\mathbf{P})}(f)\mathbf{k} + \mathcal{R}_{g(\mathbf{P})}(f, \mathbf{k}).$$

Now let us substitute in the value of **k**. We find that

$$
\begin{aligned}
(f \circ g)(\mathbf{P} + \mathbf{h}) &= f(g(\mathbf{P})) + M_{g(\mathbf{P})}(f)[M_{\mathbf{P}}(g)\mathbf{h} + \mathcal{R}_{\mathbf{P}}(g, \mathbf{h})] \\
&\quad + \mathcal{R}_{g(\mathbf{P})}(f, M_{\mathbf{P}}(g)\mathbf{h} + \mathcal{R}_{\mathbf{P}}(g, \mathbf{h})) \\
&= f(g(\mathbf{P})) + M_{g(\mathbf{P})}(f)M_{\mathbf{P}}(g)\mathbf{h} \\
&\quad + \{M_{g(\mathbf{P})}(f)\mathcal{R}_{\mathbf{P}}(g, \mathbf{h}) \\
&\quad + \mathcal{R}_{g(\mathbf{P})}(f, M_{\mathbf{P}}(g)\mathbf{h} + \mathcal{R}_{\mathbf{P}}(g, \mathbf{h}))\} \\
&\equiv f(g(\mathbf{P})) + M_{g(\mathbf{P})}(f)M_{\mathbf{P}}(g)\mathbf{h} \\
&\quad + Q_{\mathbf{P}}(f \circ g, \mathbf{h}),
\end{aligned}
$$

where the last equality *defines* Q. The term Q should be thought of as a remainder term. Since

$$
\frac{\|\mathcal{R}_{\mathbf{P}}(g, \mathbf{h})\|}{\|\mathbf{h}\|} \to 0
$$

as $\mathbf{h} \to 0$, it follows that

$$
\frac{M_{g(\mathbf{P})}(f)\mathcal{R}_{\mathbf{P}}(g, \mathbf{h})}{\|\mathbf{h}\|} \to 0.
$$

(Details of this assertion are requested of you in the exercises.) Similarly,

$$
\frac{\mathcal{R}_{g(\mathbf{P})}(f, M_{\mathbf{P}}(g)\mathbf{h} + \mathcal{R}_{\mathbf{P}}(g, \mathbf{h}))}{\|\mathbf{h}\|} \to 0
$$

as $\mathbf{h} \to 0$.

In conclusion, we see that $f \circ g$ is differentiable at $\mathbf{P}$ and that the derivative equals $M_{g(\mathbf{P})}(f)M_{\mathbf{P}}(g)$, the product of the derivatives of f and g.□

Remark 10.16: Notice that, by our hypotheses, $M_{\mathbf{P}}(g)$ is an $m \times k$ size matrix and $M_{g(\mathbf{P})}(f)$ is an $n \times m$ size matrix. Thus, their product makes sense.

In general, if g is a function from a subset of $\mathbb{R}^k$ to $\mathbb{R}^m$, then, if we want $f \circ g$ to make sense, f must be a function from a subset of $\mathbb{R}^m$ to some $\mathbb{R}^n$. In other words, the dimension of the range of g had better match the dimension of the domain of f. Then, the derivative of g at some point $\mathbf{P}$ will be an $m \times k$ matrix and the derivative of f at $g(\mathbf{P})$ will be an $n \times m$ matrix. Hence the matrix multiplication $M_{g(\mathbf{P})}(f)M_{\mathbf{P}}(g)$ will make sense.

Corollary 10.17: (The Chain Rule in Coordinates) Let $f: \mathbb{R}^m \to \mathbb{R}^n$ and $g: \mathbb{R}^k \to \mathbb{R}^m$ be vector-valued functions and assume that $h = f \circ g$ makes sense. If g

is differentiable at a point $\mathbf{P}$ *of its domain and f is differentiable at* $g(\mathbf{P})$, *then for each i and j we have*

$$\frac{\partial h_i}{\partial x_j}(\mathbf{P}) = \sum_{\ell=1}^{m} \frac{\partial f_i}{\partial s_\ell}(g(\mathbf{P})) \cdot \frac{\partial g_\ell}{\partial x_j}(\mathbf{P}).$$

Proof: The function $\partial h_i / x_j$ is the entry of $M_{\mathbf{P}}(h)$ in the ith row and jth column. However, $M_{\mathbf{P}}(h)$ is the product of $M_{g(\mathbf{P})}(f)$ with $M_{\mathbf{P}}(g)$ by Theorem 10.15. The entry in the ith row and jth column of that product is

$$\sum_{\ell=1}^{m} \frac{\partial f_i}{\partial s_\ell}(g(\mathbf{P})) \cdot \frac{\partial g_\ell}{\partial x_j}(\mathbf{P}).$$

□

Example 10.18: Let $f(x, y) = x^2 - y^2$ and let $g(s, t) = (st, -t^3)$. Then, $f \circ g$ makes sense and we may calculate the derivative of this composition at the point $P = (1, 3)$. Let us first do so according to the matrix rule given in Theorem 10.15. And, then let us follow that with the analogous calculation in coordinates (as in the corollary).

We write $g(s, t) = (st, -t^3) = (g_1(s, t), g_2(s, t))$. Now, begin by calculating

$$\frac{\partial g_1}{\partial s} = t \text{ and } \frac{\partial g_1}{\partial t} = s$$

and

$$\frac{\partial g_2}{\partial s} = 0 \text{ and } \frac{\partial g_2}{\partial t} = -3t^3.$$

Thus,

$$\frac{\partial g_1}{\partial s}(\mathbf{P}) = 3 \text{ and } \frac{\partial g_1}{\partial t}(\mathbf{P}) = 1$$

and

$$\frac{\partial g_2}{\partial s}(\mathbf{P}) = 0 \text{ and } \frac{\partial g_2}{\partial t}(\mathbf{P}) = -27.$$

Therefore,

$$M_P(g) = \begin{pmatrix} 3 & 1 \\ 0 & -27 \end{pmatrix}.$$

Next, we note that $g(P) = (3, -27)$ and

$$\frac{\partial f}{\partial x} = 2x \text{ and } \frac{\partial f}{\partial y} = -2y$$

so that

$$\frac{\partial f}{\partial x}(g(P)) = 6 \text{ and } \frac{\partial f}{\partial y}(g(P)) = 54.$$

Thus,

$$M_{g(P)}(f) = (6, 54).$$

In conclusion,

$$M_P(f \circ g) = (6, 54) \cdot \begin{pmatrix} 3 & 1 \\ 0 & -27 \end{pmatrix} = (18, -1452).$$

Now, let us perform the same calculation in coordinates. We begin by writing

$$f \circ g(s, t) = (st)^2 - (-t^3)^2 = s^2 t^2 - t^6.$$

Now, we calculate (using the fact that $x = 3$ and $y = -27$) that

$$\frac{\partial(f \circ g)}{\partial s} = \frac{\partial f}{\partial x} \cdot \frac{\partial g_1}{\partial s} + \frac{\partial f}{\partial y} \cdot \frac{\partial g_2}{\partial s} = 2x \cdot 3 + (-2y) \cdot 0 = 2 \cdot 3 \cdot 3 = 18$$

and

$$\frac{\partial(f \circ g)}{\partial t} = \frac{\partial f}{\partial x} \cdot \frac{\partial g_1}{\partial t} + \frac{\partial f}{\partial y} \cdot \frac{\partial g_2}{\partial t} = 2x \cdot 1 + (-2y) \cdot (-27) = 6 - 1458 = -1452.$$

In sum, the two methods of calculation give the same answer. □

We conclude this section by deriving a Taylor expansion for the scalar-valued functions of k real variables: this expansion for functions of several variables is derived in an interesting way from the expansion for functions of one variable. We say that a function f of several real variables is k times continuously differentiable if all partial derivatives of orders up to, and including k, exist and are continuous on the domain of f.

Theorem 10.19: (Taylor's Expansion) *For q, a nonnegative integer, let f be a $q + 1$ times continuously differentiable scalar-valued function on a neighborhood of a closed ball $\bar{B}(\mathbf{P}, r) \subseteq \mathbb{R}^k$. Then, for $x \in B(\mathbf{P}, r)$,*

$$f(x) = \sum_{0 \le j_1 + j_2 + \cdots + j_k \le q} \frac{\partial^{j_1 + j_2 + \cdots + j_k} f}{\partial x_1^{j_1} \partial x_2^{j_2} \cdots \partial x_k^{j_k}}(\mathbf{P}) \cdot \frac{(x_1 - P_1)^{j_1}(x_2 - P_2)^{j_2} \cdots (x_k - P_k)^{j_k}}{(j_1)!(j_2)! \cdots (j_k)!}$$
$$+ \mathcal{R}_{q,\mathbf{P}}(x),$$

where

$$|\mathcal{R}_{q,\mathbf{P}}(x)| \le C_0 \cdot \frac{\|x - \mathbf{P}\|^{q+1}}{(q+1)!},$$

and

$$C_0 = \sup_{\substack{s \in \bar{B}(\mathbf{P},r) \\ \ell_1 + \ell_2 + \cdots + \ell_k = q+1}} \left| \frac{\partial^{j_1 + j_2 + \cdots + j_k} f}{\partial x_1^{j_1} \partial x_2^{j_2} \cdots \partial x_k^{j_k}}(s) \right|.$$

Proof: With $\mathbf{P}$ and x fixed, define

$$\mathcal{F}(s) = f(\mathbf{P} + s(x - \mathbf{P})), \quad 0 \le s < \frac{r}{\|x - \mathbf{P}\|}.$$

We apply the one-dimensional Taylor theorem to the function $\mathcal{F}$, expanded about the point 0:

$$\mathcal{F}(s) = \sum_{\ell=0}^{q} \mathcal{F}^{(\ell)}(0) \frac{s^\ell}{\ell!} + R_{q,0}(\mathcal{F}, s).$$

Now, the Chain Rule shows that

$$\mathcal{F}^{(\ell)}(0) =$$
$$\sum_{j_1 + j_2 + \cdots + j_k = \ell} \frac{\partial^{j_1 + j_2 + \cdots + j_k} f}{\partial x_1^{j_1} \partial x_2^{j_2} \cdots \partial x_k^{j_k}}(\mathbf{P})$$
$$\cdot \frac{\ell!}{(j_1)!(j_2)! \cdots (j_k)!} \cdot (x_1 - P_1)^{j_1}(x_2 - P_2)^{j_2} \cdots (x_k - P_k)^{j_k}.$$

Substituting this last equation, for each ℓ, into the formula for $\mathcal{F}(s)$ and setting $s = 1$ (recall that $r/\|x - P\| > 1$ since $x \in B(P, r)$) yields the desired expression for $f(x)$. It remains to estimate the remainder term.

The one-variable Taylor theorem tells us that, for $s > 0$,

$$|R_{q,0}(\mathcal{F}, s)| = \left| \int_0^s \mathcal{F}^{(q+1)}(\sigma) \frac{(s-\sigma)^q}{q!} d\sigma \right|$$

$$\leq \int_0^s C_0 \cdot \|x - P\|^{q+1} \cdot \left| \frac{(s-\sigma)^q}{q!} \right| d\sigma$$

$$= C_0 \cdot \frac{\|x - P\|^{q+1}}{(q+1)!}.$$

Here, we have of course used the Chain Rule to pass from derivatives of $\mathcal{F}$ to derivatives of f. This is the desired result. □

Example 10.20: Let us determine the degree-three Taylor expansion for the function $f(x, y) = x \cos y$ expanded about the point $P = (0, 0)$.

Following the theorem, we calculate as follows:

$$f(P) = 0,$$

$$\frac{\partial f}{\partial x}(P) = \cos y \big|_{(0,0)} = 1,$$

$$\frac{\partial f}{\partial y}(P) = -x \sin y \big|_{(0,0)} = 0,$$

$$\frac{\partial^2 f}{\partial x^2}(P) = 0,$$

$$\frac{\partial^2 f}{\partial y^2}(P) = -x \cos y \big|_{(0,0)} = 0,$$

$$\frac{\partial^2 f}{\partial x \partial y}(P) = - \sin y \big|_{(0,0)} = 0,$$

$$\frac{\partial^3 f}{\partial x^3}(P) = 0,$$

$$\frac{\partial^3 f}{\partial x^2 \partial y}(P) = 0,$$

$$\frac{\partial^3 f}{\partial x \partial y^2} = - \cos y \big|_{(0,0)} = -1,$$

$$\frac{\partial^3 f}{\partial y^3} = x \sin y \big|_{(0,0)} = 0.$$

We find, then, that the Taylor expansion is

$$f(x, y) = 0 + 1(x - 0) + 0(y - 0)$$

$$+ \frac{1}{2!}(0(x - 0)^2 + 2 \cdot 0(x - 0)(y - 0) + 0(y - 0)^2)$$

$$+ \frac{1}{3!}(0(x - 0)^3 + 3 \cdot 0(x - 0)^2(y - 0) + 3 \cdot (-1)(x - 0)(y - 0)^2$$

$$+ 0(y - 0)^3) + \mathcal{R}_{3,\mathbf{P}}(x).$$

This of course simplifies to

$$f(x, y) = x - \frac{3}{3!}xy^2 + \mathcal{R}_{3,\mathbf{P}}(x).$$

□

Exercises

1. Formulate a sum, product, and quotient rule for derivatives of functions of two real variables taking values in $\mathbb{R}$.

2. Use the Chain Rule for a function $f \colon \mathbb{R}^n \to \mathbb{R}^n$ to find a formula for the derivative of the inverse of f in terms of the derivative of f itself.

3. Formulate a definition of second derivative parallel to the definition of first derivative given in Section 10.1. Your definition should involve a matrix. What does this matrix tell us about the second partial derivatives of the function?

*4. If f and g are vector-valued functions with domain $\mathbb{R}^k$, both taking values in $\mathbb{R}^m$ and both having the same domain, then we can define the dot product function $h(x) = f(x) \cdot g(x)$. Formulate and prove a derivative Product Rule for this type of product.

5. Prove that if a function with domain an open subset of $\mathbb{R}^k$ is differentiable at a point P then it is continuous at P.

6. Let f be a function defined on a ball $B(\mathbf{P}, r)$. Let $\mathbf{u} = (u_1, u_2, \ldots, u_k)$ be a vector of unit length. If f is differentiable at $\mathbf{P}$, then give a definition of the directional derivative $D_{\mathbf{u}}f(\mathbf{P})$ of f in the direction $\mathbf{u}$ at P in terms of $M_{\mathbf{P}}$.

7. If f is differentiable on a ball $B(\mathbf{P}, r)$ and if M_x is the zero matrix for every $x \in B(\mathbf{P}, r)$, then prove that f is constant on $B(\mathbf{P}, r)$.

8. Refer to Exercise **6** for notation. For which collections of vectors $\mathbf{u}_1, \mathbf{u}_2, \ldots, \mathbf{u}_k$ in $\mathbb{R}^k$ is it true that if $D_{\mathbf{u}_j}f(x) = 0$ for all $x \in B(\mathbf{P}, r)$ and all $j = 1, 2, \ldots, k$, then f is identically constant?

9. Prove that an $\mathbb{R}^m$-valued function $\mathbf{f}$ is differentiable at a point $\mathbf{P} \in \mathbb{R}^k$ if and only if there is an $m \times k$ matrix (where k is the

dimension of the domain and m the dimension of the range) $M_P(f)$ such that

$$f(P + h) = f(P) + M_P(f)h + \mathcal{R}_P(f, h);$$

here, h is a k-column vector and the remainder term $\mathcal{R}_P$ is a column vector satisfying

$$\frac{\|\mathcal{R}_P(f, h)\|}{\|h\|} \to 0$$

as $h \to 0$.

10. Verify the last assertion in the proof of Theorem 10.15.

11. Prove, in the context of two real variables, that the composition of two continuously differentiable mappings is continuously differentiable.

12. Prove that the product of continuously differentiable functions is continuously differentiable.

*13. There is no mean value theorem as such in the theory of functions of several real variables. For example, if $\gamma: [0, 1] \to \mathbb{R}^k$ is a differentiable function on $(0, 1)$, continuous on $[0, 1]$, then it is not necessarily the case that there is a point $\xi \in (0, 1)$ such that $\dot{\gamma}(\xi) = \gamma(1) - \gamma(0)$. Provide a counterexample to substantiate this claim.

However, there is a serviceable substitute for the mean value theorem: if we assume that $\gamma: [a, b] \to \mathbb{R}^N$ is continuously differentiable on an open interval that contains $[a, b]$ and if $M = \max_{t \in [a,b]} |\dot{\gamma}(t)|$, then

$$|\gamma(b) - \gamma(a)| \le M \cdot |b - a|.$$

Prove this statement.

14. Refer to Exercise 13. Another alternative to the mean value theorem is this. Let $U \subseteq \mathbb{R}^N$ be a convex, open set. Let $f: U \to \mathbb{R}$ be a continuously differentiable function. Assume that $|\nabla f| \le M$ on U. Let $P, Q \in U$. Then,

$$|f(P) - f(Q)| \le M \cdot \|P - Q\|.$$

Prove this result.

*15. Let f be a twice continuously differentiable function on a neighborhood U of the origin in $\mathbb{R}^2$. Define

$$Hf(x, y) = \begin{pmatrix} \frac{\partial^2}{\partial x \partial x}(x, y) & \frac{\partial^2}{\partial x \partial y}(x, y) \\ \frac{\partial^2}{\partial y \partial x}(x, y) & \frac{\partial^2}{\partial y \partial y}(x, y) \end{pmatrix}.$$

We call Hf the *Hessian matrix* of f. If P is a point at which f has a local minimum, then what characteristics does $Hf(P)$ have? If Q is a point at which f has a local maximum, then what characteristics does $Hf(Q)$ have? If R is a saddle point for f, then what characteristics does $Hf(R)$ have?

16. Let U be an open set in $\mathbb{R}^N$. Let $f: U \to \mathbb{R}$. Define what it means for f to be Lipschitz on U. Show that a Lipschitz function must be uniformly continuous.

10.3 The Inverse and Implicit Function Theorems

It is easy to tell whether a continuous function of one real variable is invertible. If the function is strictly monotone increasing or strictly monotone decreasing on an interval then the restriction of the function to that interval is invertible. The converse is true as well. It is more difficult to tell whether a function of several variables, when restricted to a neighborhood of a point, is invertible. The reason, of course, is that such a function will in general have different monotonicity behavior in different directions. Also, the domain of the function could have a strange shape.

However, if we look at the one-variable situation in a new way, it can be used to give us an idea for analyzing functions of several variables. Suppose that f is continuously differentiable on an open interval I and that $P \in I$. If $f'(P) > 0$, then the continuity of f' tells us that, for x near P, $f'(x) > 0$. Thus, f is strictly increasing on some (possibly smaller) open interval J centered at P. Such a function, when restricted to J, is an invertible function. The same analysis applies when $f'(P) < 0$.

Now, the hypothesis that $f'(P) > 0$ or $f'(P) < 0$ has an important geometric interpretation—the positivity of $f'(P)$ means that the tangent line to the graph of f at P has positive slope; hence, that the tangent line is the graph of an invertible function (Figure 10.8); likewise, the negativity of $f'(P)$ means that the tangent line to the graph of f at P has negative slope; hence, that the tangent line is the graph of an invertible function (Figure 10.9). Since the tangent line is a very close approximation at P to the graph of f, our geometric intuition suggests that the local invertibility of f is closely linked to the invertibility of the function describing the tangent line. This guess is in fact borne out in the discussion in the last paragraph.

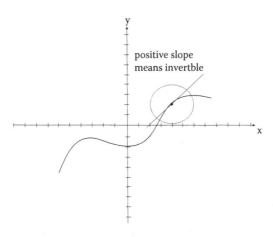

FIGURE 10.8
An invertible function.

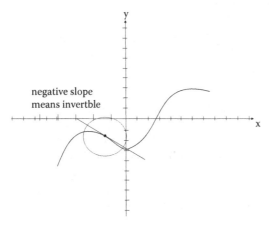

FIGURE 10.9
Negative slope implies invertible.

We would like to carry out an analysis of this kind for a function f from a subset of $\mathbb{R}^k$ into $\mathbb{R}^k$. If P is in the domain of f and if a certain derivative of f at P (to be discussed later) does not vanish, then we would like to conclude that there is a neighborhood U of P such that the restriction of f to U is invertible. That is the content of the Inverse Function Theorem.

Before we formulate and prove this important theorem, we first discuss the kind of derivative of f at P that we shall need to examine.

Definition 10.21: Let $f = (f_1, f_2, ..., f_k)$ be a differentiable function from an open subset U of $\mathbb{R}^k$ into $\mathbb{R}^k$. The *Jacobian matrix* of f at a point $\mathbf{P} \in U$ is the matrix

$$Jf(\mathbf{P}) = \begin{pmatrix} \frac{\partial f_1}{\partial x_1}(\mathbf{P}) & \frac{\partial f_1}{\partial x_2}(\mathbf{P}) & \cdots & \frac{\partial f_1}{\partial x_k}(\mathbf{P}) \\ \frac{\partial f_2}{\partial x_1}(\mathbf{P}) & \frac{\partial f_2}{\partial x_2}(\mathbf{P}) & \cdots & \frac{\partial f_2}{\partial x_k}(\mathbf{P}) \\ & & \cdots & \\ \frac{\partial f_k}{\partial x_1}(\mathbf{P}) & \frac{\partial f_k}{\partial x_2}(\mathbf{P}) & \cdots & \frac{\partial f_k}{\partial x_k}(\mathbf{P}) \end{pmatrix}.$$

We have seen the matrix in the definition before: it is just the derivative of f at P.

Example 10.22: Let $f(x, y) = (x^2 - y, y^2 - x)$. Let $\mathbf{P} = (1, 1)$. Then,

$$Jf(\mathbf{P}) = \begin{pmatrix} 2x & -1 \\ -1 & 2y \end{pmatrix}\Big|_{\mathbf{P}} = \begin{pmatrix} 2 & -1 \\ -1 & 2 \end{pmatrix}.$$

This is the Jacobian matrix for f at $\mathbf{P}$.   □

Notice that if we were to expand the function f in a Taylor series about $\mathbf{P}$ (this would be in fact a k-tuple of expansions, since $f = (f_1, f_2, \dots, f_k)$), then the expansion would be

$$f(\mathbf{P} + \mathbf{h}) = f(\mathbf{P}) + Jf(P)\mathbf{h} + \dots.$$

Thus, the Jacobian matrix is a natural object to study. Moreover, we see that the expression $f(\mathbf{P} + \mathbf{h}) - f(\mathbf{P})$ is well approximated by the expression $Jf(\mathbf{P})\mathbf{h}$. Thus, in analogy with one-variable analysis, we might expect that the invertibility of the matrix $Jf(\mathbf{P})$ would imply the existence of a neighborhood of $\mathbf{P}$ on which the function f is invertible. This is indeed the case, shown as follows:

Theorem 10.23: (The Inverse Function Theorem) *Let f be a continuously differentiable function from an open set $U \subseteq \mathbb{R}^k$ into $\mathbb{R}^k$. Suppose that $\mathbf{P} \in U$ and that the matrix $Jf(\mathbf{P})$ is invertible. Then, there is a neighborhood V of P such that the restriction of f to V is invertible.*

Proof: The proof of the theorem as stated is rather difficult. Therefore, we shall content ourselves with the proof of a special case: we shall make the additional hypothesis that the function f is twice continuously differentiable in a neighborhood of $\mathbf{P}$.

Choose $s > 0$ such that $\bar{B}(\mathbf{P}, s) \subseteq U$ and so that $\det Jf(x) \neq 0$ for all $x \in \bar{B}(P, s)$. Thus, the Jacobian matrix $Jf(x)$ is invertible for all $x \in \bar{B}(P, s)$. With the extra hypothesis, Taylor's theorem tells us that there is a constant C such that if $\|\mathbf{h}\| < s/2$, then

$$f(\mathbf{Q} + \mathbf{h}) - f(\mathbf{Q}) = Jf(\mathbf{Q})\mathbf{h} + \mathcal{R}_{1,\mathbf{Q}}(f, \mathbf{h}), \qquad (10.23.1)$$

where

$$|\mathcal{R}_{1,\mathbf{Q}}(\mathbf{h})| \leq C \cdot \frac{\|\mathbf{h}\|^2}{2!},$$

and

$$C = \sup_{\substack{t \in B(\mathbf{Q},r) \\ j_1+j_2+\cdots+j_k=2}} \left| \frac{\partial^{j_1+j_2+\cdots+j_k} f}{\partial x_1^{j_1} \partial x_2^{j_2} \cdots \partial x_k^{j_k}} \right|.$$

However, all the derivatives in the sum specifying C are, by hypothesis, continuous functions. Since all the balls $B(\mathbf{Q}, s/2)$ are contained in the compact subset $\bar{B}(P, s)$ of U, it follows that we may choose C to be a finite number *independent of* $\mathbf{Q}$.

Now, the matrix $Jf(\mathbf{Q})^{-1}$ exists by hypothesis. The coefficients of this matrix will be continuous functions of $\mathbf{Q}$ because those of Jf are. Thus, these coefficients will be bounded above on $\bar{B}(P, s)$. We conclude that there is a constant $K > 0$ *independent of* $\mathbf{Q}$ such that for every $\mathbf{k} \in \mathbb{R}^k$ we have

$$\|Jf(\mathbf{Q})^{-1}\mathbf{k}\| \leq K\|\mathbf{k}\|.$$

Taking $\mathbf{k} = Jf(\mathbf{Q})\mathbf{h}$ yields

$$\|\mathbf{h}\| \leq K\|Jf(\mathbf{Q})\mathbf{h}\|. \qquad (10.23.2)$$

Now, set

$$r = \min \{s/2, 1/(KC)\}.$$

Line (10.23.1) tells us that, for $\mathbf{Q} \in B(\mathbf{P}, r)$ and $\|\mathbf{h}\| < r$,

$$\|f(\mathbf{Q} + \mathbf{h}) - f(\mathbf{Q})\| \geq \|Jf(\mathbf{Q})\mathbf{h}\| - \|\mathcal{R}_{1,\mathbf{Q}}(\mathbf{h})\|.$$

But estimate (10.23.2), together with our estimate from the aforementioned equation on the error term $\mathcal{R}$, yields that the right side of this equation is

$$\geq \frac{\|\mathbf{h}\|}{K} - \frac{C}{2}\|\mathbf{h}\|^2.$$

The choice of r tells us that $\|\mathbf{h}\| \le 1/(KC)$; hence, the last line majorizes $(K/2)\|\mathbf{h}\|$.

But, this tells us that, for any $\mathbf{Q} \in B(\mathbf{P}, r)$ and any $\mathbf{h}$ satisfying $\|\mathbf{h}\| < r$, it holds that $f(\mathbf{Q} + \mathbf{h}) \ne f(\mathbf{Q})$. In particular, the function f is one-to-one when restricted to the ball $B(\mathbf{P}, r/2)$. Thus, $f|_{B(P,s/2)}$ is invertible.   □

In fact, the estimate

$$\|f(\mathbf{Q} + \mathbf{h}) - f(\mathbf{Q})\| \ge \frac{K}{2}\|\mathbf{h}\|$$

that we derived easily implies that the image of every $B(\mathbf{Q}, s)$ contains an open ball $B(f(\mathbf{Q}), s')$, some $s' > 0$. This means that f is an *open mapping*. You will be asked in the exercises to provide the details of this assertion.

Example 10.24: Let $f(x, y) = (xy - y^3, y + x^2)$. Notice that the Jacobian matrix of this function is

$$Jf = \begin{pmatrix} y & x - 3y^2 \\ 2x & 1 \end{pmatrix}.$$

At the point $(1, 1)$, the Jacobian is

$$Jf(1, 1) = \begin{pmatrix} 1 & -2 \\ 2 & 1 \end{pmatrix}.$$

The determinant of $Jf(1, 1)$ is $5 \ne 0$. Thus, the Inverse Function Theorem guarantees that f is invertible in a neighborhood of the point $(1, 1)$.   □

Example 10.25: Define

$$f(x, y) = (xy + y, y - x).$$

It is easy to calculate that the Jacobian determinant at the point $(1, 1)$ is $3 \ne 0$. So, the Inverse Function Theorem applies and we know that f is invertible in a neighborhood.

In this example, it is actually possible to calculate f^{-1}, and we ask you to perform this calculation as an exercise.   □

With some additional effort it can be shown that f^{-1} is continuously differentiable in a neighborhood of $f(\mathbf{P})$. However, the details of this matter are beyond the scope of this book. We refer the interested reader to Ref. [RUD1].

Next, we turn to the implicit function theorem. This result addresses the question of when we can solve an equation

$$f(x_1, x_2, \ldots, x_k) = 0$$

for one of the variables in terms of the other $(k - 1)$ variables. It is illustrative to first consider a simple example. Look at the equation

$$f(x_1, x_2) = (x_1)^2 + (x_2)^2 = 1.$$

We may restrict attention to $-1 \le x_1 \le 1, -1 \le x_2 \le 1$. As a glance at the graph shows, we can solve this equation for x_2, uniquely in terms of x_1, in a neighborhood of any point *except* for the points $(\pm 1, 0)$. At these two exceptional points, it is impossible to avoid the ambiguity in the square root process, even by restricting to a very small neighborhood. At other points, we may write

$$t_2 = +\sqrt{1 - (t_1)^2}$$

for points (t_1, t_2) near (x_1, x_2) when $x_2 > 0$ and

$$t_2 = -\sqrt{1 - (t_1)^2}$$

for points (t_1, t_2) near (x_1, x_2) when $x_2 < 0$.

What distinguishes the two exceptional points from the others is that the tangent line to the locus (a circle) is vertical at each of these points. Another way of saying this is that

$$\frac{\partial f}{x_2} = 0$$

at these points (Figure 10.10). These preliminary considerations motivate the following theorem.

Theorem 10.26: (The Implicit Function Theorem) *Let f be a function of k real variables, taking scalar values, whose domain contains a neighborhood of a point* **P**. *Assume that f is continuously differentiable and that* $f(\mathbf{P}) = 0$. *If* $(\partial f / \partial x_k)(\mathbf{P}) \ne 0$, *then there are numbers* $\delta > 0$, $\eta > 0$ *such that if* $|x_1 - P_1| < \delta, |x_2 - P_2| < \delta, \ldots,$ $|x_{k-1} - P_{k-1}| < \delta$, *then there is a unique* x_k *with* $|x_k - P_k| < \eta$ *and*

$$f(x_1, x_2, \ldots, x_k) = 0. \tag{10.26.1}$$

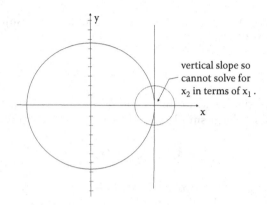

FIGURE 10.10
Vertical tangents.

In other words, in a neighborhood of **P**, the equation (10.26.1) uniquely determines x_k in terms of $x_1, x_2, \ldots, x_{k-1}$.

Proof: We consider the function

$$T: (x_1, x_2, \ldots, x_k) \mapsto (x_1, x_2, \ldots, x_{k-1}, f(x_1, x_2, \ldots, x_k)).$$

The Jacobian matrix of T at **P** is

$$\begin{pmatrix} 1 & 0 & \cdots & 0 \\ 0 & 1 & \cdots & 0 \\ & \cdots & & \\ 0 & \cdots & 1 & 0 \\ \frac{\partial f}{\partial x_1}(\mathbf{P}) & \frac{\partial f}{\partial x_2}(\mathbf{P}) & \cdots & \frac{\partial f}{\partial x_k}(\mathbf{P}) \end{pmatrix}.$$

Of course, the determinant of this matrix is $\partial f / \partial x_k(\mathbf{P})$, which we hypothesized to be nonzero. Thus, the Inverse Function Theorem applies to T. We conclude that T is invertible in a neighborhood of **P**. That is, there is a number $\eta > 0$ and a neighborhood W of the point $(P_1, P_2, \ldots, P_{k-1}, 0)$ such that

$$T: B(\mathbf{P}, \eta) \mapsto W$$

is a one-to-one, onto, continuously differentiable function, which is invertible. Select $\delta > 0$ such that if $|x_1 - P_1| < \delta$, $|x_2 - P_2| < \delta$, ..., $|x_{k-1} - P_{k-1}| < \delta$, then the point $(x_1, x_2, \ldots, x_{k-1}, 0) \in W$. Such a point $(x_1, x_2, \ldots, x_{k-1}, 0)$ then has a unique inverse image under T that lies in $B(\mathbf{P}, \eta)$. But this just says that there is a unique x_k such that

$f(x_1, x_2, \ldots, x_k) = 0$. We have established the existence of δ and η as required; hence, the proof is complete. □

Example 10.27: Let $f(x, y) = yx^2 - x + y$. Observe that $f(0, 0) = 0$. Also, note that

$$\frac{\partial f}{\partial y}(0, 0) = 1 \neq 0.$$

Thus, the implicit function theorem guarantees that we can solve for y in terms of x in a neighborhood of the origin. And in fact, in this simple instance, we can solve explicitly:

$$y = \frac{x}{x^2 + 1}.$$
□

Exercises

1. Prove that a function satisfying the hypotheses of the Inverse Function Theorem is an open mapping in a neighborhood of the point **P**.

2. Prove that the Implicit Function Theorem is still true if the equation $f(x_1, x_2, \ldots, x_k) = 0$ is replaced by $f(x_1, x_2, \ldots, x_k) = c$. (**Hint:** Do *not* repeat the proof of the Implicit Function Theorem.)

3. Let $y = \varphi(x)$ be a twice continuously differentiable function on $[0, 1]$ with nonvanishing first derivative. Let U be the graph of φ. Show that there is an open neighborhood W of U so that, if $P \in W$, then there is a unique point $X \in U$ which is nearest to P.

4. Give an example of a curve that is not twice continuously differentiable for which the result of Exercise **3** fails.

5. Use the Implicit Function Theorem to show that the natural logarithm function can have only one zero.

6. Use the Implicit Function Theorem to show that the exponential function can have no zeros.

7. It is not true that if a function f from $\mathbb{R}^k$ to $\mathbb{R}^k$ is invertible in a neighborhood of a point P in the domain then the Jacobian determinant at P is nonzero. Provide an example to illustrate this point.

8. It is not true that if the equation $f(x, y) = 0$ can be solved for y in terms of x near the point P then $\partial f/\partial y(P) \neq 0$. Provide an example to illustrate this point.

*9. Use the Implicit Function Theorem to give a proof of the Fundamental Theorem of Algebra.

10. Let U, V be open subsets of $\mathbb{R}^N$. Let X be the collection of functions $f: U \to V$, which are continuously differentiable and have continuously differentiable inverse. Show that X is closed under addition and multiplication and scalar multiplication. We call X an *algebra*.

*11. Show that the Inverse Function Theorem implies the Implicit Function Theorem.

*12. Show that the Implicit Function Theorem implies the Inverse Function Theorem.

11

Advanced Topics

11.1 Metric Spaces

As you studied Chapter 10, and did the exercises developing the basic properties of functions of several variables, you should have noticed that many of the proofs were identical to those in Chapters 5 and 6. The arguments generally involved clever use of the triangle inequality. For functions of one variable, the inequality was for $|\quad|$. For functions of several variables, the inequality was for $\|\quad\|$.

This section formalizes a general context in which we may do analysis any time we have a reasonable notion of calculating distance. Such a structure will be called a metric:

Definition 11.1: Metric space A *metric space* is a pair (X, ρ), where X is a set and

$$\rho\colon X \times X \to \{t \in \mathbb{R}\colon t \geq 0\}$$

is atriangle inequality function satisfying

1. $\forall\, x, y \in X,\quad \rho(x, y) = \rho(y, x)$;
2. $\rho(x, y) = 0$ if and only if $x = y$;
3. $\forall\, x, y, z \in X,\quad \rho(x, y) \leq \rho(x, z) + \rho(z, y)$.

The function ρ is called a *metric* on X.

Example 11.2: The pair $(\mathbb{R}, \rho)$, where $\rho(x, y) = |x - y|$, is a metric space. Each of the properties required of a metric is in this case a restatement of familiar facts from the analysis of one dimension.

The pair $(\mathbb{R}^k, \rho)$, where $\rho(x, y) = \|x - y\|$, is a metric space. Each of the properties required of a metric is in this case a restatement of familiar facts from the analysis of k dimensions. □

DOI: 10.1201/9781003222682-12

The first example presented familiar metrics on two familiar spaces. Now we look at some new ones.

Example 11.3: The pair $(\mathbb{R}^2, \rho)$, where $\rho(x, y) = \max\{|x_1 - y_1|, |x_2 - y_2|\}$, is a metric space. Only the triangle inequality is not trivial to verify, but that reduces to the triangle inequality of one variable.

The pair $(\mathbb{R}, \mu)$, where $\mu(x, y) = 1$ if $x \neq y$ and 0 otherwise, is a metric space. Checking the triangle inequality reduces to seeing that if $x \neq y$ then either $x \neq z$ or $y \neq z$.   □

Example 11.4: Let X denote the space of continuous functions on the interval $[0, 1]$. If $f, g \in X$ then let $\rho(f, g) = \sup_{t \in [0,1]} |f(t) - g(t)|$. Then the pair (X, ρ) is a metric space. The first two properties of a metric are obvious and the triangle inequality reduces to the triangle inequality for real numbers.

This example is a dramatic new departure from the analysis we have done in the previous ten chapters. For X is a very large space—infinite dimensional in a certain sense. Using the ideas that we are about to develop, it is nonetheless possible to study convergence, continuity, compactness, and the other basic concepts of analysis in this more general context. We shall see applications of these new techniques in later sections.   □

Now we begin to develop the tools of analysis in metric spaces.

Definition 11.5: Let (X, ρ) be a metric space. A sequence $\{x_j\}$ of elements of X is said to *converge* to a point $\alpha \in X$ if, for each $\varepsilon > 0$, there is an $N > 0$ such that if $j > N$ then $\rho(x_j, \alpha) < \varepsilon$. We call α the *limit* of the sequence $\{x_j\}$. We sometimes write $x_j \to \alpha$.

Compare this definition of convergence with the corresponding definition for convergence in the real line in Section 2.1. Notice that it is identical, except that the sense in which distance is measured is now more general.

Example 11.6: Let (X, ρ) be the metric space from Example 11.4, consisting of the continuous functions on the unit interval with the indicated metric function ρ. Then $f = \sin x$ is an element of this space, and so are the functions

$$f_j = \sum_{\ell=0}^{j} (-1)^\ell \frac{x^{2\ell+1}}{(2\ell + 1)!}.$$

Observe that the functions f_j are the partial sums for the Taylor series of $\sin x$. We can check from simple estimates on the error term of Taylor's theorem that the functions f_j converge uniformly to f. Thus, in the language of metric spaces, $f_j \to f$ in the metric space notion of convergence.   □

Definition 11.7: Let (X, ρ) be a metric space. A sequence $\{x_j\}$ of elements of X is said to be *Cauchy* if, for each $\varepsilon > 0$ there is an $N > 0$ such that if $j, k > N$ then $\rho(x_j, x_k) < \varepsilon$.

Now the Cauchy criterion and convergence are connected in the expected fashion:

Proposition 11.8: *Let $\{x_j\}$ be a convergent sequence, with limit α, in the metric space (X, ρ). Then the sequence $\{x_j\}$ is Cauchy.*

Proof: Let $\varepsilon > 0$. Choose an N so large that, if $j > N$, then $\rho(x_j, \alpha) < \varepsilon/2$. If $j, k > N$ then

$$\rho(x_j, x_k) \leq \rho(x_j, \alpha) + \rho(\alpha, x_k) < \frac{\varepsilon}{2} + \frac{\varepsilon}{2} = \varepsilon.$$

That completes the proof. □

The converse of the proposition is true in the real numbers (with the usual metric), as we proved in Section 1.1. However, it is not true in every metric space. For example, the rationals $\mathbb{Q}$ with the usual metric $\rho(s, t) = |s - t|$ is a metric space; but the sequence

$$3, 2.1, 2.14, 2.141, 2.1415, 2.14159, \ldots,$$

while certainly Cauchy, *does not converge to a rational number*. Thus we are led to a definition:

Definition 11.9: We say that a metric space (X, ρ) is *complete* if every Cauchy sequence converges to an element of the metric space.

Thus the real numbers, with the usual metric, form a complete metric space. The rational numbers do not.

Example 11.10: Consider the metric space (X, ρ) from Example 11.4 above, consisting of the continuous functions on the closed unit interval with the indicated metric function ρ. If $\{g_j\}$ is a Cauchy sequence in this metric space then each g_j is a continuous function on the unit interval and this sequence of continuous functions is Cauchy in the uniform sense (see Chapter 8). Therefore they converge uniformly to a limit function g that must be continuous. We conclude that the metric space (X, ρ) is complete. □

Example 11.11: Consider the metric space (X, ρ) consisting of the polynomials, taken to have domain the interval $[0, 1]$, with the distance function $\rho(f, g) = \sup_{t \in [0,1]} |f(t) - g(t)|$. This metric space is *not* complete. For if h is any continuous function on $[0, 1]$ that is not a polynomial, such as

$h(x) = \sin x$, then by the Weierstrass Approximation Theorem there is a sequence $\{p_j\}$ of polynomials that converges uniformly on $[0, 1]$ to h. Thus this sequence $\{p_j\}$ will be Cauchy in the metric space, but it *does not converge to an element of the metric space*. We conclude that the metric space (X, ρ) is not complete.   □

If (X, ρ) is a metric space then an *(open) ball* with center $P \in X$ and radius r is the set

$$B(P, r) = \{x \in X : \rho(x, P) < r\}.$$

The *closed ball* with center P and radius r is the set

$$\bar{B}(P, r) = \{x \in X : \rho(x, P) \le r\}.$$

Definition 11.12: Let (X, ρ) be a metric space and E a subset of X. A point $P \in E$ is called an *isolated point* of E if there is an $r > 0$ such that $E \cap B(P, r) = \{P\}$. If a point of E is not isolated then it is called *nonisolated*.

We see that the notion of "isolated" has intuitive appeal: an isolated point is one that is spaced apart—at least distance r—from the other points of the space. A nonisolated point, by contrast, has neighbors that are arbitrarily close.

Definition 11.13: Let (X, ρ) be a metric space and $f : X \to \mathbb{R}$. If $P \in X$ is a nonisolated point and $\ell \in \mathbb{R}$ we say that *the limit of f at P is ℓ*, and write

$$\lim_{x \to P} f(x) = \ell,$$

if for any $\varepsilon > 0$ there is a $\delta > 0$ such that if $0 < \rho(x, P) < \delta$ then $|f(x) - \ell| < \varepsilon$.

Notice in this definition that we use ρ to measure distance in X—that is the natural notion of distance with which X comes equipped—but we use absolute values to measure distance in $\mathbb{R}$.

The following lemma will prove useful.

Lemma 11.14: *Let (X, ρ) be a metric space and $P \in X$ a nonisolated point. Let f be a function from X to $\mathbb{R}$. Then $\lim_{x \to P} f(x) = \ell$ if and only if, for every sequence $\{x_j\} \subseteq X$ satisfying $x_j \to P$, it holds that $f(x_j) \to f(P)$.*

Proof: This is straightforward and is treated in the exercises.   □

Definition 11.15: Let (X, ρ) be a metric space and E a subset of X. Suppose that $P \in E$ is a nonisolated point. We say that a function $f : E \to \mathbb{R}$ is *continuous* at P if

$$\lim_{x \to P} f(x) = f(P).$$

Example 11.16: Let (X, ρ) be the space of continuous functions on the interval $[0, 1]$ equipped with the supremum metric as in Example 11.4 above. Define the function $\mathcal{F}: X \to \mathbb{R}$ by the formula

$$\mathcal{F}(f) = \int_0^1 f(t)dt \ .$$

Then $\mathcal{F}$ takes an element of X, namely a continuous function, to a real number, namely its integral over $[0, 1]$. We claim that $\mathcal{F}$ is continuous at every point of X.

For fix a point $f \in X$. If $\{f_j\}$ is a sequence of elements of X converging in the metric space sense to the limit f, then (in the language of classical analysis as in Chapters 5–8) the f_j are continuous functions converging uniformly to the continuous function f on the interval $[0, 1]$. But, by Theorem 8.8, it follows that

$$\int_0^1 f_j(t)dt \ \to \ \int_0^1 f(t)dt.$$

But this just says that $\mathcal{F}(f_j) \to \mathcal{F}(f)$. Using the lemma, we conclude that

$$\lim_{g \to f} \mathcal{F}(g) = \mathcal{F}(f).$$

Therefore $\mathcal{F}$ is continuous at f.

Since $f \in X$ was chosen arbitrarily, we conclude that the function $\mathcal{F}$ is continuous at every point of X. □

In the next section we shall develop some topological properties of metric spaces.

Exercises

1. Let (X, ρ) be a metric space. Prove that the function

$$\sigma(s, t) = \frac{\rho(s, t)}{1 + \rho(s, t)}$$

is also a metric on X and that the open sets defined by the metric ρ are the same as the open sets defined by σ. Finally prove that $\sigma(s, t) < 1$ for all $s, t \in X$. This is a method for constructing a bounded metric from an arbitrary metric.

2. Let (X, ρ) be a metric space and E a subset of X. Define the *boundary* boundary of a set of E to be those elements $x \in X$ with the property that every ball $B(x, r)$ contains both points of E and points of ${}^{c}E$. Prove that the boundary of E must be closed. Prove that the interior of E (define!) is disjoint from the boundary of E.

3. Let (X, ρ) and (Y, σ) be metric spaces. Describe a method for equipping the set $X \times Y$ with a metric manufactured from ρ and σ.

4. Let X be the collection of all continuously differentiable functions on the interval $[0, 1]$. If $f, g \in X$ then define

$$\rho(f, g) = \sup_{x \in [0,1]} |f'(x) - g'(x)|.$$

   Is ρ a metric? Why or why not?

5. Prove Lemma 11.14.

6. Consider the set of all polynomials of one variable of degree not exceeding 4, and define

$$\rho(p, q) = \max\{|p(1) - q(1)|, |p(2) - q(2)|, |p(3) - q(3)|, \\ |p(4) - q(4)|, |p(5) - q(5)|, |p(6) - q(6)|\}.$$

   Prove that ρ is a metric on this space of polynomials. Why does this not work if we consider polynomials of degree not exceeding 10?

7. How many different metrics are there on the space with three points?

8. Define a metric on the real numbers $\mathbb{R}$ so that the space becomes discrete.

9. Give an example of a metric space which is discrete.

*10. Can there be a countable, complete metric space?

11.2 Topology in a Metric Space

Fix a metric space (X, ρ). A set $U \subseteq X$ is called *open* if for each $u \in U$ there is an $r > 0$ such that $B(u, r) \subseteq U$. A set $E \subseteq X$ is called *closed* if its complement in X is open. Notice that these definitions are analogous to those that we gave in the topology chapter (Chapter 4) for subsets of $\mathbb{R}$.

Example 11.17: Consider the set of real numbers $\mathbb{R}$ equipped with the metric $\rho(s, t) = 1$ if $s \neq t$ and $\rho(s, t) = 0$ otherwise. Then each singleton $U = \{x\}$ is an open set. For let P be a point of U. Then $P = x$ and the ball $B(P, 1/2)$ lies in U.

However, each singleton is also closed. For the complement of the singleton $U = \{x\}$ is the set $S = \mathbb{R}\setminus\{x\}$. If $s \in S$ then $B(s, 1/2) \subseteq S$ as in the preceding paragraph.   □

Example 11.18: Let (X, ρ) be the metric space of continuous functions on the interval $[0, 1]$ equipped with the metric $\rho(f, g) = \sup_{x \in [0,1]} |f(x) - g(x)|$. Define

$$U = \{f \in X : f(1/2) > 5\}.$$

Then U is an open set in the metric space. To verify this assertion, fix an element $f \in U$. Let $\varepsilon = f(1/2) - 5 > 0$. We claim that the metric ball $B(f, \varepsilon)$ lies in U. For let $g \in B(f, \varepsilon)$. Then

$$\begin{aligned} g(1/2) &= f(1/2) - |f(1/2) - g(1/2)| \\ &\geq f(1/2) - \rho(f, g) \\ &> f(1/2) - \varepsilon \\ &= 5. \end{aligned}$$

It follows that $g \in U$. Since $g \in B(f, \varepsilon)$ was chosen arbitrarily, we may conclude that $B(f, \varepsilon) \subseteq U$. But this says that U is open.

We may also conclude from this calculation that

$$^c U = \{f \in X : f(1/2) \leq 5\}$$

is closed.   □

Definition 11.19: Let (X, ρ) be a metric space and $S \subseteq X$. A point $x \in X$ is called an *accumulation point* of S (also called a limit point or a cluster point) if every $B(x, r)$ contains infinitely many elements of S.

Proposition 11.20: *Let (X, ρ) be a metric space. A set $S \subseteq X$ is closed if and only if every accumulation point of S lies in S.*

Proof: The proof is similar to the corresponding result in Section 4.1 and we leave it to the exercises.   □

Definition 11.21: Let (X, ρ) be a metric space. A subset $S \subseteq X$ is said to be *bounded* if S lies in some ball $B(P, r)$.

Definition 11.22: Let (X, ρ) be a metric space. A set $S \subseteq X$ is said to be *compact* if every sequence in S has a subsequence that converges *to an element of* S.

Example 11.23: In Chapter 4 we learned that, in the real number system, compact sets are closed and bounded, and conversely. Such is not the case in general metric spaces.

As an example, consider the metric space (X, ρ) consisting of all continuous functions on the interval $[0, 1]$ with the supremum metric as in previous examples. Let

$$S = \{f_j(x) = x^j \colon j = 1, 2, \ldots\}.$$

This set is bounded since it lies in the ball $B(0, 2)$ (here 0 denotes the identically zero function). We claim that S contains no Cauchy sequences. This follows (see the discussion of uniform convergence in Chapter 8) because, no matter how large N is, if $k > j > N$ then we may write

$$|f_j(x) - f_k(x)| = |x^j| \, |(x^{k-j} - 1)|.$$

Fix j. If x is sufficiently near to 1 then $|x^j| > 3/4$. But then we may pick k so large that $|x^{k-j}| < 1/4$. Thus

$$|f_k(x) - f_j(x)| \geq 9/16.$$

So there is no Cauchy subsequence. We may conclude (for vacuous reasons) that S is closed.

But S is not compact. For, as just noted, the sequence $\{f_j\}$ consists of infinitely many distinct elements of S which do not have a convergent subsequence (indeed not even a Cauchy subsequence).   □

In spite of the last example, half of the Heine-Borel theorem is true:

Proposition 11.24: *Let (X, ρ) be a metric space and S a subset of X. If S is compact then S is closed and bounded.*

Proof: Let $\{s_j\}$ be a Cauchy sequence in S. By compactness, this sequence must contain a subsequence converging to some limit P. But since the full sequence is Cauchy, the full sequence must converge to P (exercise). Thus S is closed.

If S is not bounded, we derive a contradiction as follows. Fix a point $P_1 \in S$. Since S is not bounded we may find a point P_2 that has distance at least 1 from P_1. Since S is unbounded, we may find a point P_3 of S that is distance at least 2 from both P_1 and P_2. Continuing in this fashion, we select $P_j \in S$ which is distance at least j from $P_1, P_2, \ldots P_{j-1}$. Such a sequence $\{P_j\}$ can have no Cauchy subsequence, contradicting compactness. Therefore S is bounded.   □

Definition 11.25: Let S be a subset of a metric space (X, ρ). A collection of open sets $\{O_\alpha\}_{\alpha \in A}$ (each O_α is an open set in X) is called an *open covering* of S if

$$\cup_{\alpha \in A} O_\alpha \supseteq S.$$

Definition 11.26: If C is an open covering of a set S and if $\mathcal{D}$ is another open covering of S such that each element of $\mathcal{D}$ is also an element of C then we call $\mathcal{D}$ a *subcovering* of C.

We call $\mathcal{D}$ a *finite subcovering* if $\mathcal{D}$ has just finitely many elements.

Theorem 11.27: *A subset S of a metric space (X, ρ) is compact if and only if every open covering $C = \{O_\alpha\}_{\alpha \in A}$ of S has a finite subcovering.*

Proof: The forward direction is beyond the scope of this book and we shall not discuss it.

The proof of the reverse direction is similar in spirit to the proof in Section 4.3 (Theorem 4.32). We leave the details for the exercises. ☐

Proposition 11.28: *Let S be a compact subset of a metric space (X, ρ). If E is a closed subset of S then E is compact.*

Proof: Let C be an open covering of E. The set $U = X \backslash E$ is open and the covering C' consisting of all the open sets in C together with the open set U covers S. Since S is compact we may find a finite subcovering

$$O_1, O_2, \ldots O_k$$

that covers S. If one of these sets is U then discard it. The remaining $k - 1$ open sets cover E. ☐

The exercises will ask you to find an alternative proof of this last fact.

Definition 11.29: If (X, ρ) is a metric space and $E \subseteq X$ then the *closure* of E is defined to be the union of E with the set of its accumulation points.

Exercises
1. Let (X, ρ) be a metric space. Prove that the closure of any set in X is closed. Prove that the closure of any E equals the union of the interior and the boundary.
2. Let (X, ρ) be a metric space. Let $K_1 \supseteq K_2 \ldots$ be a nested family of countably many nonempty compact sets. Prove that $\cap_j K_j$ is a nonempty set.

3. Give an example of a metric space (X, ρ), a point $P \in X$, and a positive number r such that $\bar{B}(P, r)$ is *not* the closure of the ball $B(P, r)$.

4. Let (X, ρ) be a compact metric space. Prove that X has a countable dense subset. [We call such a space *separable*.]

5. Let K be a compact subset of a metric space (X, ρ). Let $P \in X$ not lie in K. Prove that there is an element $k_0 \in K$ such that

$$\rho(k_0, P) = \inf_{x \in K} \rho(x, P).$$

6. Consider the metric space Q equipped with the Euclidean metric. Describe all the open sets in this metric space.

7. In $\mathbb{R}$, if I is an open interval then every element of I is a limit point of I. Is the analogous statement true in an arbitrary metric space, with "interval" replaced by "ball"?

8. The Bolzano-Weierstrass Theorem tells us that, in $\mathbb{R}^1$, a bounded infinite set must have a limit point. Show by example that the analogous statement is false in an arbitrary metric space. But it is true in $\mathbb{R}^N$.

9. Let E be a subset of a metric space. Is the interior of E equal to the interior of the closure of E? Is the closure of the interior of E equal to the closure of E itself?

10. Let (X, ρ) be a metric space. Call a subset E of X *connected* if there do not exist open sets U and V in X such that $U \cap E$ and $V \cap E$ are nonempty, disjoint, and $(U \cap E) \cup (V \cap E) = E$. connected set Is the closure of a connected set connected? Is the product of two connected sets connected? Is the interior of a connected set connected?

11. Refer to Exercise **10** for terminology. Give exact conditions that will guarantee that the union of two connected sets is connected.

12. Let (X, ρ) be the metric space of continuously differentiable functions on the interval $[0, 1]$ equipped with the metric

$$\rho(f, g) = \sup_{x \in [0,1]} |f(x) - g(x)|.$$

Consider the function

$$T(f) = f'(1/2).$$

Is T continuous? Is there some metric with which we can equip X that will make T continuous?

13. Let (X, ρ) be a metric space and let $\{x_j\}$ be a Cauchy sequence in X. If a subsequence $\{x_{j_k}\}$ converges to a point $P \in X$ then prove that the full sequence $\{x_j\}$ converges to P.

14. Prove the converse direction of Theorem 11.27.

15. Give a proof of Proposition 11.28 that uses the sequential definition of compactness.

*16. Let (X, ρ) be any metric space. Consider the space $\hat{X}$ of all Cauchy sequences of elements of X, subject to the equivalence relation that $\{x_j\}$ and $\{y_j\}$ are equivalent if $\rho(x_j, y_j) \to 0$ completion of a metric space as $j \to \infty$. Explain why, in a natural way, this space of equivalence class of Cauchy sequences may be thought of as *the completion* of X, that is, explain in what sense $\hat{X} \supseteq X$ and $\hat{X}$ is complete. Prove that $\hat{X}$ is minimal in a certain sense. Prove that if X is already complete then this space of equivalence classes can be identified in a natural way with X.

17. It is a theorem (fairly tricky to prove) that a metric space always has a countable dense subset. What is the countable dense subset of the space of continuous functions on the interval $[0, 1]$? What is the countable dense subset of the space of polynomials of degree not exceeding 10 on the interval $[0, 1]$ (equipped with the supremum norm)?

18. Prove Proposition 11.20.

11.3 The Baire Category Theorem

Let (X, ρ) be a metric space and $S \subseteq X$ a subset. A set $E \subseteq X$ is said to be *dense* density in S if every element of S is the limit of some sequence of elements of E.

Example 11.30: The set of rational numbers $\mathbb{Q}$ is dense in any nontrivial interval of $\mathbb{R}$. □

Example 11.31: Let (X, ρ) be the metric space of continuous functions on the interval $[0, 1]$ equipped with the supremum metric as usual. Let $E \subseteq X$ be the polynomial functions. Then the Weierstrass Approximation Theorem tells us that E is dense in X. □

Example 11.32: Consider the real numbers $\mathbb{R}$ with the metric $\rho(s, t) = 1$ if $s \neq t$ and $\rho(s, t) = 0$ otherwise. Then no proper subset of $\mathbb{R}$ is dense in $\mathbb{R}$. To see this, notice that if E were dense and were not all of $\mathbb{R}$ and if $P \in \mathbb{R}\backslash E$

then $\rho(P, e) > 1/2$ for all $e \in E$. So elements of E do not get close to P. Thus E is not dense in $\mathbb{R}$.   □

Refer to Definition 11.29 for the concept of closure of a set.

Example 11.33: Let (X, ρ) be the set of real numbers with the usual metric and set $E = Q \cap (-2, 2)$. Then the closure of E is $[-2, 2]$.

Let (Y, σ) be the continuous functions on $[0, 1]$ equipped with the supremum metric as in Example 11.4. Take $E \subseteq Y$ to be the polynomials. Then the closure of E is Y.   □

We note in passing that, if $B(P, r)$ is a ball in a metric space (X, ρ), then $\bar{B}(P, r)$ will contain but need not be equal to the closure of $B(P, r)$ (for which see Exercise 3 of the last section).

Definition 11.34: Let (X, ρ) be a metric space. We say that $E \subseteq X$ is *nowhere dense* in X if the closure of E contains no ball $B(x, r)$ for any $x \in X, r > 0$.

Example 11.35: Let us consider the integers $\mathbb{Z}$ as a subset of the metric space $\mathbb{R}$ equipped with the standard metric. Then the closure of $\mathbb{Z}$ is $\mathbb{Z}$ itself. And of course $\mathbb{Z}$ contains no metric balls. Therefore $\mathbb{Z}$ is nowhere dense in $\mathbb{R}$. □

Example 11.36: Consider the metric space X of all continuous functions on the unit interval $[0, 1]$, equipped with the usual supremum metric. Fix $k > 0$ and consider

$$E \equiv \{p(x): p \text{ is a polynomial of degree not exceeding } k\}.$$

Then the closure of E is E itself (that is, the limit of a sequence of polynomials of degree not exceeeding k is still a polynomial of degree not exceeding k—details are requested of you in the exercises). And E contains no metric balls. For if $p \in E$ and $r > 0$ then $p(x) + (r/2) \cdot x^{k+1} \in B(p, r)$ but $p(x) + (r/2) \cdot x^{k+1} \notin E$.

We recall, as noted in Example 11.31 above, that the set of *all* polynomials is dense in X; but if we restrict attention to polynomials of degree not exceeding a fixed number k then the resulting set is nowhere dense.   □

Theorem 11.37: (The Baire Category Theorem) *Let (X, ρ) be Baire category theorem a complete metric space. Then X cannot be written as the union of countably many nowhere dense sets.*

Proof: This proof is quite similar to the proof that we presented in Chapter 4 that a perfect set must be uncountable. You may wish to review that proof at this time.

Seeking a contradiction, suppose that X may be written as a countable union of nowhere dense sets $Y_1, Y_2, \ldots$. Choose a point $x_1 \in {}^c \bar{Y}_1$. Since Y_1 is nowhere dense we may select an $r_1 > 0$ such that $\bar{B}_1 \equiv \bar{B}(x_1, r_1)$ satisfies $\bar{B}_1 \cap \bar{Y}_1 = \emptyset$. Assume without loss of generality that $r_1 < 1$.

Next, since Y_2 is nowhere dense, we may choose $x_2 \in \bar{B}_1 \cap {}^c \bar{Y}_2$ and an $r_2 > 0$ such that $\bar{B}_2 = \bar{B}(x_2, r_2) \subseteq \bar{B}_1 \cap {}^c \bar{Y}_2$. Shrinking B_2 if necessary, we may assume that $r_2 < \frac{1}{2} r_1$. Continuing in this fashion, we select at the jth step a point $x_j \in \bar{B}_{j-1} \cap {}^c \bar{Y}_j$ and a number $r_j > 0$ such that $r_j < \frac{1}{2} r_{j-1}$ and $\bar{B}_j = \bar{B}(x_j, r_j) \subseteq \bar{B}_{j-1} \cap {}^c \bar{Y}_j$.

Now the sequence $\{x_j\}$ is Cauchy since all the terms x_j for $j > N$ are contained in a ball of radius $r_N < 2^{-N}$ hence are not more than distance 2^{-N} apart. Since (X, ρ) is a complete metric space, we conclude that the sequence converges to a limit point P. Moreover, by construction, $P \in \bar{B}_j$ for every j hence is in the complement of *every* $\bar{Y}_j$. Thus $\cup_j Y_j \neq X$. That is a contradiction, and the proof is complete. □

There is quite a lot of terminology associated with the Baire theorem, and we shall not detail it all here. We do note that a G_δ is the countable intersection of open sets.

Before we apply the Baire Category Theorem, let us formulate some re-statements, or corollaries, of the theorem which follow immediately from the definitions.

Corollary 11.38: *Let (X, ρ) be a complete metric space. Let $Y_1, Y_2, \ldots$ be countably many closed subsets of X, each of which contains no nontrivial open ball. Then $\cup_j Y_j$ also has the property that it contains no nontrivial open ball.*

Corollary 11.39: *Let (X, ρ) be a complete metric space. Let $O_1, O_2, \ldots$ be countably many dense open subsets of X. Then $\cap_j O_j$ is dense in X.*

Note that the result of the second corollary follows from the first corollary by complementation. The set $\cap_j O_j$, while dense, need not be open.

Example 11.40: The metric space $\mathbb{R}$, equipped with the standard Euclidean metric, cannot be written as a countable union of nowhere dense sets. □

By contrast, $\mathbb{Q}$ *can* be written as the union of the singletons $\{q_j\}$ where the q_j represent an enumeration of the rationals. Each singleton is of course no-where dense since it is the limit of other rationals in the set. However, $\mathbb{Q}$ is not complete.

Example 11.41: Baire's theorem contains the fact that a perfect set of real numbers must be uncountable. For if P were perfect and countable we could write $P = \{p_1, p_2, \ldots\}$. Therefore

$$P = \bigcup_{j=1}^{\infty} \{p_j\}.$$

But each of the singletons $\{p_j\}$ is a nowhere dense set in the metric space P. And P is complete. (You should verify both these assertions for yourself.) This contradicts the Baire Category Theorem. So P cannot be countable. $\square$

A set that can be written as a countable union of nowhere dense sets is said to be of *first category*. If a set is not of first category, then it is said to be of *second category*. The Baire Category Theorem says that a complete metric space must be of second category. We should think of a set of first category as being "thin" and a set of second category as being "fat" or "robust." (This is one of many ways that we have in mathematics of distinguishing "fat" sets. Countability and uncountability is another. Lebesgue's measure theory is a third.)

One of the most striking applications of the Baire Category Theorem is the following result to the effect that "most" continuous functions are nowhere differentiable. This explodes the myth that most of us mistakenly derive from calculus class that a typical continuous function is differentiable at all points except perhaps at a discrete set of bad points.

Theorem 11.42: *Genericity of nowhere differentiable functions Let (X, ρ) be the metric space of continuous functions on the unit interval $[0, 1]$ equipped with the metric*

$$\rho(f, g) = \sup_{x \in [0,1]} |f(x) - g(x)|.$$

Define a subset of E of X as follows: $f \in E$ if there exists one point at which f is differentiable. Then E is of first category in the complete metric space (X, ρ).

Proof: For each pair of positive integers m, n we let

$$A_{m,n} = \{f \in X: \exists x \in [0, 1] \text{ such that } |f(x) - f(t)| \leq n|x - t|$$
$$\forall\ t \in [0, 1] \text{ that satisfy } |x - t| \leq 1/m\}.$$

Fix m and n. We claim that $A_{m,n}$ is nowhere dense in X. In fact, if $f \in A_{m,n}$ set

$$K_f = \max_{x \in [0,1]} \left| \frac{f(x \pm 1/m) - f(x)}{1/m} \right|.$$

Let $h(x)$ be a continuous piecewise linear function, bounded by 1, consisting of linear pieces having slope $3K_f$. Then for every $\varepsilon > 0$ it holds that $f + \varepsilon \cdot h$

has metric distance less than ε from f and is not a member of $A_{m,n}$. This proves that $A_{m,n}$ is nowhere dense.

We conclude from Baire's theorem that $\cup_{m,n} A_{m,n}$ is nowhere dense in X. Therefore $S = X \backslash \cup_{m,n} A_{m,n}$ is of second category. But if $f \in S$ then for every $x \in [0, 1]$ and every $n > 0$ there are points t arbitrarily close to x (that is, at distance $\leq 1/m$ from x) such that

$$\left| \frac{f(x) - f(t)}{t - x} \right| > n.$$

It follows that f is differentiable at no $x \in [0, 1]$. That proves the assertion. $\qquad \square$

Exercises

1. Let (X, ρ) be the collection of continuous functions on the interval $[0, 1]$ equipped with the usual supremum metric. For j a positive integer, let

$$E_j = \{p(x) : p \text{ is a polynomial of degree not exceeding } j\}.$$

Then, as noted in the text, each E_j is nowhere dense in X. Yet $\cup_j E_j$ *is* dense in X. Explain why these assertions do not contradict Baire's theorem.

*2. Assume f_j is a sequence of continuous, real-valued functions on $\mathbb{R}$ with the property that $\{f_j(x)\}$ is unbounded whenever $x \in \mathbb{Q}$. Use the Category Theorem to prove that it cannot then be true that whenever t is irrational then the sequence $\{f_j(t)\}$ is bounded.

3. Even if we did not know the transcendental functions $\sin x$, $\cos x$, $\ln x$, e^x, etc. explicitly, the Baire Category Theorem demonstrates that transcendental functions must exist. Explain why this assertion is true.

4. Fix a positive integer k. Let $\{p(x)\}$ be a sequence of polynomial functions on the real line, each of degree not exceeding k. Assume that this sequence converges pointwise to a limit function f. Prove that f is a polynomial of degree not exceeding k.

5. Give an example of a perfect set with empty interior. Show that, in the reals, there is no perfect set with countable interior.

6. Show that the set of polynomials is of first category in the space of continuous functions on the interval $[0, 1]$.

7. Show that the rational numbers $\mathbb{Q}$ are of first category in the reals $\mathbb{R}$.

8. Is the set $\mathbb{Z}$ of integers of first category in the rationals $\mathbb{Q}$?

9. Show that any compact metric space is of second category.

10. Give two examples of sets of first category that are dense in the reals $\mathbb{R}$.

11.4 The Ascoli-Arzela Theorem

Let $\mathcal{F} = \{f_\alpha\}_{\alpha \in A}$ be a family, not necessarily countable, of functions on a metric space (X, ρ). We say that the family $\mathcal{F}$ is *equicontinuous* equicontinuous family on X if for every $\varepsilon > 0$ there is a $\delta > 0$ such that when $\rho(s, t) < \delta$ then $|f_\alpha(s) - f_\alpha(t)| < \varepsilon$. Notice that equicontinuity mandates not only uniform continuity of each f_α but also that the uniformity occur simultaneously, and at the same rate, for all the f_α.

Example 11.43: Let (X, ρ) be the unit interval $[0, 1]$ with the usual Euclidean metric. Let $\mathcal{F}$ consist of all functions f on X that satisfy the Lipschitz condition

$$|f(s) - f(t)| \leq 2 \cdot |s - t|$$

for all s, t. Then $\mathcal{F}$ is an equicontinuous family of functions. For if $\varepsilon > 0$ then we may take $\delta = \varepsilon/2$. Then if $|s - t| < \delta$ and $f \in \mathcal{F}$ we have

$$|f(s) - f(t)| \leq 2 \cdot |s - t| < 2 \cdot \delta = \varepsilon.$$

Observe, for instance, that the Mean Value Theorem tells us that $\sin x$, $\cos x$, $2x$, x^2 are elements of $\mathcal{F}$. $\quad\square$

If $\mathcal{F}$ is a family of functions on X, then we call $\mathcal{F}$ *equibounded* if there is a number $M > 0$ such that equibounded family

$$|f(x)| \leq M$$

for all $x \in X$ and all $f \in \mathcal{F}$. Thus we are not only mandating that each f be bounded, but also that the entire family be uniformly bounded. For example, the functions $f_j(x) = \sin jx$ on $[0, 1]$ form an equibounded family.

One of the cornerstones of classical analysis is the following result of Ascoli and Arzela: Ascoli-Arzela Theorem

Theorem 11.44: (The Ascoli-Arzela Theorem) *Let* (Y, σ) *be a metric space and assume that Y is compact. Let* $\mathcal{F}$ *be an equibounded, equicontinuous family of functions on Y. Then there is a sequence* $\{f_j\} \subseteq \mathcal{F}$ *that converges uniformly to a continuous function on Y.*

Before we prove this theorem, let us comment on it. Let (X, ρ) be the metric space consisting of the continuous functions on the unit interval $[0, 1]$ equipped with the usual supremum norm. Let $\mathcal{F}$ be an equicontinuous, equibounded family of functions on $[0, 1]$. Then the theorem says that $\mathcal{F}$ is a compact set in this metric space. For any infinite subset of $\mathcal{F}$ is guaranteed to have a convergent subsequence. As a result, we may interpret the Ascoli-Arzela theorem as identifying certain compact collections of continuous functions.

Proof of the Ascoli-Arzela Theorem: We divide the proof into a sequence of lemmas.

Lemma 11.45: *Let* $\eta > 0$. *There exist finitely many points* $y_1, y_2, \ldots y_k \in Y$ *such that every ball* $B(s, \eta) \subseteq Y$ *contains one of the* y_j. *We call* $y_1, \ldots, y_k$ *an* η-net for Y.

Proof: Consider the collection of balls $\{B(y, \eta/2): y \in Y\}$. This is an open covering of Y hence, by compactness, has a finite subcovering $B(y_1, \eta/2), \ldots, B(y_k, \eta/2)$. The centers $y_1, \ldots, y_k$ are the points we seek. For if $B(s, \eta)$ is *any* ball in Y then its center s must be contained in some ball $B(y_j, \eta/2)$. But then $B(y_j, \eta/2) \subseteq B(s, \eta)$ hence, in particular, $y_j \in B(s, \eta)$. □

Lemma 11.46: *Let* $\varepsilon > 0$. *There is an* $\eta > 0$, *a corresponding* η-net $y_1, \ldots y_k$, *and a sequence* $\{f_m\} \subseteq \mathcal{F}$ *such that*

- The sequence $\{f_m(y_\ell)\}_{m=1}^{\infty}$ converges for each y_ℓ;
- For any $y \in Y$ the sequence $\{f_m(y)\}_{j=1}^{\infty}$ is contained in an interval in the real line of length at most ε.

Proof: By equicontinuity there is an $\eta > 0$ such that if $\rho(s, t) < \eta$ then $|f(s) - f(t)| < \varepsilon/3$ for every $f \in \mathcal{F}$. Let $y_1, \ldots, y_k$ be an η-net. Since the family $\mathcal{F}$ is equibounded, the set of numbers $\{f(y_1): f \in \mathcal{F}\}$ is bounded. Thus there is a subsequence f_j such that $\{f_j(y_1)\}$ converges. But then, by similar reasoning, we may choose a subsequence f_{j_k} such that $\{f_{j_k}(y_2)\}$ converges. Continuing in this fashion, we may find a sequence, which we call $\{f_m\}$, which converges at each point y_ℓ. The first assertion is proved. Discarding finitely many of the f_ms, we may suppose that for every m, n and every j it holds that $|f_m(y_j) - f_n(y_j)| < \varepsilon/3$.

Now if y is *any* point of Y then there is an element y_t of the η-net such that $\rho(y, y_t) < \eta$. But then, for any m, n, we have

$$|f_m(y) - f_n(y)| \leq |f_m(y) - f_m(y_t)|$$
$$+ \ |f_m(y_t) - f_n(y_t)|$$
$$+ \ |f_n(y_t) - f_n(y)|$$
$$< \frac{\varepsilon}{3} + \frac{\varepsilon}{3} + \frac{\varepsilon}{3}$$
$$= \varepsilon.$$

That proves the second assertion.   □

Proof of the Theorem: With $\varepsilon = 2^{-1}$ apply Lemma 11.46 to obtain a sequence f_m. Apply Lemma 11.46 again, with $\varepsilon = 2^{-2}$ and the role of $\mathcal{F}$ being played by the sequence $\{f_m\}$. This yields a new sequence $\{f_{m_r}\}$. Apply Lemma 11.46 once again with $\varepsilon = 2^{-3}$ and the role of $\mathcal{F}$ being played by the second sequence $\{f_{m_r}\}$. Keep going to produce a countable list of sequences.

Now produce the final sequence by selecting the first element of the first sequence, the second element of the second sequence, the third element of the third sequence, and so forth.[1] This sequence, which we call $\{f_w\}$, will satisfy the conclusion of the theorem.

For, if $\varepsilon > 0$, then there is a j such that $2^{-j} < \varepsilon$. After j terms, the sequence $\{f_w\}$ is a subsequence of the jth sequence constructed above. Hence, at every $y \in Y$, all the terms $f_w(y)$, $w > j$, lie in an interval of length ε. But that just verifies convergence at the point y. Note moreover that the choice of j in this last argument was independent of $y \in Y$. That shows that the convergence is uniform. The proof is complete.   □

Exercises

*1. Consider a collection $\mathcal{F}$ of differentiable functions on the interval $[a, b]$ that satisfy the conditions $|f(x)| \leq K$ and $|f'(x)| \leq C$ for all $x \in [a, b]$. Demonstrate that the Ascoli-Arzela theorem applies to $\mathcal{F}$ and describe the resulting conclusion.

2. A function on the interval $[0, 1]$ is Lipschitz if it satisfies the condition

$$|f(s) - f(t)| \leq C|s - t|$$

for some positive constant C. Use the Ascoli-Arzela theorem to show that the set of Lipschitz functions with constant C less than or equal to 1 and uniform bound less than or equal to 1 is a compact subset of the continuous functions on $[0, 1]$.

3. Explain in detail why the Ascoli-Arzela theorem is a compactness theorem.

4. Why is there no Ascoli-Arzela theorem (without any additional hypotheses) for the continuous functions on a compact interval?

5. A version of the Rellich lemma says that if $\beta > \alpha$ then the Lipschitz space of order β is compact in the Lipschitz space of order α. Explain exactly what this means, and why it is true.

6. Let X be a finite set and Y a finite set and let $\mathcal{F}$ be the set of functions from X to Y. Then, no matter what topology we put on X and Y, $\mathcal{F}$ will be compact. Why is that so?

*7. Give an example of a space that is compact inside the space of integrable functions on the unit interval.

8. Is the space of twice continuously differentiable functions compact inside the space of once continuously differentiable functions? Why or why not?

9. On the domain the unit interval $[0, 1]$, consider the set S of all polynomials of degree not exceeding 10 with coefficients of absolute value not larger than 1. Show that the Ascoli-Arzela theorem applies to S.

10. On the domain the interval $[0, 2\pi]$, consider the set T of all trigonometric polynomials of degree not exceeding 50 with coefficients of absolute value not larger than 5. Show that the Ascoli-Arzela theorem applies to T.

Note

1 This very standard type of construction is called a "diagonalization argument."

12

Applications of Analysis to Differential Equations

Differential equations are the heart and soul of analysis. Virtuallydifferential equations any law of physics or engineering or biology or chemistry can be expressed as a differential equation—and frequently as a first-order equation (i.e., an equation involving only first derivatives). Much of mathematical analysis has been developed in order to find techniques for solving differential equations.

Most introductory books on differential equations ([COL], [KRA7], and [BIR] are three examples) devote themselves to elementary techniques for finding solutions to a very limited selection of equations. In the present book we take a different point of view. We instead explore certain central and broadly applicable principles which apply to virtually any differential equation. These principles, in particular, illustrate some of the key ideas of the book.

12.1 Picard's Existence and Uniqueness Theorem
Picard's Theorem

12.1.1 The Form of a Differential Equation

A fairly general first-order differential equation will have the formdifferential equations, first order

$$\frac{dy}{dx} = F(x, y). \tag{12.1.1}$$

We say that the equation is "first order" because the highest derivative that appears is the first derivative.

In equation (12.1.1), F is a continuously differentiable function on some domain $(a, b) \times (c, d)$. We think of y as the dependent variable (that is, the function that we seek) and x as the independent variable. That is to say, $y = y(x)$. For technical reasons, we assume that the function F is bounded,

$$|F(x, y)| \le M, \tag{12.1.2}$$

DOI: 10.1201/9781003222682-13

and in addition that F satisfies a *Lipschitz condition*:

$$|F(x, s) - F(x, t)| \leq C \cdot |s - t|. \tag{12.1.3}$$

[In many treatments it is standard to assume that F is bounded and $\partial F / \partial y$ is bounded. It is easy to see, using the Mean Value Theorem, that these two conditions imply (12.1.2), (12.1.3).]

Example 12.1: Consider the equation

$$\frac{dy}{dx} = x^2 \sin y - y \ln x.$$

Then this equation fits the paradigm of equation (12.1.1) with $F(x, y) = x^2 \sin y - y \ln x$ provided that $1 \leq x \leq 2$ and $0 \leq y \leq 3$ (for instance). □

In fact the most standard, and physically appealing, setup for a first-order equation such as (12.1.1) is to adjoin to it an *initial condition*. For usinitial condition this condition will have the form

$$y(x_0) = y_0. \tag{12.1.4}$$

Thus the problem we wish to solve is (12.1.1) and (12.1.4) together.

Picard's idea is to set up an iterative scheme for doing so. The most remarkable fact about Picard's technique is that it always works: As long as F is bounded and satisfies the Lipschitz condition, then the problem will possess one and only one solution.

12.1.2 Picard's Iteration Technique

While we will not actually give a complete proof that Picard's technique works, we will set it up and indicate the sequence of functions it produces that converges uniformly to the solution of our problem.

Picard's approach is inspired by the fact that the differential equation (12.1.1) and initial condition (12.1.4), taken together, are equivalent to the single integral equation

$$y(x) = y_0 + \int_{x_0}^{x} F(t, y(t)) dt. \tag{12.1.5}$$

We invite the reader to differentiate both sides of thisintegral equation equation, using the Fundamental Theorem of Calculus, to derive the original differential equation (12.1.1). Of course the initial condition (12.1.4) is built into (12.1.5). This integral equation inspires the iteration scheme that we now describe.

We assume that $x_0 \in (a, b)$ and that $y_0 \in (c, d)$. We set

$$y_1(x) = y_0 + \int_{x_0}^{x} F(t, y_0)dt.$$

For x near to x_0, this definition makes sense.

Next we define

$$y_2(x) = y_0 + \int_{x_0}^{x} F(t, y_1(t))dt$$

and, more generally,

$$y_{j+1}(x) = y_0 + \int_{x_0}^{x} F(t, y_j(t))dt \qquad (12.1.6)$$

for $j = 2, 3, \dots$.

It turns out that the sequence of functions $\{y_1, y_2, \dots\}$ will converge uniformly on an interval of the form $[x_0 - h, x_0 + h] \subseteq (a, b)$ to a solution of (12.1.1) that satisfies (12.1.4).

12.1.3 Some Illustrative Examples

Picard's iteration method is best apprehended by way of some examples that show how the iterates arise and how they converge to a solution. We now proceed to develop such illustrations.

Example 12.2: Consider the initial value problem

$$y' = 2y, \quad y(0) = 1.$$

Of course this could easily be solved by the method of first order linear equations, or by separation of variables (see [KRA7] for a description of these methods). Our purpose here is to illustrate how the Picard method works.

First notice that the stated initial value problem is equivalent to the integral equation

$$y(x) = 1 + \int_{0}^{x} 2y(t)dt.$$

Following the paradigm (12.1.6), we thus find that

$$y_{j+1}(x) = 1 + \int_0^x 2y_j(x)\,dx.$$

Using $x_0 = 0$, $y_0 = 1$, we then find that

$$y_1(x) = 1 + \int_0^x 2\,dt = 1 + 2x,$$

$$y_2(x) = 1 + \int_0^x 2(1 + 2t)\,dt = 1 + 2x + 2x^2,$$

$$y_3(x) = 1 + \int_0^x 2(1 + 2t + 2t^2)\,dt = 1 + 2x + 2x^2 + \frac{4x^3}{3}.$$

In general, we find that

$$y_j(x) = 1 + \frac{2x}{1!} + \frac{(2x)^2}{2!} + \frac{(2x)^3}{3!} + \cdots + \frac{(2x)^j}{j!} = \sum_{\ell=0}^{j} \frac{(2x)^\ell}{\ell!}.$$

It is plain that these are the partial sums for the power series expansion of $y = e^{2x}$. We conclude that the solution of our initial value problem is $y = e^{2x}$. You are encouraged to check that $y = e^{2x}$ does indeed solve the differential equation and initial condition stated at the beginning of the example. □

Example 12.3: Let us use Picard's method to solve the initial value problem

$$y' = 2x - y, \quad y(0) = 1.$$

The equivalent integral equation is

$$y(x) = 1 + \int_0^x [2t - y(t)]\,dt$$

and (12.1.6) tells us that

$$y_{j+1}(x) = 1 + \int_0^x [2t - y_j(t)]\,dt.$$

Taking $x_0 = 0$, $y_0 = 1$, we then find that

$$y_1(x) = 1 + \int_0^x (2t - 1)dt = 1 + x^2 - x,$$

$$y_2(x) = 1 + \int_0^x (2t - [1 + t^2 - t])dt$$

$$= 1 + \frac{3x^2}{2} - x - \frac{x^3}{3},$$

$$y_3(x) = 1 + \int_0^x (2t - [1 + 3t^2/2 - t - t^3/3])dt$$

$$= 1 + \frac{3x^2}{2} - x - \frac{x^3}{2} + \frac{x^4}{4 \cdot 3},$$

$$y_4(x) = 1 + \int_0^x (2t - [1 + 3t^2/2 - t - t^3/2 + t^4/4 \cdot 3])dt$$

$$= 1 + \frac{3x^2}{2} - x - \frac{x^3}{2} + \frac{x^4}{4 \cdot 2} - \frac{x^5}{5 \cdot 4 \cdot 3}.$$

In general, we find that

$$y_j(x) = 1 - x + \frac{3x^2}{2!} - \frac{3x^3}{3!} + \frac{3x^4}{4!} - + \cdots$$

$$+ (-1)^j \frac{3x^j}{j!} + (-1)^{j+1} \frac{2x^{j+1}}{(j+1)!}$$

$$= [2x - 2] + 3 \cdot \sum_{\ell=0}^j \frac{(-x)^j}{j!} + (-1)^{j+1} \frac{2x^{j+1}}{(j+1)!}.$$

Notice that the $2x - 2$ terms cancel with the first two terms of the infinite sum to give $1 - x$.

Of course the last term tends to 0 as $j \to +\infty$. Thus we see that the iterates $y_j(x)$ converge to the solution $y(x) = [2x - 2] + 3e^{-x}$ for the initial value problem. Check that this function does indeed satisfy the given differential equation and initial condition. □

12.1.4 Estimation of the Picard Iterates

To get an idea of why the assertion at the end of Subsection 12.1.2—that the functions y_j converge uniformly—is true, let us do some elementary estimations. Choose $h > 0$ so small that $h \cdot C < 1$, where C is the constant from the Lipschitz condition (12.1.3). We will assume in the following calculations that $|x - x_0| < h$.

Now we proceed with the iteration. Let y_0 be the initial value at x_0 as usual. Then

$$|y_0 - y_1(t)| = |\int_{x_0}^{x} F(t, y_0)dt|$$

$$\leq \int_{x_0}^{x} |F(t, y_0)|\, dt$$

$$\leq M \cdot |x - x_0|$$

$$\leq M \cdot h.$$

We have of course used the boundedness condition (12.1.2).
  Next we have

$$|y_1(x) - y_2(x)| = |\int_{x_0}^{x} F(t, y_0(t))dt - \int_{x_0}^{x} F(t, y_1(t))dt|$$

$$\leq \int_{x_0}^{x} |F(t, y_0(t)) - F(t, y_1(t))|\, dt$$

$$\leq \int_{x_0}^{x} C \cdot |y_0(t) - y_1(t)|\, dt$$

$$\leq C \cdot M \cdot h \cdot h$$

$$= M \cdot C \cdot h^2.$$

One can continue this procedure to find that

$$|y_2(x) - y_3(x)| \leq M \cdot C^2 \cdot h^3 = M \cdot h \cdot (Ch)^2.$$

and, more generally,

$$|y_j(x) - y_{j+1}(x)| \leq M \cdot C^j \cdot h^{j+1} = M \cdot h \cdot (Ch)^j.$$

Now, if $0 < K < L$ are integers, then

$$|y_K(x) - y_L(x)| \leq |y_K(x) - y_{K+1}(x)| + |y_{K+1}(x) - y_{K+2}(x)|$$

$$+ \cdots + |y_{L-1}(x) - y_L(x)|$$

$$\leq M \cdot h \cdot ([Ch]^K + [Ch]^{K+1} + \cdots [Ch]^{L-1}).$$

Since $|Ch| < 1$ by design, the geometric series $\sum_j [Ch]^j$ converges. As a result, the expression on the right of our last display is as small as we please, for K and L large, just by the Cauchy criterion for convergent series. It follows that the sequence $\{y_j\}$ of approximate solutions converges uniformly to a function $y = y(x)$. In particular, y is continuous.

Furthermore, we know that

$$y_{j+1}(x) = y_0 + \int_{x_0}^{x} F(t, y_j(t))dt.$$

Letting $j \to \infty$, and invoking the uniform convergence of the y_j, we may pass to the limit and find that

$$y(x) = y_0 + \int_{x_0}^{x} F(t, y(x))dt.$$

This says that y satisfies the integral equation that is equivalent to our original initial value problem. This equation also shows that y is continuously differentiable. Thus y is the function that we seek.

It can be shown that this y is in fact the *unique* solution to our initial value problem. We shall not provide the details of the proof of that assertion.

In case F is not Lipschitz—say that F is only continuous—then it is still possible to show that a solution y exists. But it will no longer be unique.

****

CHARLES ÉMILE PICARD

Charles Émile Picard (1856–1941) was a French mathematician. He was elected the fifteenth member to occupy seat 1 of the Académie française in 1924.

He was born in Paris on 24 July 1856 and educated there at the Lycée Henri IV. He then studied Mathematics at the École Normale Supérieure.

Picard's mathematical papers, textbooks, and many popular writings exhibit an extraordinary range of interests, as well as an impressive mastery of the mathematics of his time. Picard's Little Theorem and Great Theorem are important cornerstones in the study of singularities of a holomorphic function. Picard made important contributions in the theory of differential equations, including work on Picard–Vessiot theory, Painlevé transcendents and his introduction of a kind of symmetry group for a linear differential equation. He also introduced the Picard group in the theory of algebraic surfaces, which describes the classes of algebraic curves on the surface modulo linear equivalence. In connection with his work on function theory, he was one of the first mathematicians to use the emerging ideas of algebraic topology. Picard studied elasticity and telegraphy and made numerous significant contributions to applied mathematics. His collected papers run to four volumes.

Louis Couturat studied integral calculus with Picard in 1891 and 1892, taking detailed notes of the lectures. These notes were preserved and now are available in three volumes from the Internet Archive.

Like his contemporary, Henri Poincaré, Picard was much concerned with the training of mathematics, physics, and engineering students. He dedicated himself to teaching and wrote some significant textbooks. He wrote a classic textbook on analysis and one of the first textbooks on the theory of relativity. People were skeptical of relativity theory for many years, so Picard's text was an important early step. Picard's popular writings include biographies of many leading French mathematicians, including his father in law, Charles Hermite. Hermite made important contributions to analysis and differential equations.

In 1881 he married Marie, the daughter of Charles Hermite.

****

Exercises

1. Use the method of Picard iteration to solve the initial value problem $y' = y + x$, $y(0) = 1$.

2. Verify that the function $y = 1/\sqrt{2(x + 1)}$ is a solution of the differential equation

$$y' + y^3 = 0 . \tag{*}$$

Can you use separation of variables to find the general solution? This means to write the equation as

$$\frac{dy}{dx} = -y^3$$

and then do some algebra to have all x terms on one side of the equation and all y terms on the other side of the equation. Then integrate. [**Hint:** It is $y = 1/\sqrt{2(x + c)}$.] Now find the solution to the initial value problem (*) with initial condition $y(1) = 4$.

3. Check that the function

$$y = \sqrt{\frac{2}{3} \ln(1 + x^2) + C}$$

solves the differential equation

$$\frac{dy}{dx} = \frac{2x}{3y + 3yx^2} .$$

Find the particular solution that satisfies the initial condition $y(0) = 2$.

4. In the method of Picard, suppose that the function F is given by a power series. Formulate a version of the Picard iteration technique in the language of power series.

5. Explain why the initial value problem

$$y' = e^y$$
$$y(0) = 1$$

has a solution in a neighborhood of the origin.

6. For each differential equation, sketch the family of solutions on a set of axes. This means, since each equation is *not* equipped with an initial condition, that the solution to each equation will have an unspecified constant in it.

   a. $y' - xy = 0$

   b. $y' + y = e^x$

   c. $y' = x$

   d. $y' = 1 - y$

*7. Formulate a version of the Picard theorem for vector-valued functions. Indicate how its proof differs, if at all, from the proof for scalar-valued functions. Now explain how one can use this vector-valued version of Picard to obtain an existence and uniqueness theorem for k th-order ordinary differential equations.

*8. Does the Picard theorem apply to the initial value problem

$$e^{dy/dx} + \frac{dy}{dx} = x^2, \quad y(1) = 2?$$

Why or why not? [**Hint:** Think in terms of the Implicit Function Theorem.]

*9. A *vector field* is a function

$$F(x, y) = \langle \alpha(x, y), \beta(x, y) \rangle$$

that assigns to each point in the plane $\mathbb{R}^2$ a vector. We call a curve $\gamma : (a, b) \to \mathbb{R}^2$ (here $\gamma(t) = (\gamma_1(t), \gamma_2(t))$) an *integral curve* of the vector field if

$$\gamma'(t) = F(\gamma(t))$$

for each t. Thus γ "flows along" the vector field, and the tangent to the curve at each point is given by the value of the vector field at that point.

Put suitable conditions on F that will guarantee that if $P \in \mathbb{R}^2$ then there will be an integral curve for F through the point P. [**Hint:** Of course use the Picard theorem to obtain your result. What is the correct initial value problem?]

*10. Give an example which illustrates that the integral curve that you found in Exercise **9** will only, in general, be defined in a small neighborhood of P. [**Hint:** Think of a vector field that "dies out."]

*11. Refer to Exercises **9** and **10**. Find integral curves for each of the following vector fields:

   a. $F(x, y) = \langle -y, x \rangle$
   b. $F(x, y) = \langle x + 1, y - 2 \rangle$
   c. $F(x, y) = \langle 2xy, x^2 \rangle$
   d. $F(x, y) = \langle -x, 2y \rangle$

*12. Solve the differential equation $y'' - xy' + y = 0$ by using a change of variable to reduce it to a first-order equation and then applying Picard's theorem.

13. Apply the first three iterations of Picard's technique to the differential equation

$$y' = y^2 + y, \quad y(0) = 1 .$$

14. Apply the first three iterations of Picard's technique to the differential equation

$$y' = xy^2, \quad y(1) = 2.$$

12.2 Power Series Methods

One of the techniques of broadest applicability in the subject of differential equations is that of power series, or real analytic functions. The philosophy is to *guess* that a given problem has a solution that may be represented by a power series, and then to endeavor to solve for the coefficients of that series. Along the way, one uses (at least tacitly) fundamental properties of these series—that they may be differentiated and integrated term by term, for instance. And that their intervals of convergence are preserved under standard arithmetic operations.

Example 12.4: Let p be an arbitrary real constant. Let us use a differential equation to derive the power series expansion for the function

$$y = (1 + x)^p.$$

Of course the given y is a solution of the initial value problem

$$(1 + x) \cdot y' = py, \quad y(0) = 1.$$

We assume that the equation has a power series solution

$$y = \sum_{j=0}^{\infty} a_j x^j = a_0 + a_1 x + a_2 x^2 + \cdots$$

with positive radius of convergence R. Then

$$y' = \sum_{j=1}^{\infty} j \cdot a_j x^{j-1} = a_1 + 2a_2 x + 3a_3 x^2 + \cdots ;$$

$$xy' = \sum_{j=1}^{\infty} j \cdot a_j x^j = a_1 x + 2a_2 x^2 + 3a_3 x^3 + \cdots ;$$

$$py = \sum_{j=0}^{\infty} pa_j x^j = pa_0 + pa_1 x + pa_2 x^2 + \cdots .$$

By the differential equation, we see that the sum of the first two of these series equals the third. Thus

$$\sum_{j=1}^{\infty} j a_j x^{j-1} + \sum_{j=1}^{\infty} j a_j x^j = \sum_{j=0}^{\infty} pa_j x^j.$$

We immediately see two interesting anomalies: the powers of x on the left-hand side do not match up, so the two series cannot be immediately added. Also the summations do not all begin in the same place. We address these two concerns as follows.

First, we can change the index of summation in the first sum on the left to obtain

$$\sum_{j=0}^{\infty} (j + 1)a_{j+1} x^j + \sum_{j=1}^{\infty} j a_j x^j = \sum_{j=0}^{\infty} pa_j x^j.$$

Write out the first few terms of the new sum, and the original sum, to see that they are just the same.

Now every one of our series has x^j in it, but they begin at different places. So we break off the extra terms as follows:

$$\sum_{j=1}^{\infty} (j + 1)a_{j+1}x^j + \sum_{j=1}^{\infty} ja_j x^j - \sum_{j=1}^{\infty} pa_j x^j = -a_1 x^0 + pa_0 x^0. \qquad (12.4.1)$$

Notice that all we have done is to break off the zeroeth terms of the first and third series, and put them on the right.

The three series on the left-hand side of (12.4.1) are begging to be put together: they have the same form, they all involve powers of x, and they all begin at the same index. Let us do so:

$$\sum_{j=1}^{\infty} [(j + 1)a_{j+1} + ja_j - pa_j]x^j = -a_1 + pa_0.$$

Now the powers of x that appear on the left are 1, 2, ..., and there are none of these on the right. We conclude that each of the coefficients on the left is zero; by the same reasoning, the coefficient $(-a_1 + pa_0)$ on the right (i.e., the constant term) equals zero. So we have the equations[1]

$$- a_1 + pa_0 = 0$$
$$(j + 1)a_{j+1} + (j - p)a_j = 0 \quad \text{for } j \geq 1.$$

Our initial condition tells us that $a_0 = 1$. Then our first equation implies that $a_1 = p$. The next equation, with $j = 1$, says that

$$2a_2 + (1 - p)a_1 = 0.$$

Hence $a_2 = (p - 1)a_1/2 = (p - 1)p/2$. Continuing, we take $j = 2$ in the second equation to get

$$3a_3 + (2 - p)a_2 = 0$$

so $a_3 = (p - 2)a_2/3 = (p - 2)(p - 1)p/(3 \cdot 2)$.
We may continue in this manner to obtain that

$$a_j = \frac{p(p - 1)(p - 2) \cdots (p - j + 1)}{j!} \quad \text{for } j \geq 1.$$

Thus the power series expansion for our solution y is

$$y = 1 + px + \frac{p(p - 1)}{2!}x^2 + \frac{p(p - 1)(p - 2)}{3!}x^3 + \cdots$$
$$+ \frac{p(p - 1)(p - 2) \cdots (p - j + 1)}{j!}x^j + \cdots .$$

Since we knew in advance that the solution of our initial value problem was

$$y = (1 + x)^p,$$

we find that we have derived Isaac Newton's general binomial theorem (or binomial series):

$$(1 + x)^p = 1 + px + \frac{p(p-1)}{2!}x^2 + \frac{p(p-1)(p-2)}{3!}x^3 + \cdots$$
$$+ \frac{p(p-1)(p-2)\cdots(p-j+1)}{j!}x^j + \cdots .$$

□

Example 12.5: Let us consider the differential equation

$$y' = y.$$

Of course we know from elementary considerations that the solution to this equation is $y = C \cdot e^x$, but let us pretend that we do not know this. Our goal is to instead use power series to *discover* the solution. We proceed by *guessing* that the equation has a solution given by a power series, and we proceed to solve for the coefficients of that power series.

So our guess is a solution of the form

$$y = a_0 + a_1 x + a_2 x^2 + a_3 x^3 + \cdots .$$

Then

$$y' = a_1 + 2a_2 x + 3a_3 x^2 + \cdots ,$$

and we may substitute these two expressions into the differential equation. Thus

$$a_1 + 2a_2 x + 3a_3 x^2 + \cdots = a_0 + a_1 x + a_2 x^2 + \cdots .$$

Now the powers of x must match up (i.e., the coefficients must be equal). We conclude that

$$a_1 = a_0$$
$$2a_2 = a_1$$
$$3a_3 = a_2$$

and so forth. Let us take a_0 to be an unknown constant C. Then we see that

$$a_1 = C;$$
$$a_2 = \frac{C}{2};$$
$$a_3 = \frac{C}{3 \cdot 2};$$
$$\text{etc.}$$

In general,

$$a_j = \frac{C}{j!}.$$

In summary, our power series solution of the original differential equation is

$$y = \sum_{j=0}^{\infty} \frac{C}{j!} x^j = C \cdot \sum_{j=0}^{\infty} \frac{x^j}{j!} = C \cdot e^x.$$

Thus we have a new way, using power series, of discovering the general solution of the differential equation $y' = y$.   □

Example 12.6: Let us use the method of power series to solve the differential equation

$$(1 - x^2)y'' - 2xy'' + p(p + 1)y = 0. \tag{12.6.1}$$

Here p is an arbitrary real constant. This is called *Legendre's equation*.
    We therefore guess a solution of the form

$$y = \sum_{j=0}^{\infty} a_j x^j = a_0 + a_1 x + a_2 x^2 + \cdots$$

and calculate

$$y' = \sum_{j=1}^{\infty} j a_j x^{j-1} = a_1 + 2a_2 x + 3a_3 x^2 + \cdots$$

and

$$y'' = \sum_{j=2}^{\infty} j(j - 1)a_j x^{j-2} = 2a_2 + 3 \cdot 2 \cdot a_3 x + \cdots .$$

It is most convenient to treat the differential equation in the form (12.6.1). We calculate

$$-x^2 y'' = -\sum_{j=2}^{\infty} j(j-1)a_j x^j$$

and

$$-2xy' = -\sum_{j=1}^{\infty} 2ja_j x^j.$$

Substituting into the differential equation now yields

$$\sum_{j=2}^{\infty} j(j-1)a_j x^{j-2} - \sum_{j=2}^{\infty} j(j-1)a_j x^j - \sum_{j=1}^{\infty} 2ja_j x^j + p(p+1)\sum_{j=0}^{\infty} a_j x^j = 0.$$

We adjust the index of summation in the first sum so that it contains x^j rather than x^{j-2} and we break off spare terms and collect them on the right. We also break off terms from the third and fourth power series and move them to the right. The result is

$$\sum_{j=2}^{\infty} (j+2)(j+1)a_{j+2} x^j - \sum_{j=2}^{\infty} j(j-1)a_j x^j$$

$$- \sum_{j=2}^{\infty} 2ja_j x^j + p(p+1)\sum_{j=2}^{\infty} a_j x^j$$

$$= -2a_2 - 6a_3 x + 2a_1 x - p(p+1)a_0 - p(p+1)a_1 x.$$

In other words,

$$\sum_{j=2}^{\infty} [(j+2)(j+1)a_{j+2} - j(j-1)a_j - 2ja_j + p(p+1)a_j]x^j$$

$$= -2a_2 - 6a_3 x + 2a_1 x - p(p+1)a_0 - p(p+1)a_1 x.$$

As a result,

$$[(j+2)(j+1)a_{j+2} - j(j-1)a_j - 2ja_j + p(p+1)a_j] = 0 \quad \text{for } j = 2, 3, \ldots$$

together with

$$-2a_2 - p(p+1)a_0 = 0$$

and

$$-6a_3 + 2a_1 - p(p+1)a_1 = 0.$$

We have arrived at the recursion

$$a_2 = -\frac{p(p+1)}{1\cdot 2}\cdot a_0,$$

$$a_3 = -\frac{(p-1)(p+2)}{2\cdot 3}\cdot a_1, \qquad\qquad (12.6.2)$$

$$a_{j+2} = -\frac{(p-j)(p+j+1)}{(j+2)(j+1)}\cdot a_j \quad \text{for } j = 2, 3, \ldots.$$

We recognize a familiar pattern: The coefficients a_0 and a_1 are unspecified, so we set $a_0 = A$ and $a_1 = B$. Then we may proceed to solve for the rest of the coefficients. Now

$$a_2 = -\frac{p(p+1)}{2}\cdot A,$$

$$a_3 = -\frac{(p-1)(p+2)}{2\cdot 3}\cdot B,$$

$$a_4 = -\frac{(p-2)(p+3)}{3\cdot 4}a_2 = \frac{p(p-2)(p+1)(p+3)}{4!}\cdot A,$$

$$a_5 = -\frac{(p-3)(p+4)}{4\cdot 5}a_3$$

$$= \frac{(p-1)(p-3)(p+2)(p+4)}{5!}\cdot B,$$

$$a_6 = -\frac{(p-4)(p+5)}{5\cdot 6}a_4$$

$$= -\frac{p(p-2)(p-4)(p+1)(p+3)(p+5)}{6!}\cdot A,$$

$$a_7 = -\frac{(p-5)(p+6)}{6\cdot 7}a_5$$

$$= -\frac{(p-1)(p-3)(p-5)(p+2)(p+4)(p+6)}{7!}\cdot B,$$

and so forth. Putting these coefficient values into our supposed power series solution we find that the general solution of our differential equation is

$$
y = A\left[1 - \frac{p(p+1)}{2!}x^2 + \frac{p(p-2)(p+1)(p+3)}{4!}x^4\right.
$$

$$
\left. - \frac{p(p-2)(p-4)(p+1)(p+3)(p+5)}{6!}x^6 + -\cdots \right]
$$

$$
+ B\left[x - \frac{(p-1)(p+2)}{3!}x^3 + \frac{(p-1)(p-3)(p+2)(p+4)}{5!}x^5\right.
$$

$$
\left. - \frac{(p-1)(p-3)(p-5)(p+2)(p+4)(p+6)}{7!}x^7 + -\cdots \right].
$$

We assure the reader that, when p is not an integer, then these are *not* familiar elementary transcendental functions. They are what we call *Legendre functions*. In the special circumstance that p is a positive even integer, the first function (that which is multiplied by A) terminates as a polynomial. In the special circumstance that p is a positive odd integer, the second function (that which is multiplied by B) terminates as a polynomial. These are called *Legendre polynomials*, and they play an important role in mathematical physics, representation theory, and interpolation theory. □

Some differential equations have singularities. In the present context, this means that the higher order terms have coefficients that vanish to high degree. As a result, one must make a slightly more general guess as to the solution of the equation. This more general guess allows for a corresponding singularity to be built into the solution. Rather than develop the full theory of these Frobenius series, we merely give one example.

Example 12.7: We use the method of Frobenius series to solve the differential equation

$$
2x^2y'' + x(2x+1)y' - y = 0 \tag{12.7.1}
$$

about the regular singular point 0.
    We guess a solution of the form

$$
y = x^m \cdot \sum_{j=0}^{\infty} a_j x^j = \sum_{j=0}^{\infty} a_j x^{m+j}
$$

and therefore calculate that

$$y' = \sum_{j=0}^{\infty} (m + j)a_j x^{m+j-1}$$

and

$$y'' = \sum_{j=0}^{\infty} (m + j)(m + j - 1)a_j x^{m+j-2}.$$

Substituting these calculations into the differential equation yields

$$2 \sum_{j=0}^{\infty} (m + j)(m + j - 1)a_j x^{m+j}$$

$$+ 2 \sum_{j=0}^{\infty} (m + j)a_j x^{m+j+1}$$

$$+ \sum_{j=0}^{\infty} (m + j)a_j x^{m+j} - \sum_{j=0}^{\infty} a_j x^{m+j}$$

$$= 0.$$

We make the usual adjustments in the indices so that all powers of x are x^{m+j}, and break off the odd terms to put on the right-hand side of the equation. We obtain

$$2 \sum_{j=1}^{\infty} (m + j)(m + j - 1)a_j x^{m+j}$$

$$+ 2 \sum_{j=1}^{\infty} (m + j - 1)a_{j-1} x^{m+j}$$

$$+ \sum_{j=1}^{\infty} (m + j)a_j x^{m+j} - \sum_{j=1}^{\infty} a_j x^{m+j}$$

$$= -2m(m - 1)a_0 x^m - ma_0 x^m + a_0 x^m.$$

The result is

$$[2(m + j)(m + j - 1)a_j + 2(m + j - 1)a_{j-1}$$
$$+ (m + j)a_j - a_j] = 0 \qquad (12.7.2)$$
$$\text{for} \quad j = 1, 2, 3, \ldots$$

together with

$$[-2m(m - 1) - m + 1]a_0 = 0.$$

It is clearly not to our advantage to let $a_0 = 0$. Thus

$$- 2m(m-1) - m + 1 = 0.$$

This is the *indicial equation*.

The roots of this quadratic equation are $m = -1/2, 1$. We put each of these values into (12.7.2) and solve the resulting recursion.

Now (12.7.2) says that

$$(2m^2 + 2j^2 + 4mj - j - m - 1)a_j = (-2m - 2j + 2)a_{j-1}.$$

For $m = -1/2$ this is

$$a_j = \frac{3 - 2j}{-3j + 2j^2} a_{j-1}$$

so

$$a_1 = -a_0, \quad a_2 = -\frac{1}{2}a_1 = \frac{1}{2}a_0, \quad \text{etc.}$$

For $m = 1$ we have

$$a_j = \frac{-2j}{3j + 2j^2} a_{j-1}$$

so

$$a_1 = -\frac{2}{5}a_0, \quad a_2 = -\frac{4}{14}a_1 = \frac{4}{35}a_0, \quad \text{etc.}$$

Thus we have found the linearly independent solutions

$$a_0 x^{-1/2} \cdot \left(1 - x + \frac{1}{2}x^2 - + \cdots \right)$$

and

$$a_0 x \cdot \left(1 - \frac{2}{5}x + \frac{4}{35}x^2 - + \cdots \right).$$

The general solution of our differential equation is then

$$y = Ax^{-1/2} \cdot \left(1 - x + \frac{1}{2}x^2 - + \cdots \right) + Bx \cdot \left(1 - \frac{2}{5}x + \frac{4}{35}x^2 - + \cdots \right).$$

□

Exercises

1. Explain why the method of power series would not work very well to solve the differential equation

$$y' - |x|y = \sin x.$$

Note here that the coefficient of y is $|x|$, and $|x|$ is not a differentiable function.

2. Solve the initial value problem

$$y'' - xy = x^2, \quad y(0) = 2, \quad y'(0) = 1$$

by the method of power series.

3. Solve the initial value problem

$$y' - xy = x, \quad y(0) = 2$$

by the method of power series.

4. Solve the differential equation

$$y''' - xy' = x$$

by the method of power series. Since there are no initial conditions, you should obtain a general solution with three free parameters.

5. Solve the initial value problem

$$y' - y = x, \quad y(0) = 1$$

both by Picard's method and by the method of power series. Verify that you get the same solution by both means.

6. When you solve a differential equation by the method of power series, you cannot in general expect the power series to converge on the entire real line. As an example, solve the differential equation

$$y' - \frac{1}{1 - x}y = 0$$

by the method of power series. What is the radius of convergence of the power series? Can you suggest why that is so?

7. Consider the differential equation

$$y'' - y = x^2 .$$

The function x^2 is even. If the function y is even, then y'' will be even also. Thus it makes sense to suppose that there is a power series solution with only even powers of x. Find it.

8. Consider the differential equation

$$y'' + y = x^3 .$$

The function x^3 is odd. If the function y is odd, then y'' will also be odd. Thus it makes sense to suppose that there is a power series solution with only odd powers of x. Find it.

9. Find all solutions of the differential equation

$$y' = xy .$$

10. Find all solutions of the differential equation

$$y' = \frac{y}{x} .$$

11. Use power series methods to solve the differential equation

$$y'' + 4y = 0 .$$

12. Solve the differential equation

$$y' = y^2 .$$

*13. What are sufficient conditions on the function F so that the differential equation

$$y' = F(x, y)$$

has the property that its solution y is continuously differentiable?

*14. Find a solution of the partial differential equation

$$\left(\frac{\partial^2}{\partial x^2} + \frac{\partial^2}{\partial y^2} \right) u(x, y) = x + y$$

using the method of power series in two variables.

15. The Cauchy-Kowalewski theorem states that any differential equation (ordinary or partial) with real analytic coefficients has a (local) real analytic solution. This is one of the only really general theorems in the theory of differential equations. Discuss what the Cauchy-Kowalewski theorem says about the differential equation

$$y'' - \frac{x}{x + 1} y = 0.$$

16. Let $\mathcal{L}$ define a differential equation

$$\mathcal{L}y = 0$$

which is a linear, constant coefficent ordinary differential equation. If y is a solution of this equation then each derivative $y^{(k)}$ a solution. What does this tell you about what y must be? What sort of power series defines a solution?

17. Give an example of a sequence of real analytic functions f_j which converge uniformly on the interval $[0, 1]$ to a non-real-analytic function.

Note

1 A set of equations like this is called a recursion. It expresses a_js with later indices in terms of a_js with earlier indices.

13

Introduction to Harmonic Analysis

13.1 The Idea of Harmonic Analysis

Fourier analysis first arose historically in the context of the study of a certain partial differential equation (we shall describe this equation in detail in the discussion below) of mathematical physics. The equation could be solved explicitly when the input (i.e., the right-hand side of the equation) was a function of the form $\sin jx$ or $\cos jx$ for j an integer. The question arose whether an *arbitrary* input could be realized as the superposition of sine functions and cosine functions.

In the late eighteenth century, debate raged over this question. It was fueled by the fact that there was no solid understanding of just what constituted a function. The important treatise [FOU] of Joseph Fourier gave a somewhat dreamy but nevertheless precise method for expanding virtually any function as a series in sines and cosines. It took almost a century, and the concerted efforts of Dirichlet, Cauchy, Riemann, Weierstrass, and many other important analysts, to put the so-called theory of "Fourier series" on a rigorous footing.

We now know, and can prove exactly, that if f is a continuously differentiable function on the interval $[0, 2\pi]$ then the coefficients

$$c_n = \frac{1}{2\pi} \int_0^{2\pi} f(t)e^{-int}dt$$

give rise to a series expansion

$$f(t) = \sum_{n=0}^{\infty} c_n e^{int}$$

that is valid (i.e., convergent) at every point, and converges back to f. [Notice that the convenient notation e^{ijt} given to us by Euler's formula carries information both about the sine and the cosine.] This expansion validates the vague but aggressive ruminations in [FOU] and lays the

DOI: 10.1201/9781003222682-14

foundations for a powerful and deep method of analysis that today has wide applicability in physics, engineering, differential equations, and harmonic analysis.[1]

Certainly harmonic analysis is one of the most vigorous and active areas of modern mathematics. New ideas are continually in development. One of the most exciting new directions in the subject is the theory of wavelets due to Yves Myer. What is remarkable about wavelet theory is that it is a "custom" harmonic analysis that allows one to design building blocks (which replace the traditional sines and cosines) that are adapted to a particular problem. The result is a theory with better convergence results, and that allows localization in both the space variable and the phase variable. See [WAL] for an accessible introduction to wavelet theory.

In the present chapter we shall explore the foundations of Fourier series and also learn some of their applications. All of our discussions will of course be rigorous and precise. They will take advantage of all the tools of analysis that we have developed thus far in the present book.

****

JEAN-BAPTISTE JOSEPH FOURIER

Jean-Baptiste Joseph Fourier (1768–1830) was a French mathematician and physicist born in Auxerre and best known for initiating the investigation of Fourier series, which eventually developed into Fourier analysis and harmonic analysis, and their applications to problems of heat transfer and vibrations. The Fourier transform and Fourier's law of conduction are also named in his honor.

Fourier was the son of a tailor. He was orphaned at the age of nine. Fourier was recommended to the Bishop of Auxerre and, through this introduction, he was educated by the Benedictine Order of the Convent of St. Mark. He took a prominent part in his own district in promoting the French Revolution, serving on the local Revolutionary Committee. He was imprisoned briefly during the Terror but, in 1795, was appointed to the École Normale and subsequently succeeded Joseph-Louis Lagrange at the École Polytechnique.

Fourier accompanied Napoleon Bonaparte on his Egyptian expedition in 1798, as scientific adviser, and was appointed secretary of the Institut d'Égypte. After the British victories and the capitulation of the French under General Menou in 1801, Fourier returned to France.

In 1801, Napoleon appointed Fourier Prefect (Governor) of the Department of Isére in Grenoble, where he oversaw road construction and other projects. It was while at Grenoble that he began to experiment on the propagation of heat. He presented his paper On the Propagation of Heat in Solid Bodies to the Paris Institute on December 21, 1807. He also contributed to the monumental Description de l'Égypte.

In 1822, Fourier succeeded Jean Baptiste Joseph Delambre as Permanent Secretary of the French Academy of Sciences. In 1830, he was elected a foreign member of the Royal Swedish Academy of Sciences.

In 1830, his diminished health began to take its toll: Fourier had already experienced, in Egypt and Grenoble, some attacks of aneurism of the heart. A fall, however, which he sustained on the 4th of May 1830, while descending a flight of stairs, aggravated the malady to an extent beyond what could have been ever feared.

Shortly after this event, he died in his bed on 16 May 1830.

Fourier was buried in the Pére Lachaise Cemetery in Paris, a tomb decorated with an Egyptian motif to reflect his position as secretary of the Cairo Institute, and his collation of Description de l'Égypte. Singer Edith Piaf, playwright Oscar Wilde, and rock star Jim Morrison are also buried there. Fourier's name is one of the 72 names inscribed on the Eiffel Tower.

A bronze statue was erected in Auxerre in 1849, but it was melted down for armaments during World War II. Joseph Fourier University in Grenoble is named after him.

****

Exercises

1. The function $f(\theta) = \cos^4 \theta$ is a nice smooth function, so will have a Fourier series expansion. That is, it will have an expansion as a sum of functions $\cos j\theta$ and $\sin j\theta$ with real coefficients. Determine what that expansion is.

*2. Explain why the only continuous multiplicative homomorphisms from the circle group $\mathbb{T}$, which is just the set of all $e^{i\theta}$ in the plane, into $\mathbb{C} \setminus \{0\}$ are given by

$$e^{i\theta} \mapsto e^{ik\theta}$$

for some integer k. Here a homomorphism φ in this context is a function φ that satisfies $\varphi(a \cdot b) = \varphi(a) \cdot \varphi(b)$.

*3. Answer Exercise 2 with the circle group replaced by the real line.

4. Classical harmonic analysis is done on a space with a group action—such as the circle group, or the line, or N-dimensional Euclidean space. Explain what this assertion means, and supply some detail.

*5. It can be proved, using elementary Fourier series (see Section 13.2), that

$$\sum_{j=1}^{\infty} \frac{1}{j^2} = \frac{\pi^2}{6}.$$

This fact was established by Leonhard Euler in 1735. It is a matter of great interest to find similar formulas for

$$\sum_{j=1}^{\infty} \frac{1}{j^k}$$

when $k = 3, 4, \ldots$. Apery has shown that, when $k = 3$, then the sum is irrational. This set of ideas has to do with the Riemann zeta function and the distribution of primes. Do some experiments on your computer to determine what this might mean.

6. Refer to Exercise 5. Use your symbol manipulation software to calculate the partial sums S_{100}, S_{1000}, and S_{10000} for the series

$$\sum_{j=1}^{\infty} \frac{1}{j^2}.$$

Compare your answers with the value of $\pi^2/6$.

7. It is counterintuitive that the function $f(x) = x$ on the interval $[0, 2\pi]$ can be uniformly approximated by trigonometric polynomials. But it is true. Write down a trigonometric polynomial that approximates f within distance $1/10$.

8. If the function $f(\theta)$ has Fourier series expansion

$$\sum_{j=-\infty}^{\infty} a_j e^{ij\theta},$$

then what can you say about the Fourier series of f^2? Can you write the first six terms?

*9. Every continuously differentiable function $f(\theta)$ has a convergent Fourier series expansion $\sum_{j=-\infty}^{\infty} a_j e^{ij\theta}$. And each of the $e^{ij\theta}$ has a convergent power series expansion. Yet f itself may not have a convergent power series expansion. Explain.

10. What is the Fourier series expansion of the function $f(\theta) = \cos 2\theta$?

13.2 The Elements of Fourier Series

In this section it will be convenient for us to work on the interval $[0, 2\pi]$. We will perform arithmetic operations on this interval *modulo* 2π: for example, $3\pi/2 + 3\pi/2$ is understood to equal π because we subtract from the answer the largest multiple of 2π that it exceeds. When we refer to a function f being continuous on $[0, 2\pi]$, we require that it be right continuous at 0, left continuous at 2π, and that $f(0) = f(2\pi)$. Similarly for continuous differentiability and so forth.

If f is a (either real- or complex-valued) Riemann integrable function on this interval and if $n \in \mathbb{Z}$, then we define

$$\hat{f}(n) = \frac{1}{2\pi} \int_0^{2\pi} f(t) e^{-int} dt.$$

We call $\hat{f}(n)$ the nth *Fourier coefficient* of f. The formal expression

$$Sf(x) \sim \sum_{n=-\infty}^{\infty} \hat{f}(n) e^{inx}$$

is called the *Fourier series* of the function f. Notice that we are *not* claiming that Sf converges, nor that it converges to f. Right now it is just a formal expression.

In circumstances where the Fourier series converges to the function f, some of which we shall discuss below, the series provides a decomposition of f into simple component functions. This type of analysis is of importance in the theory of differential equations, in signal and image processing, and in scattering theory. There is a rich theory of Fourier series which is of interest in its own right.

It is important that we say right away how we sum Fourier series. Define the Nth *partial sum* of the Fourier series of f to be

$$S_N f(x) = \sum_{j=-N}^{N} \hat{f}(j) e^{ijx}.$$

We say that the Fourier series Sf *converges* to f at x if $S_N f(x) \to f(x)$.

Observe that, in case f has the special form

$$f(x) = \sum_{j=-N}^{N} a_j e^{ijt}, \tag{13.2.1}$$

then we may calculate that

$$\frac{1}{2\pi}\int_0^{2\pi} f(t)e^{-int}dt = \frac{1}{2\pi}\sum_{j=-N}^{N} a_j \int_0^{2\pi} e^{i(j-n)t}dt$$

Now the integral equals 0 if $j \neq n$ (this is so because $\int_0^{2\pi} e^{ikt}dt = 0$ when k is a nonzero integer). And the term with $j = n$ gives rise to $a_n \cdot 1$. Thus we find that

$$a_n = \frac{1}{2\pi}\int_0^{2\pi} f(t)e^{-int}dt. \tag{13.2.2}$$

Since, in Exercise 5 of Section 9.3, we showed that functions of the form (13.2.1) are dense in the continuous functions, we might hope that a formula like (13.2.2) will give a method for calculating the coefficients of a trigonometric expansion in considerable generality. In any event, this calculation helps to justify (after the fact) our formula for $\hat{f}(n)$.

Example 13.1: Let $f(x) = x$. Then

$$a_n = \frac{1}{2\pi}\int_0^{2\pi} te^{-int}dt.$$

This is easily calculated to equal

$$a_n = -\frac{1}{in}.$$

Therefore the Fourier expansion of f is

$$\sum_{n=-\infty}^{\infty} \frac{-1}{in}e^{int}. \qquad\qquad \square$$

The other theory that you know for decomposing a function into simple components is the theory of Taylor series. However, in order for a function to have a Taylor series it must be infinitely differentiable. Even then, as we have learned, the Taylor series of a function usually does not converge, and if it does converge its limit may not be the original function—see Section 9.2. The Fourier series of f converges to the original function f under fairly mild hypotheses on f, and thus provides a useful tool in analysis.

The first result we shall prove about Fourier series gives a growth condition on the coefficients $\hat{f}(n)$:

Proposition 13.2: (Bessel's Inequality) *If f^2 is integrable then*

$$\sum_{n=-N}^{N} |\hat{f}_n|^2 \leq \int_0^{2\pi} |f(t)|^2 \, dt.$$

Proof: Recall that $\overline{e^{ijt}} = e^{-ijt}$ and $|a|^2 = a \cdot \bar{a}$ for $a \in \mathbb{C}$. We calculate

$$\frac{1}{2\pi} \int_0^{2\pi} |f(t) - S_N f(t)|^2 \, dt$$

$$= \frac{1}{2\pi} \int_0^{2\pi} \left(f(t) - \sum_{n=-N}^{N} \hat{f}(n) e^{int} \right) \cdot \overline{\left(f(t) - \sum_{n=-N}^{N} \hat{f}(n) e^{int} \right)} dt$$

$$= \frac{1}{2\pi} \int_0^{2\pi} |f(t)|^2 \, dt - \sum_{n=-N}^{N} \frac{1}{2\pi} \int_0^{2\pi} f(t) e^{-int} dt \cdot \overline{\hat{f}(n)}$$

$$- \sum_{n=-N}^{N} \frac{1}{2\pi} \int_0^{2\pi} \overline{f(t) e^{-int} dt} \cdot \hat{f}(n)$$

$$+ \sum_{m,n} \hat{f}(m) \overline{\hat{f}}(n) \frac{1}{2\pi} \int_0^{2\pi} e^{imt} \cdot e^{-int} dt.$$

Now each of the first two sums equals $\sum_{n=-N}^{N} |\hat{f}(n)|^2$. In the last sum, any summand with $m \neq n$ equals 0. The summands with $m = n$ equal $|\hat{f}(n)|^2$. Thus our equation simplifies to

$$\frac{1}{2\pi} \int_0^{2\pi} \left|f(t) - S_N f(t)\right|^2 \, dt = \frac{1}{2\pi} \int_0^{2\pi} \left|f(t)\right|^2 \, dt - \sum_{n=-N}^{N} |\hat{f}(n)|^2.$$

Since the left side is nonnegative, it follows that

$$\sum_{n=-N}^{N} |\hat{f}(n)|^2 \leq \frac{1}{2\pi} \int_0^{2\pi} |f(t)|^2 \, dt,$$

as desired. □

Corollary 13.3: *If f^2 is integrable then the Fourier coefficients $\hat{f}(n)$ satisfy*

$$\hat{f}(n) \to 0 \quad \text{as} \quad n \to \infty.$$

Proof: Since $\sum |\hat{f}(n)|^2 < \infty$ we know that $|\hat{f}(n)|^2 \to 0$. This implies the result. □

Remark 13.4: In fact, with a little extra effort, one can show that the conclusion of the corollary holds if only f is integrable. This entire matter is addressed from a slightly different point of view in Proposition 13.16 below. □

Since the coefficients of the Fourier series, at least for a square integrable function, tend to zero, we might hope that the Fourier series will converge in some sense. Of course the best circumstance would be that $S_N f \to f$ (pointwise, or in some other manner). We now turn our attention to this problem.

Proposition 13.5: (The Dirichlet Kernel) *If f is integrable then*

$$S_N f(x) = \frac{1}{2\pi} \int_0^{2\pi} D_N(x - t) f(t) dt,$$

where

$$D_N(t) = \frac{\sin\left(N + \frac{1}{2}\right)t}{\sin \frac{1}{2}t}.$$

Proof: Observe that

$$\begin{aligned}
S_N f(x) &= \sum_{n=-N}^{N} \hat{f}(n) e^{inx} \\
&= \sum_{n=-N}^{N} \frac{1}{2\pi} \int_0^{2\pi} f(t) e^{-int} dt \cdot e^{inx} \\
&= \sum_{n=-N}^{N} \frac{1}{2\pi} \int_0^{2\pi} f(t) e^{in(x-t)} dt \\
&= \frac{1}{2\pi} \int_0^{2\pi} f(t) \left[\sum_{n=-N}^{N} e^{in(x-t)} \right] dt.
\end{aligned}$$

Thus we are finished if we can show that the sum in [ ] equals $D_N(x - t)$. Rewrite the sum as

$$\sum_{n=0}^{N} (e^{i(x-t)})^n + \sum_{n=0}^{N} (e^{-i(x-t)})^n - 1.$$

Then each of these last two sums is the partial sum of a geometric series. Thus we use the formula from Proposition 3.15 to write the last line as

$$\frac{e^{i(x-t)(N+1)} - 1}{e^{i(x-t)} - 1} + \frac{e^{-i(x-t)(N+1)} - 1}{e^{-i(x-t)} - 1} - 1.$$

We put everything over a common denominator to obtain

$$\frac{\cos N(x-t) - \cos(N+1)(x-t)}{1 - \cos(x-t)}.$$

We write

$$N(x-t) = \left(\left(N + \frac{1}{2}\right)(x-t) - \frac{1}{2}(x-t)\right),$$

$$(N+1)(x-t) = \left(\left(N + \frac{1}{2}\right)(x-t) + \frac{1}{2}(x-t)\right),$$

$$(x-t) = \frac{1}{2}(x-t) + \frac{1}{2}(x-t)$$

and use the sum formula for the cosine function to find that the last line equals

$$\frac{2\sin\left(\left(N + \frac{1}{2}\right)(x-t)\right)\sin\left(\frac{1}{2}(x-t)\right)}{2\sin^2\left(\frac{1}{2}(x-t)\right)}$$

$$= \frac{\sin\left(N + \frac{1}{2}\right)(x-t)}{\sin\frac{1}{2}(x-t)}$$

$$= D_N(x-t).$$

$\square$

That is the desired conclusion.

Remark 13.6: We have presented this particular proof of the formula for D_N because it is the most natural. It is by no means the shortest. Another proof is explored in the exercises.

Note also that, by a change of variable, the formula for $S_N f$ presented in the proposition can also be written as

$$S_N f(x) = \frac{1}{2\pi}\int_0^{2\pi} D_N(t) f(x-t)\,dt$$

provided we adhere to the convention of doing all arithmetic modulo multiples of 2π. $\square$

Lemma 13.7: *For any N it holds that*

$$\frac{1}{2\pi}\int_0^{2\pi} D_N(t)\,dt = 1.$$

Proof: It would be quite difficult to prove this property of D_N from the formula that we just derived. However, if we look at the proof of the proposition we notice that

$$D_N(t) = \sum_{n=-N}^{N} e^{int}.$$

Hence

$$\frac{1}{2\pi} \int_0^{2\pi} D_N(t)\,dt = \frac{1}{2\pi} \int_0^{2\pi} \sum_{n=-N}^{N} e^{int}\,dt$$

$$= \sum_{n=-N}^{N} \frac{1}{2\pi} \int_0^{2\pi} e^{int}\,dt$$

$$= 1$$

because any power of e^{it}, except the zeroeth power, integrates to zero. This completes the proof. □

Next we prove that, for a large class of functions, the Fourier series converges back to the function at every point.

Theorem 13.8: *Let f be a functinn on $[0, 2\pi]$ that satisfies a Lipschitz condition: there is a constant $C > 0$ such that if $s, t \in [0, 2\pi]$, then*

$$|f(s) - f(t)| \le C \cdot |s - t|. \tag{13.8.1}$$

[Note that at 0 and 2π this condition is required to hold modulo 2π —see the remarks at the beginning of the section.] Then, for every $x \in [0, 2\pi]$, it holds that

$$S_N f(x) \to f(x) \quad \text{as} \quad N \to \infty.$$

Indeed, the convercence is uniform in x.

Proof: Fix $x \in [0, 2\pi]$. We calculate that

$$|S_N f(x)' - f(x)| = \left| \frac{1}{2\pi} \int_0^{2\pi} f(x - t) D_N(t)\,dt - f(x) \right|$$

$$= \left| \frac{1}{2\pi} \int_0^{2\pi} f(x - t) D_N(t)\,dt \right.$$

$$\left. - \frac{1}{2\pi} \int_0^{2\pi} f(x) D_N(t)\,dt \right|,$$

where we have made use of the lemma. It is convenient here to exploit periodicity and write our integrals as $\int_{-\pi}^{\pi}$ instead of $\int_0^{2\pi}$. Now we combine the integrals to write

$$|(S_N f(x) - f(x)|$$

$$= \left| \frac{1}{2\pi} \int_{-\pi}^{\pi} [f(x-t) - f(x)] D_N(t) dt \right|$$

$$= \left| \frac{1}{2\pi} \int_{-\pi}^{\pi} \left[\frac{f(x-t) - f(x)}{\sin t/2} \right] \cdot \sin\left(\left(N + \frac{1}{2}\right) t \right) dt \right|$$

$$\leq \left| \frac{1}{2\pi} \int_{-\pi}^{\pi} \left[\frac{f(x-t) - f(x)}{\sin t/2} \cdot \cos \frac{t}{2} \right] \sin Nt \, dt \right|$$

$$+ \left| \frac{1}{2\pi} \int_{-\pi}^{\pi} \left[\frac{f(x-t) - f(x)}{\sin t/2} \cdot \sin \frac{t}{2} \right] \cos Nt \, dt \right|$$

$$\leq \left| \frac{1}{2\pi} \int_{-\pi}^{\pi} h(t) \sin Nt \, dt \right| + \left| \frac{1}{2\pi} \int_{-\pi}^{\pi} k(t) \cos Nt \, dt \right|,$$

where we have denoted the first expression in [ ] by $h_x(t) = h(t)$ and the second expression in [ ] by $k_x(t) = k(t)$. We use our hypothesis (13.8.1) about f to see that

$$|h(t)| = \left| \frac{f(x-t) - f(t)}{t} \right| \cdot \left| \frac{t}{\sin(t/2)} \right| \cdot \left| \cos \frac{t}{2} \right| \leq C \cdot 3.$$

[Here we have used the elementary fact that $2/\pi \leq |\sin u/u| \leq 1$ for $-\pi/2 \leq u \leq \pi/2$.] Thus h is a bounded function. It is obviously continuous, because f is, except perhaps at $t = 0$. So h is integrable—since it is bounded it is even square integrable. An even easier discussion shows that k is square integrable. Therefore Corollary 13.3 applies and we may conclude that the Fourier coefficients of h and of k tend to zero. However, the integral involving h is nothing other than $(\hat{h}(N) - \hat{h}(-N))/(2i)$ and the integral involving k is precisely $(\hat{k}(N) + \hat{k}(-N))/2$. We conclude that these integrals tend to zero as $N \to \infty$; in other words,

$$|S_N f(x) - f(x)| \to 0 \quad \text{as} \quad N \to \infty.$$

Since the relevant estimates are independent of x, we see that the convergence is uniform. □

Corollary 13.9: *If $f \in C^1([0, 2\pi])$ (that is, f is continuously differentiable) then $S_N f \to f$ uniformly.*

Proof: A C^1 function, by the Mean Value Theorem, satisfies a Lipschitz condition. □

In fact the proof of the theorem suffices to show that, if f is a Riemann square-integrable function on $[0, 2\pi]$ and if f is differentiable at x, then $S_N f(x) \to f(x)$.

In the exercises we shall explore other methods of summing Fourier series that allow us to realize even discontinuous functions as the limits of certain Fourier expressions.

It is natural to ask whether the Fourier series of a function characterizes that function. We can now give a partial answer to this question:

Corollary 13.10: *If f is a function on $[0, 2\pi]$ that satisfies a Lipschitz condition and if the Fourier series of f is identically zero then $f \equiv 0$.*

Proof: By the preceding corollary, the Fourier series converges uniformly to f. But the Fourier series is 0. □

Corollary 13.11: *If f and g are functions on $[0, 2\pi]$ that satisfy a Lipschitz condition and if the Fourier coefficients of f are the same as the Fourier coefficients of g then $f \equiv g$.*

Proof: Apply the preceding corollary to $f - g$. □

Example 13.12: Let $f(t) = t^2 - 2\pi t$, $0 \leq t \leq 2\pi$. Then $f(0) = f(2\pi) = 0$ and f is Lipschitz modulo 2π. Calculating the Fourier series of f, setting $t = 0$, and using the theorem reveals that

$$\sum_{j=1}^{\infty} \frac{1}{j^2} = \frac{\pi^2}{6}.$$

You are requested to provide the details. □

Exercises

1. Find the Fourier series for the function

$$f(x) = \begin{cases} 0 & \text{if } -\pi \leq x < 0 \\ 1 & \text{if } 0 \leq x \leq \frac{\pi}{2} \\ 0 & \text{if } \frac{\pi}{2} < x \leq \pi. \end{cases}$$

2. Find the Fourier series of the function

$$f(x) = \begin{cases} 0 & \text{if } -\pi \leq x < 0 \\ \sin x & \text{if } 0 \leq x \leq \pi \end{cases}$$

3. Find the Fourier series for each of these functions. Pay special attention to the reasoning used to establish your conclusions; consider alternative lines of thought.

   a. $f(x) = \pi, \; -\pi \leq x \leq \pi$

   b. $f(x) = \sin x, \; -\pi \leq x \leq \pi$

   c. $f(x) = \cos x, \; -\pi \leq x \leq \pi$

   d. $f(x) = \pi + \sin x + \cos x, \; -\pi \leq x \leq \pi$

4. Find the Fourier series for the function given by

$$f(x) = \begin{cases} -a & \text{if} \;\; -\pi \leq x < 0 \\ a & \text{if} \;\; 0 \leq x \leq \pi \end{cases} \text{ for } a \text{ a positive real number.}$$

$$f(x) = \begin{cases} -1 & \text{if} \;\; -\pi \leq x < 0 \\ 1 & \text{if} \;\; 0 \leq x \leq \pi \end{cases}$$

$$f(x) = \begin{cases} -\frac{\pi}{4} & \text{if} \;\; -\pi \leq x < 0 \\ \frac{\pi}{4} & \text{if} \;\; 0 \leq x \leq \pi \end{cases}$$

$$f(x) = \begin{cases} -1 & \text{if} \;\; -\pi \leq x < 0 \\ 2 & \text{if} \;\; 0 \leq x \leq \pi \end{cases}$$

$$f(x) = \begin{cases} 1 & \text{if} \;\; -\pi \leq x < 0 \\ 2 & \text{if} \;\; 0 \leq x \leq \pi \end{cases}$$

5. The functions $\sin^2 x$ and $\cos^2 x$ are both even. Show, without using any calculations, that the identities

$$\sin^2 x = \frac{1}{2}(1 - \cos 2x) = \frac{1}{2} - \frac{1}{2}\cos 2x$$

and

$$\cos^2 x = \frac{1}{2}(1 + \cos 2x) = \frac{1}{2} + \frac{1}{2}\cos 2x$$

are actually the Fourier series expansions of these functions.

6. Prove the trigonometric identities

$$\sin^3 x = \frac{3}{4}\sin x - \frac{1}{4}\sin 3x \quad \text{and} \quad \cos^3 x = \frac{3}{4}x + \frac{1}{4}\cos 3x$$

and show briefly, without calculation, that these are the Fourier series expansions of the functions $\sin^3 x$ and $\cos^3 x$.

7. Give another proof for the formula for $D_N(t)$ by completing the following outline:

    a. $D_N(t) = \sum_{n=-N}^{N} e^{int}$;

    b. $(e^{it} - 1) \cdot D_N(t) = e^{i(N+1)t} - e^{-iNt}$;

    c. Multiply both sides of the last equation by $e^{-it/2}$.

    d. Conclude that $D_N(t) = \frac{\sin(N+1/2)t}{\sin(t/2)}$.

8. Complete the details of Example 13.12.

*9. If f is integrable on the interval $[0, 2\pi]$ and if N is a nonnegative integer then define

$$\sigma_N f(x) = \frac{1}{N+1} \sum_{n=0}^{N} S_N f(x).$$

This is called the Nth *Cesaro mean* for the Fourier series of f. Prove that

$$\sigma_N f(x) = \frac{1}{2\pi} \int_0^{2\pi} K_N(x-t) f(t)\,dt,$$

where

$$K_N(x-t) = \frac{1}{N+1} \left\{ \frac{\sin \frac{N+1}{2}(x-t)}{\sin \frac{1}{2}t} \right\}^2.$$

10. Refer to Exercise **9** for notation. Prove that if $\delta > 0$ then $\lim_{N \to \infty} K_N(t) = 0$ with the limit being uniform for all $|t| \geq \delta$.

*11. Refer to Exercise **9** for notation. Prove that $\frac{1}{2\pi} \int_0^{2\pi} K_N(t)\,dt = 1$.

12. What is the Fourier series for the function $f(x) = x$ on the interval $[-\pi, \pi]$? Why does it only involve sine terms?

13. What is the Fourier series for the function $f(x) = |x|$ on the interval $[-\pi, \pi]$? Why does it only involve cosine terms?

14. What is the Fourier series for the function $f(x) = \sin 3x$ on the interval $[-\pi, \pi]$. Why does it only involve sine terms?

13.3 An Introduction to the Fourier Transform

It turns out that Fourier analysis on the interval $[0, 2\pi]$ and Fourier analysis on the entire real line $\mathbb{R}$ are analogous; but they differ in certain particulars that are well worth recording. In the present section we present an outline of the theory of he Fourier transform on the line. A thorough treatment of Fourier analysis on Euclidean space may be found in [STG]. See also [KRA2].

We define the *Fourier transform* of an integrable function f on $\mathbb{R}$ by

$$\hat{f}(\xi) = \int_{\mathbb{R}} f(t)e^{it\cdot\xi}dt .$$

Many references will insert a factor of 2π in the exponential or in the measure. Others will insert a minus sign in the exponent. There is no agreement on this matter. We have opted for this particular definition because of its simplicity.

We note that the significance of the exponentials $e^{it\xi}$ is that the only continuous multiplicative homomorphisms of $\mathbb{R}$ intocharactev group of $\mathbb{R}$ the circle group are the functions $\phi_{\xi}(t) = e^{it\xi}$, $\xi \in \mathbb{R}$. These functions are called the *characters* of the additive group $\mathbb{R}$. We refer the reader to [KRA2] for more on this matter.

Proposition 13.13: *If f is an integrable function, then*

$$|\hat{f}(\xi)| \le \int_{\mathbb{R}} |f(t)|\,dt .$$

Proof: Observe that, for any $\xi \in \mathbb{R}$,

$$|\hat{f}(\xi)| = \left| \int_{\mathbb{R}} f(t)e^{it\cdot\xi}dt \right| \le \int_{\mathbb{R}} |f(t)e^{it\cdot\xi}|\,dt \le \int |f(t)|\,dt.$$

Proposition 13.14: *If f is integrable, f is differentiable, and f' is integrable, then*

$$(\hat{f}')(\xi) = -i\xi\hat{f}(\xi).$$

Proof: Integrate by parts: if f is an infinitely differentiable function that vanishes outside a compact set, then

$$\begin{aligned}
(\hat{f}')(\xi) &= \int f'(t)e^{it\cdot\xi}dt \ \ dt \\
&= -\int f(t)[e^{it\cdot\xi}]'dt \\
&= -i\xi\int f(t)e^{it\cdot\xi}dt \\
&= -i\xi\hat{f}(\xi).
\end{aligned}$$

[Of course the "boundary terms" in the integration by parts vanish since f vanishes outside a compact set.] The general case follows from a limiting argument (see the Appendix at the end of this section).   □

Proposition 13.15: If f is integrable and $it f$ is integrable, then

$$(it\,\hat{f}\,) = \frac{d}{d\xi}\hat{f}.$$

Proof: Differentiate under the integral sign:

$$\frac{d}{d\xi}\hat{f}\,(\xi) = \frac{d}{d\xi}\int_{\mathbb{R}} f(t)e^{it\xi}dt$$
$$= \int_{\mathbb{R}} f(t)\frac{d}{d\xi}(e^{it\xi})dt$$
$$= \int_{\mathbb{R}} f(t)ite^{it\xi}dt$$
$$= (it\hat{f}\,).$$

□

Proposition 13.16: (The Riemann–Lebesgue Lemma) *If f is integrable, then*

$$\lim_{\xi \to \infty}|\hat{f}\,(\xi)| = 0.$$

Proof: First assume that $g \in C^2(\mathbb{R})$ and vanishes outside a compact set. We know that $|\hat{g}\,|$ is bounded. Also

$$|\xi^2\hat{g}\,(\xi)| = |\,[\hat{g}'']\,\hat{}\,| \le \int_{\mathbb{R}} |g''(x)|\,dx = C'.$$

Then $(1 + |\xi|^2)\hat{g}$ is bounded. Thus

$$|\hat{g}\,(\xi)| \le \frac{C''}{1 + |\xi|^2} \stackrel{|\xi|\to\infty}{\longrightarrow} 0.$$

This proves the result for $g \in C_c^2$. [Notice that the argument also shows that, if $f \in C^2(\mathbb{R})$ and vanishes outside a compact set, then $\hat{g}$ is integrable.]

Now let f be an arbitrary integrable function. Then there is a function $\psi \in C^2(\mathbb{R})$, vanishing outside a compqct set, such that

$$\int_{\mathbb{R}} |f(x) - \psi(x)|\,dx < \varepsilon/2.$$

[See the appendix to this section for the details of this assertion.] Choose M so large that, when $|\xi| > M$, then $|\hat{\psi}(\xi)| < \varepsilon/2$. Then, for $|\xi| > M$, we have

$$
\begin{aligned}
|\hat{f}(\xi)| &= |(f - \hat{\psi})(\xi) + \widehat{psi}(\xi)| \\
&\leq |(f - \hat{\psi})(\xi)| + |\hat{\psi}(\xi)| \\
&\leq \int_{\mathbb{R}} |f(x) - \psi(x)| \, dx + \frac{\varepsilon}{2} \\
&< \frac{\varepsilon}{2} + \frac{\varepsilon}{2} = \varepsilon.
\end{aligned}
$$

$\square$

This proves the result.

Example 13.17: The Riemann–Lebesgue lemma is intuitively clear when viewed in the following way. Fix an integrable function f. An integrable function is of well-approximated by a continuous function, so we may as well suppose that f is continuous. But a continuous function is well-approximated by a smooth function (see the Appendix to this section), so we may as well suppose that f is smooth. On a small interval I—say of length $1/M$—a smooth function is nearly constant. So, if we let $|\xi| >> 2\pi M^2$, then the character $e^{i\xi \cdot x}$ will oscillate at least M times on I, and will therefore integrate against a constant to a value that is very nearly zero. As M becomes larger, this statement becomes more and more accurate. That is the Riemann–Lebesgue lemma. $\square$

Proposition 13.18: *Let f be integrable on $\mathbb{R}$. Then $\hat{f}$ is uniformly continuous.*

Proof: Let us first assume that f is continuous and vanishes outside a compact set. Then

$$
\lim_{\xi \to \xi_0} \hat{f}(\xi) = \lim_{\xi \to \xi_0} \int f(x) e^{ix \cdot \xi} dx = \int \lim_{\xi \to \xi_0} f(x) e^{ix \cdot \xi} dx = \hat{f}(\xi_0).
$$

[*Exercise:* Justify passing the limit under the integral sign.] Since $\hat{f}$ also vanishes at ∞, the result is immediate when f is continuous and vanishing outside a compact set. The general result follows from an approximation argument (see the Appendix to this section). $\square$

Let $C_0(\mathbb{R})$ denote the continuous functions on $\mathbb{R}$ that vanish at ∞. Equip this space with the supremum norm. Then our results show that the Fourier transform maps the integrable functions to C_0 continuously.

It is natural to ask whether the Fourier transform is univalent; put in other words, can we recover a function from its Fourier transform? If so, can we do so with an explicit integral formula? The answer to all these questions is "yes," but advanced techniques are required for the proofs. We cannot treat them here, but see [KRA2] for the details. We content ourselves with the formulation of a sample result and its consequences.

Theorem 13.19: *Let f be a continuous, integrable function on $\mathbb{R}$ and suppose also $\hat{f}$ is integrable. Then*

$$f(x) = \frac{1}{2\pi} \int_{\mathbb{R}} \hat{f}(\xi) e^{-ix \cdot \xi} d\xi$$

for every x.

Corollary 13.20: *If f is continuous and integrable and $\hat{f} \equiv 0$ then $f \equiv 0$.*

Corollary 13.21: *If f, g are continuous and integrable and $\hat{f}(\xi) = \hat{g}(\xi)$ for all ξ, then $f \equiv g$.*

We refer to the circle of ideas in this theorem and the two corollaries as "Fourier inversion." See [KRA2] for the details of all these assertions.

Appendix: Approximation by Smooth Functions

At several junctures in this section we have used the idea that an integrable function may be approximated by smooth functions. We take a moment now to discuss this notion. Not all of the details appear here, but the interested reader may supply them as an exercise.

Let f be any integrable function on the interval $[0, 1]$. Then f may be approximated by its Riemann sums in the following sense. Let

$$0 = x_0 < x_1 < \cdots < x_k = 1$$

be a partition of the interval. For $j = 1, \ldots, k$ define

$$h_j(x) = \begin{cases} 0 & \text{if } 0 \le x < x_{j-1} \\ 1 & \text{if } x_{j-1} \le x \le x_j \\ 0 & \text{if } x_j < x \le 1. \end{cases}$$

Then the function

$$\mathcal{R}f(x) = \sum_{j=1}^{k} f(x_j) \cdot h_j(x)$$

is a piecewise constant approximation for f and the expression

$$\int_{\mathbb{R}} |f(x) - \mathcal{R}f(x)| \, dx \qquad (13.22)$$

will be small if the mesh of the partition is sufficiently fine. In fact the expression (13.22) is a standard "distance between functions" that is used in mathematical analysis. We often denote this quantity by $\|f - \mathcal{R}f\|_{L^1}$ and we call it "the L^1 norm" or "L^1 distance." More generally, we call the expression

$$\int_{\mathbb{R}} |g(x)| \, dx \equiv \|g\|_{L^1}$$

the L^1 norm of the function g.

Now our strategy is to approximate each of the functions h_j by a "smooth" function. Let $f(x) = 10x^3 - 15x^4 + 6x^5$. Notice that $f(0) = 0$, $f(1) = 1$, and both f' and f'' vanish at 0 and at 1.

The model for the sort of smooth function we are looking for is

$$\psi(x) = \begin{cases} 0 & \text{if} & x < -2 \\ f(x+2) & \text{if} & -2 \le x \le -1 \\ 1 & \text{if} & -1 < x < 1 \\ f(2-x) & \text{if} & 1 \le x \le 2 \\ 0 & \text{if} & 2 < x. \end{cases}$$

Refer to Figure 13.1. You may calculate that this function is twice continuously differentiable. It vanishes outside the interval $[-2, 2]$. And it is identically equal to 1 on the interval $[-1, 1]$.

More generally, we will consider the functions

$$\psi_\delta(x) = \begin{cases} 0 & \text{if} & x < -1 - \delta \\ f\left(\frac{x + (1 + \delta)}{\delta}\right) & \text{if} & -1 - \delta \le x \le -1 \\ 1 & \text{if} & -1 < x < 1 \\ f\left(\frac{(1 + \delta) - x}{\delta}\right) & \text{if} & 1 \le x \le 1 + \delta \\ 0 & \text{if} & 1 + \delta < x. \end{cases}$$

for $\delta > 0$ and

$$\psi_\delta^{[a,b]}(x) = \psi_\delta\left(\frac{2x - b - a}{b - a}\right)$$

for $\delta > 0$ and $a < b$. Figure 13.2 shows that ψ_δ is similar to the function ψ, but its sides are contracted so that it climbs from 0 to 1 over the interval

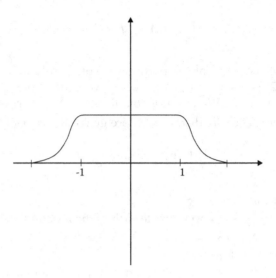

FIGURE 13.1
A compactly supported, smooth function.

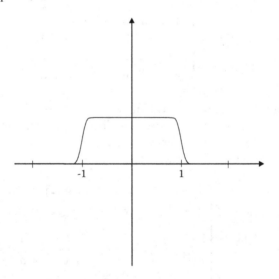

FIGURE 13.2
Another compactly supported, smooth function.

$[-1 - \delta, -1]$ of length δ and then descends from 1 to 0 over the interval $[1, 1 + \delta]$ of length δ. The function $\psi_\delta^{[a,b]}$ is simply the function ψ_δ adapted to the interval $[a, b]$ (Figure 13.3). The function $\psi_\delta^{[a,b]}$ climbs from 0 to 1 over the interval $[a - (\delta(b - a))/2, a]$ of length $\delta(b - a)/2$ and descends from 1 to 0 over the interval $[b, b + (\delta(b - a)/2)]$ of length $\delta(b - a)/2$.

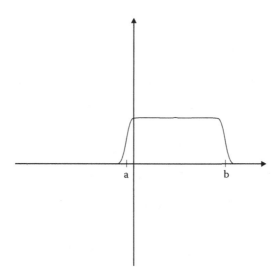

FIGURE 13.3
The compactly supported, smooth function translated and dilated.

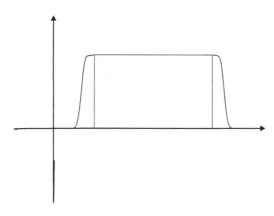

FIGURE 13.4
Unit for approximation.

Finally, we approximate the function h_j by $k_j(x) \equiv \psi_\delta^{[x_{j-1}, x_j]}$ for $j = 1, \ldots, k$. See Figure 13.4. Then the function f is approximated in L^1 norm by

$$Sf(x) = \sum_{j=1}^{k} f(x_j) \cdot k_j(x).$$

See Figure 13.5. If $\delta > 0$ is sufficiently small, then we can make $\|f - Sf\|_{L^1}$ as small as we please.

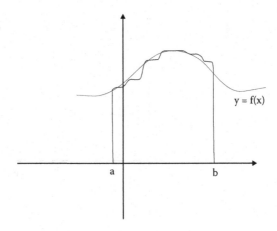

FIGURE 13.5
Approximation by a smooth function.

The approximation by twice continuously differentiable (or C^2) functions that we have constructed here is easily modified to achieve approximation by C^k functions for any k. One merely replaces the polynomial f by a polynomial that vanishes to higher order (order at least k) at 0 and at 1.

Exercises

1. Determine whether each of the following functions is even, odd, or neither:

$$x^5 \sin x, \ x^2 \sin 2x, \ e^x, \ (\sin x)^3, \ \sin x^2,$$

$$\cos(x + x^3), \ x + x^2 + x^3, \ \ln \frac{1 + x}{1 - x}.$$

2. Show that any function f defined on an interval symmetric about the origin can be written as the sum of an even function and an odd function.

3. Calculate the Fourier transform of $f(x) = x \cdot \chi_{[0,1]}$, where $\chi_{[0,1]}(x)$ equals 1 if $x \in [0, 1]$ and equals 0 otherwise.

4. See Exercise **3** for notation. Calculate the Fourier transform of $g(x) = \cos x \cdot \chi_{[0,2]}$.

5. If f, g are integrable functions on $\mathbb{R}$, then define their *convolution* to be

$$h(x) = f * g(x) = \int_{\mathbb{R}} f(x - t)g(t)dt.$$

Prove that

$$\hat{h}(\xi) = \hat{f}(\xi) \cdot \hat{g}(\xi).$$

6. Refer to Exercise **5** for notation and terminology. Fix an integrable function g on $\mathbb{R}$. Define a linear operator by

$$T: f \rightarrow f * g.$$

Prove that

$$\|Tf\|_{L^1} \leq C\|f\|_{L^1},$$

where

$$\|f\|_{L^1} = \int_{\mathbb{R}} |f(x)| \, dx$$

and $C = \int |g| \, dx = \|g\|_{L^1}$.

*7. Let f be a function on $\mathbb{R}$ that vanishes outside a compact set. Prove that $\hat{f}$ does *not* vanish outside any compact set.

*8. Calculate the Fourier transform of the function $f(x) = e^{-x^2}$.

*9. Use the calculation from Exercise **8** to discover an eigenfunction of the Fourier transform.

*10. Refer to Exercises **8**, **9**. What are the eigenvalues of the Fourier transform?

*11. A version of the Poisson summation formula says that, if f is a suitable function on the real line, then

$$\sum_{n=-\infty}^{\infty} f(n) = \sum_{k=-\infty}^{\infty} \hat{f}(k).$$

Find a proof of this assertion.

*12. Plancherel's theorem says that, if f is a continuous function with compact support, then

$$(2\pi)^{-1} \int |\hat{f}(\xi)|^2 \, d\xi = \int |f(x)|^2 \, dx.$$

Find a proof of this result. [**Hint:** Consider the Fourier transform of $f * \tilde{f}$, where $\tilde{g}(x) \equiv g(-x)$.]

*13. Extend the result of Exercise 12 to all square integrable functions f.

*14. Refer to Exercises 12 and 13. The Fourier transform maps integrable functions to bounded functions and square integrable functions to square integrable functions. One can use interpolation of operators to conclude how the Fourier transform acts on pth-power integrable functions, $1 < p < 2$. Discuss.

15. Let

$$\varphi_N(x) = \begin{cases} 2N & \text{if } |x| \leq 1/N \\ 0 & \text{otherwise .} \end{cases}$$

Calculate the Fourier transform ψ of φ_N. What is the limit of ψ as $N \to \infty$? This will be the Fourier transform of the Dirac delta mass.

13.4 Fourier Methods in the Theory of Differential Equations

In fact an entire separate book could be written about the applications of Fourier analysis to differential equations and to other parts of mathematical analysis. The subject of Fourier series grew up hand in hand with the analytical areas to which it is applied. In the present brief section we merely indicate a couple of examples.

13.4.1 Remarks on Different Fourier Notations

In Section 13.2, we found it convenient to define the Fourier coefficients of an integrable function on the interval $[0, 2\pi]$ to be

$$\hat{f}(n) = \frac{1}{2\pi} \int_0^{2\pi} f(x)e^{-inx}dx.$$

From the point of view of pure mathematics, this complex notation has proved to be useful, and it has become standardized.

But, in applications, there are other Fourier paradigms. They are easily seen to be equivalent to the one we have already introduced. The reader who wants to be conversant in this subject should be aware of these different ways of writing the basic ideas of Fourier series. We will introduce one of them now, and use it in the ensuing discussion.

If f is integrable on the interval $[-\pi, \pi]$ (note that, by 2π-periodicity, this is not essentially different from $[0, 2\pi]$), then we define the Fourier coefficients

$$a_0 = \frac{1}{2\pi} \int_{-\pi}^{\pi} f(x)dx,$$
$$a_n = \frac{1}{\pi} \int_{-\pi}^{\pi} f(x)\cos nx \ dx \quad \text{for } n \geq 1,$$
$$b_n = \frac{1}{\pi} \int_{-\pi}^{\pi} f(x)\sin nx \ dx \quad \text{for } n \geq 1.$$

This new notation is not essentially different from the old, for

$$\hat{f}(n) = \frac{1}{2}[a_n + ib_n]$$

for $n \geq 1$. The change in normalization (i.e., whether the constant before the integral is $1/\pi$ or $1/2\pi$) is dictated by the observation that we want to exploit the fact (so that our formulas come out in a neat and elegant fashion) that

$$\frac{1}{2\pi} \int_0^{2\pi} |e^{-int}|^2 \ dt = 1,$$

in the theory from Section 13.2 and that

$$\frac{1}{2\pi} \int_{-\pi}^{\pi} 1^2 dx = 1,$$
$$\frac{1}{\pi} \int_{-\pi}^{\pi} |\cos nt|^2 \ dt = 1 \quad \text{for } n \geq 1,$$
$$\frac{1}{\pi} \int_{-\pi}^{\pi} |\sin nt|^2 \ dt = 1 \quad \text{for } n \geq 1$$

in the theory that we are about to develop.

It is clear that any statement (as in Section 13.2) that is formulated in the language of $\hat{f}(n)$ is easily translated into the language of a_n and b_n and vice versa. In the present discussion we shall use a_n and b_n just because that is the custom, and because it is convenient for the points that we want to make.

13.4.2 The Dirichlet Problem on the Disc

We now study the two-dimensional Laplace equation, which is

$$\Delta = \frac{\partial^2 u}{\partial x^2} + \frac{\partial^2 u}{\partial y^2} = 0. \tag{13.4.1}$$

This is probably the most important differential equation of mathematical physics. It describes a steady-state heat distribution, electrical fields, and many other important phenomena of nature.

It will be useful for us to write this equation in polar coordinates. To do so, recall that

$$r^2 = x^2 + y^2, \quad x = r \cos \theta, \quad y = r \sin \theta.$$

Thus

$$\frac{\partial}{\partial r} = \frac{\partial x}{\partial r} \frac{\partial}{\partial x} + \frac{\partial y}{\partial r} \frac{\partial}{\partial y} = \cos \theta \frac{\partial}{\partial x} + \sin \theta \frac{\partial}{\partial y}$$

$$\frac{\partial}{\partial \theta} = \frac{\partial x}{\partial \theta} \frac{\partial}{\partial x} + \frac{\partial y}{\partial \theta} \frac{\partial}{\partial y} = -r \sin \theta \frac{\partial}{\partial x} + r \cos \theta \frac{\partial}{\partial y}$$

We may solve these two equations for the unknowns $\partial/\partial x$ and $\partial/\partial y$. The result is

$$\frac{\partial}{\partial x} = \cos \theta \frac{\partial}{\partial r} - \frac{\sin \theta}{r} \frac{\partial}{\partial \theta} \quad \text{and} \quad \frac{\partial}{\partial y} = \sin \theta \frac{\partial}{\partial r} - \frac{\cos \theta}{r} \frac{\partial}{\partial \theta}.$$

A tedious calculation now reveals that

$$\begin{aligned}
\Delta = \frac{\partial^2}{\partial x^2} + \frac{\partial^2}{\partial y^2} &= \left(\cos \theta \frac{\partial}{\partial r} - \frac{\sin \theta}{r} \frac{\partial}{\partial \theta} \right) \left(\cos \theta \frac{\partial}{\partial r} - \frac{\sin \theta}{r} \frac{\partial}{\partial \theta} \right) \\
&\quad + \left(\sin \theta \frac{\partial}{\partial r} - \frac{\cos \theta}{r} \frac{\partial}{\partial \theta} \right) \left(\sin \theta \frac{\partial}{\partial r} - \frac{\cos \theta}{r} \frac{\partial}{\partial \theta} \right) \\
&= \frac{\partial^2}{\partial r^2} + \frac{1}{r} \frac{\partial}{\partial r} + \frac{1}{r^2} \frac{\partial^2}{\partial \theta^2}.
\end{aligned}$$

Let us use the so-called separation of variables methodseparation of variables method to analyze our partial differential equation (13.4.1). We will seek a solution $w = w(r, \theta) = u(r) \cdot v(\theta)$ of the Laplace equation. Using the polar form, we find that this leads to the equation

$$u''(r) \cdot v(\theta) + \frac{1}{r} u'(r) \cdot v(\theta) + \frac{1}{r^2} u(r) \cdot v''(\theta) = 0.$$

Thus

$$\frac{r^2 u''(r) + r u'(r)}{u(r)} = -\frac{v''(\theta)}{v(\theta)}.$$

Since the left-hand side depends only on r, and the right-hand side only on θ, both sides must be constant. Denote the common constant value by λ. Then we have

$$v''(\theta) + \lambda v(\theta) = 0 \tag{13.4.2}$$

and

$$r^2 u''(r) + r u'(r) - \lambda u(r) = 0. \tag{13.4.3}$$

If we demand that v be continuous and periodic, then we must insist that $\lambda > 0$ and in fact that $\lambda = n^2$ for some nonnegative integer $n.^2$ For $n = 0$ the only suitable solution is $v \equiv$ constant and for $n > 0$ the general solution (with $\lambda = n^2$) is

$$v = A \cos n\theta + B \sin n\theta,$$

as you can verify directly.

We set $\lambda = n^2$ in equation (13.4.3), and obtain

$$r^2 u'' + r u' - n^2 u = 0, \tag{13.4.4}$$

which is Euler's equidimensional equation. The change of variables $r = e^z$ transforms this equation to a linear equation with constant coefficients, and that can in turn be solved with standard techniques. To wit, the equation that we now have is

$$u'' - n^2 u = 0.$$

The variable is now z. We guess a solution of the form $u(z) = e^{\alpha z}$. Thus

$$\alpha^2 e^{\alpha z} - n^2 e^{\alpha z} = 0 \tag{13.4.5}$$

so that

$$\alpha = \pm n \,.$$

Hence the solutions of (13.4.5) are

$$u(z) = e^{nz} \quad \text{and} \quad u(z) = e^{-nz}$$

provided that $n \neq 0$. It follows that the solutions of the original Euler equation (13.4.4) are

$$u(r) = r^n \quad \text{and} \quad u(r) = r^{-n} \quad \text{for } n \neq 0.$$

In case $n = 0$ the solution is readily seen to be $u = 1$ or $u = \ln r$.
The result is

$$u = A + B \ln r \quad \text{if} \quad n = 0;$$
$$u = Ar^n + Br^{-n} \quad \text{if} \quad n = 1, 2, 3, \ldots.$$

We are most interested in solutions u that are continuous at the origin; so we take $B = 0$ in all cases. The resulting solutions are

$$n = 0, \quad w = a \text{ constant } a_0/2;$$
$$n = 1, \quad w = r(a_1 \cos\theta + b_1 \sin\theta);$$
$$n = 2, \quad w = r^2(a_2 \cos 2\theta + b_2 \sin 2\theta);$$
$$n = 3, \quad w = r^3(a_3 \cos 3\theta + b_3 \sin 3\theta);$$
$$\cdots$$

Of course any finite sum of solutions of Laplace's equation is also a solution. The same is true for infinite sums. Thus we are led to consider

$$w = w(r, \theta) = \frac{1}{2}a_0 + \sum_{j=0}^{\infty} r^j \left(a_j \cos j\theta + b_j \sin j\theta \right).$$

On a formal level, letting $r \to 1^-$ in this last expression gives

$$w = \frac{1}{2}a_0 + \sum_{j=1}^{\infty} \left(a_j \cos j\theta + b_j \sin j\theta \right).$$

We draw all these ideas together with the following physical rubric. Consider a thin aluminum disc of radius 1, and imagine applying a heat distribution to the boundary of that disc. In polar coordinates, this distribution is specified by a function $f(\theta)$. We seek to understand the steady-state heat distribution on the entire disc. See Figure 13.6. So we seek a function $w(r, \theta)$, continuousheat distribution on the disc on the closure of the disc, which agrees with f on the boundary and which represents the steady-state distribution of heat inside. Some physical analysis shows that such a function w is the solution of the boundary value problem

$$\Delta w = 0,$$
$$w \big|_{\partial D} = f.$$

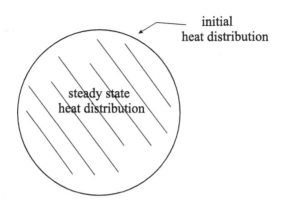

FIGURE 13.6
Steady-state heat.

According to the calculations we performed prior to this last paragraph, a natural approach to this problem is to expand the given function f in its sine/cosine series:

$$f(\theta) = \frac{1}{2}a_0 + \sum_{j=1}^{\infty} \left(a_j \cos j\theta + b_j \sin j\theta \right)$$

and then posit that the w we seek is

$$w(r, \theta) = \frac{1}{2}a_0 + \sum_{j=1}^{\infty} r^j \left(a_j \cos j\theta + b_j \sin j\theta \right).$$

This process is known as *solving the Dirichlet problem on the disc with boundary data f.*

Example 13.22: Let us follow the paradigm just sketched to solve the Dirichlet problem on the disc with $f(\theta) = 1$ on the top half of the boundary and $f(\theta) = -1$ on the bottom half of the boundary. See Figure 13.7.

It is straightforward to calculate that the Fourier series (sine series) expansion for this f is

$$f(\theta) = \frac{4}{\pi} \left(\sin \theta + \frac{\sin 3\theta}{3} + + \frac{\sin 5\theta}{5} + \cdots \right).$$

There are no cosine terms because f is an odd function.

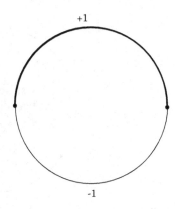

FIGURE 13.7
Boundary data.

The solution of the Dirichlet problem is therefore

$$w(r, \theta) = \frac{4}{\pi}\left(r \sin \theta + \frac{r^3 \sin 3\theta}{3} + + \frac{r^5 \sin 5\theta}{5} + \cdots \right).$$

$\square$

13.4.3 Introduction to the Heat and Wave Equations

In the middle of the eighteenth century much attention was given to the problem of determining the mathematical laws governing the motion of a vibrating string with fixed endpoints at 0 and π (Figure 13.8). An elementary analysis of tension shows that, if $y(x, t)$ denotes the ordinate of the string at time t above the point x, then $y(x, t)$ satisfies the *wave equation*

$$\frac{\partial^2 y}{\partial t^2} = a^2 \frac{\partial^2 y}{\partial x^2}.$$

Here a is a parameter that depends on the tension of the string. A change of scale will allow us to assume that $a = 1$. (A bit later we shall actually provide a formal derivation of the wave equation. See also [KRA2] for a more thorough consideration of these matters.)

FIGURE 13.8
The wave equation.

In 1747 d'Alembert showed that solutions of this equation have the form

$$y(x, t) = \frac{1}{2}(f(t + x) + g(t - x)), \tag{13.4.6}$$

where f and g are "any" functions of one variable. (The following technicality must be noted: the functions f and g are initially specified on the
interval $[0, \pi]$. We extend f and g to $[-\pi, 0]$ and to $[\pi, 2\pi]$ by odd reflection.
Continue f and g to the rest of the real line so that they are 2π-periodic.)

In fact the wave equation, when placed in a "well-posed" setting, wave
equation comes equipped with two initial conditions:

i. $y(x, 0) = \phi(x)$
ii. $\partial_t y(x, 0) = \psi(x)$.

These conditions mean **(i)** that the wave has an initial configuration that is the
graph of the function ϕ and **(ii)** that the string is released with initial velocity ψ.

If (13.4.6) is to be a solution of this initial value problem then f and g must
satisfy

$$\frac{1}{2}(f(x) + g(-x)) = \phi(x) \tag{13.4.7}$$

and

$$\frac{1}{2}(f'(x) + g'(-x)) = \psi(x). \tag{13.4.8}$$

Integration of (13.4.8) gives a formula for $f(x) - g(-x)$. That and (13.4.7)
give a system that may be solved for f and g with elementary algebra.

The converse statement holds as well: for any functions f and g, a
function y of the form (13.4.6) satisfies the wave equation (Exercise). The
work of d'Alembert brought to the fore a controversy which had been
implicit in the work of Daniel Bernoulli, Leonhard Euler, and others: what is
a "function"? (We recommend the article [LUZ] for an authoritative discussion of the controversies that grew out of classical studies of the wave
equation. See also [LAN].)

It is clear, for instance, in Euler's writings that he did not perceive a
function to be an arbitrary "rule" that assigns points of the range to points
of the domain; in particular, Euler did not thinkfunction, what is? that a
function could be specified in a fairly arbitrary fashion at different points of
the domain. Once a function was specified on some small interval, Euler
thought that it could only be extended in one way to a larger interval.

Therefore, on physical grounds, Euler objected to d'Alembert's work. He claimed that the initial position of the vibrating string could be specified by several different functions pieced together continuously, so that a single f could not generate the motion of the string.

Daniel Bernoulli solved the wave equation by a different method (separation of variables, which we treat below) and was able to show that there are infinitely many solutions of the wave equation having the form

$$\phi_j(x, t) = \sin jx \cos jt, \quad j \geq 1 \text{ an integer.}$$

Proceeding formally, he posited that all solutions of the wave equation satisfying $y(0, t) = y(\pi, t) = 0$ and $\partial_t y(x, 0) = 0$ will have the form

$$y = \sum_{j=1}^{\infty} a_j \sin jx \cos jt.$$

Setting $t = 0$ indicates that the initial form of the string is $f(x) \equiv \sum_{j=1}^{\infty} a_j \sin jx$. In d'Alembert's language, the initial form of the string is $\frac{1}{2}(f(x) - f(-x))$, for we know that

$$0 \equiv y(0, t) = f(t) + g(t)$$

(because the endpoints of the string are held stationary), hence $g(t) = -f(t)$. If we suppose that d'Alembert's function is odd (as is $\sin jx$, each j), then the initial position is given by $f(x)$. Thus the problem of reconciling Bernoulli's solution to d'Alembert's reduces to the question of whether an "arbitrary" function f on $[0, \pi]$ may be written in the form $\sum_{j=1}^{\infty} a_j \sin jx$.

Since most mathematicians contemporary with Bernoulli believed that properties such as continuity, differentiability, and periodicity were preserved under (even infinite) addition, the consensus was that arbitrary f could *not* be represented as a (even infinite) trigonometric sum. The controversy extended over some years and was fueled by further discoveries (such as Lagrange's technique for interpolation by trigonometric polynomials) and more speculations.

****

JOHANN BERNOULLI

Johann Bernoulli was born in Basel, the son of Nicolaus Bernoulli, an apothecary, and began studying medicine at Basel University. He convinced his father to allow him to study medicine. However, Johann Bernoulli did not enjoy medicine and began studying mathematics on the side with his older brother Jacob. Throughout Johann Bernoulli's education at Basel University, the Bernoulli brothers worked together studying infinitesimal calculus.

After graduating from Basel University, Johann Bernoulli taught differential equations. Later, in 1694, he married Dorothea Falkner, the daughter of an alderman of Basel, and soon after accepted a position as the Professor of Mathematics at the University of Groningen. In 1705 Bernoulli planned to return to his home town of Basel. Just after setting out on the journey he learned of his brother's death by tuberculosis. Bernoulli had planned on becoming the Professor of Greek at Basel University upon returning but instead was able to take over as Professor of Mathematics, his older brother's former position. As a student of Leibniz's calculus, Bernoulli sided with him in 1713 in the Leibniz–Newton debate over who deserved credit for the discovery of calculus. This ultimately delayed acceptance of Newton's theory in continental Europe.

In 1724, Johann Bernoulli entered a competition sponsored by the French Acad?e Royale des Sciences. It posed this question:

What are the laws according to which a perfectly hard body, put into motion, moves another body of the same nature either at rest or in motion, and which it encounters either in a vacuum or in a plenum? Bernoulli was disqualified for the prize, which was won by Maclaurin. However, Bernoulli's paper was subsequently accepted in 1726 when the Acad?e considered papers regarding elastic bodies, for which the prize was awarded to Pierre Mazière. Bernoulli received an honourable mention in both competitions.

Johann and his brother developed a jealous and competitive relationship. Johann was jealous of Jacob's position and the two often attempted to outdo each other. After Jacob's death Johann's jealousy shifted toward his own talented son, Daniel. In 1738 the father–son duo nearly simultaneously published separate works on hydrodynamics. Johann attempted to take precedence over his son by purposely and falsely predating his work two years prior to his son's.

The Bernoulli brothers often worked on the same problems, but not without friction. Their most bitter dispute concerned the brachistochrone curve problem, concerning which curve allows a particle to descend from an upper location to a lower one in the least time. Johann presented the problem in 1696, offering a reward for its solution. Entering the challenge, Johann proposed the cycloid. Jacob proposed the same solution, but Johann's derivation of the solution was incorrect, and he presented his brother Jacob's derivation as his own.

Bernoulli was hired by Guillaume de l'Hôpital for tutoring in mathematics. Bernoulli and l'Hopital signed a contract which gave l'Hôpital the right to use Bernoulli's discoveries as he pleased. L'Hôpital authored the first textbook on infinitesimal calculus, *Analyse des Infiniment Petits pour l'Intelligence des Lignes*

Courbes in 1696, which mainly consisted of the work of Bernoulli, including what is now known as l'Hôpital's rule. Subsequently, in letters to Leibniz, Varignon and others, Bernoulli complained that he had not received enough credit for his contributions, in spite of the preface of his book.

****

In the 1820s, the problem of representation of an "arbitrary" function by trigonometric series was given a satisfactory answer as a result of two events. First, there is the sequence of papers by Joseph Fourier culminating with the tract [FOU]. FourierFourier, J. B. J. gave a formal method of expanding an "arbitrary" function f into a trigonometric series. He computed some partial sums for some sampleDirichlet, P. G. L. fs and verified that they gave very good approximations to f. Second, Dirichlet proved the first theorem giving sufficient (and very general) conditions for the Fourier series series of a function f to converge pointwise to f. *Dirichlet was one of the first, in 1828, to formalize the notions of partial sum and convergence of a series;* his ideas had antecedents in the work of Gauss and Cauchy.

For all practical purposes, these events mark the beginning of the mathematical theory of Fourier series (see [LAN]).

13.4.4 Boundary Value Problems

We wish to motivate the physics of the vibrating string. We begin this discussion by seeking a nontrivial solution y of the differential equation

$$y'' + \lambda y = 0 \tag{13.4.9}$$

subject to the conditions

$$y(0) = 0 \quad \text{and} \quad y(\pi) = 0 . \tag{13.4.10}$$

Notice that this is a different situation from the one we have studied in earlier parts of the book. Ordinary differential equations generally have "initial conditions." Now we have what are called *boundary conditions*: we specify one condition (in this instance the *value*) for the function at two different points. For instance, in the discussion of the vibrating string in the last section, we wanted our string to be pinned down at the two endpoints. These are typical boundary conditions coming from a physical problem.

The situation with boundary conditions is quite different from that for initial conditions. The latter is a sophisticated variation of the fundamental theorem of calculus. The former is rather more subtle. So let us begin to analyze.

First, we can just solve the equation explicitly when $\lambda < 0$ and see that the independent solutions are a pair of exponentials, no linear combination of which can satisfy (13.4.10).

If $\lambda = 0$ then the general solution of (13.4.9) is the linear function $y = Ax + B$. Such a function cannot vanish at two points unless it is identically zero.

So the only interesting case is $\lambda > 0$. In this situation, the general solution of (13.4.9) is

$$y = A \sin \sqrt{\lambda} x + B \cos \sqrt{\lambda} x.$$

Since $y(0) = 0$, this in fact reduces to

$$y = A \sin \sqrt{\lambda} x.$$

In order for $y(\pi) = 0$, we must have $\sqrt{\lambda} \pi = n\pi$ for some positive integer n, thus $\lambda = n^2$. These values of λ are termed the *eigenvalues* of the problem, and the corresponding solutions

$$\sin x, \quad \sin 2x, \quad \sin 3x \ldots$$

are called the *eigenfunctions* of the problem (13.4.9), (13.4.10).

We note these immediate properties of the eigenvalues and eigenfunctions for our problem:

i. If ϕ is an eigenfunction for eigenvalue λ, then so is $c \cdot \phi$ for any constant c.

ii. The eigenvalues $1, 4, 9, \ldots$ form an increasing sequence that approaches $+\infty$.

iii. The nth eigenfunction $\sin nx$ vanishes at the endpoints $0, \pi$ (as we originally mandated) and has exactly $n - 1$ zeros in the interval $(0, \pi)$.

13.4.5 Derivation of the Wave Equation

Now let us re-examine the vibrating string from the last section and see how eigenfunctions and eigenvalues arise naturally in a physical problem. We consider a flexible string with negligible weight that is fixed at its ends at the points $(0, 0)$ and $(\pi, 0)$. The curve is deformed into an initial position $y = f(x)$ in the $x - y$ plane and then released.

Our analysis will ignore damping effects, such as air resistance. We assume that, in its relaxed position, the string is as in Figure 13.9. The string is plucked in the vertical direction, and is thus set in motion in a vertical

FIGURE 13.9
The string in relaxed position.

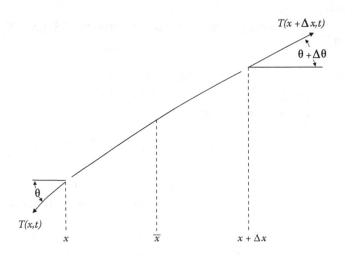

FIGURE 13.10
An element of the plucked string.

plane. We will bedamping effects supposing that the oscillation has small amplitude.

We focus attention on an "element" Δx of the string (Figure 13.10) that lies between x and $x + \Delta x$. We adopt the usual physical conceit of assuming that the displacement (motion) of this string element is *small*, so that there is only a slight error in supposing that the motion of each point of the string element is strictly vertical. We let the tension of the string, at the point x at time t, be denoted by $T(x, t)$. Note that T acts only in the tangential direction (i.e., along the string). We denote the mass density of the string by ρ.

Since *there is no horizontal component of acceleration*, we see that

$$T(x + \Delta x, t) \cdot \cos(\theta + \Delta\theta) - T(x, t) \cdot \cos(\theta) = 0. \qquad (13.4.11)$$

(Refer to Figure 13.11: The expression $T(\star) \cdot \cos(\star)$ denotes $H(\star)$, the horizontal component of the tension.) Thus equation (13.4.11) says that H is independent of x.

Now we look at the vertical component of force (acceleration):

$$T(x + \Delta x, t) \cdot \sin(\theta + \Delta\theta) - T(x, t) \cdot \sin(\theta) = \rho \cdot \Delta x \cdot u_{tt}(\bar{x}, t). \quad (13.4.12)$$

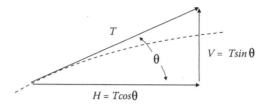

FIGURE 13.11
The horizontal component of the tension.

Here $\bar{x}$ is the mass center of the string element and we are applying Newton's second law—that the external force is the mass of the string element times the acceleration of its center of mass. We use subscripts to denote derivatives. We denote the vertical component of $T(\star)$ by $V(\star)$. Thus equation (13.4.12) can be written as

$$\frac{V(x + \Delta x, t) - V(x, t)}{\Delta x} = \rho \cdot u_{tt}(x, t).$$

Letting $\Delta x \to 0$ yields

$$V_x(x, t) = \rho \cdot u_{tt}(x, t). \tag{13.4.13}$$

We would like to express equation (13.4.13) entirely in terms of u, so we notice that

$$V(x, t) = H(t) \tan \theta = H(t) \cdot u_x(x, t).$$

(We have used the fact that the derivative in x is the slope of the tangent line, which is $\tan \theta$.) Substituting this expression for V into (13.4.13) yields

$$(Hu_x)_x = \rho \cdot u_{tt}.$$

But H is independent of x, so this last line simplifies to

$$H \cdot u_{xx} = \rho \cdot u_{tt}.$$

For small displacements of the string, θ is nearly zero, so $H = T \cos \theta$ is nearly T. We are most interested in the case where T is constant. And of course ρ is constant. Thus we finally write our equation as

$$\frac{T}{\rho} u_{xx} = u_{tt}.$$

It is traditional to denote the constant T/ρ on the left by a^2. We finally arrive at the *wave equation*

$$a^2 u_{xx} = u_{tt}.$$

13.4.6 Solution of the Wave Equation

We consider the wave equation

$$a^2 y_{xx} = y_{tt} \qquad\qquad (13.4.14)$$

with the boundary conditions

$$y(0, t) = 0$$

and

$$y(\pi, t) = 0 \,.$$

Physical considerations dictate that we also impose the initial conditions

$$\left. \frac{\partial y}{\partial t} \right|_{t=0} = 0 \qquad\qquad (13.4.15)$$

(indicating that the initial velocity of the string is 0) and

$$y(x, 0) = f(x) \qquad\qquad (13.4.16)$$

(indicating that the initial configuration of the string is the graph of the function f).

We solve the wave equation using a classical technique known as "separation of variables." For convenience, we assume that the constant $a = 1$. We guess a solution of the form $u(x, t) = u(x) \cdot v(t)$. Putting this guess into the differential equation

$$u_{xx} = u_{tt}$$

gives

$$u''(x)v(t) = u(x)v''(t).$$

We may obviously separate variables, in the sense that we may write

$$\frac{u''(x)}{u(x)} = \frac{v''(t)}{v(t)}.$$

The left-hand side depends only on x while the right-hand side depends only on t. The only way this can be true is if

$$\frac{u''(x)}{u(x)} = \lambda = \frac{v''(t)}{v(t)}$$

for some constant λ. But this gives rise to two second-order linear, ordinary differential equations that we can solve explicitly:

$$u'' = \lambda \cdot u \tag{13.4.17}$$

$$v'' = \lambda \cdot v. \tag{13.4.18}$$

Observe that this is the *same* constant λ in both of these equations. Now, as we have already discussed, we want the initial configuration of the string to pass through the points $(0, 0)$ and $(\pi, 0)$. We can achieve these conditions by solving (13.4.17) with $u(0) = 0$ and $u(\pi) = 0$. But of course this is the eigenvalue problem that we treated at the beginning of the section. The problem has a nontrivial solution if and only if $\lambda = -n^2$ for some positive integer n, and the corresponding eigenfunction is

$$u_n(x) = \sin nx.$$

For this same λ, the general solution of (13.4.15) is

$$v(t) = A \sin nt + B \cos nt.$$

If we impose the requirement that $v'(0) = 0$, so that (10) is satisfied, then $A = 0$ and we find the solution

$$v(t) = B \cos nt.$$

This means that the solution we have found of our differential equation with boundary and initial conditions is

$$y_n(x, t) = \sin nx \cos nt. \tag{13.4.19}$$

And in fact any finite sum with coefficients (or *linear combination*) of these solutions will also be a solution:

$$y = \alpha_1 \sin x \cos t + \alpha_2 \sin 2x \cos 2t + \cdots \alpha_k \sin kx \cos kt.$$

Ignoring the rather delicate issue of convergence, we may claim that any *infinite* linear combination of the solutions (13.4.19) will also be a solution:

$$y = \sum_{j=1}^{\infty} b_j \sin jx \cos jt. \tag{13.4.20}$$

Now we must examine the initial condition (13.4.16). The mandate $y(x, 0) = f(x)$ translates to

$$\sum_{j=1}^{\infty} b_j \sin jx = y(x, 0) = f(x) \tag{13.4.21}$$

or

$$\sum_{j=1}^{\infty} b_j u_j(x) = y(x, 0) = f(x). \tag{13.4.22}$$

Thus we demand that f have a valid Fourier series expansion. Such an expansion is correct for a rather broad class of functions f. Thus the wave equation is solvable in considerable generality.

Now fix $m \neq n$. We know that our eigenfunctions u_j satisfy

$$u''_m = -m^2 u_m \quad \text{and} \quad u''_n = -n^2 u_n.$$

Multiply the first equation by u_n and the second by u_m and subtract. The result is

$$u_n u''_m - u_m u''_n = (n^2 - m^2) u_n u_m$$

or

$$[u_n u'_m - u_m u'_n]' = (n^2 - m^2) u_n u_m.$$

We integrate both sides of this last equation from 0 to π and use the fact that $u_j(0) = u_j(\pi) = 0$ for every j. The result is

$$0 = [u_n u'_m - u_m u'_n]\Big|_0^{\pi} = (n^2 - m^2) \int_0^{\pi} u_m(x) u_n(x)\, dx.$$

Thus

$$\int_0^\pi \sin mx \sin nx \, dx = 0 \quad \text{for } n \neq m \tag{13.4.23}$$

or

$$\int_0^\pi u_m(x)u_n(x)dx = 0 \quad \text{for } n \neq m. \tag{13.4.24}$$

Of course this is a standard fact from calculus. But now we understand it as an orthogonality condition, and we see how the condition arises naturally from the differential equation.

In view of the orthogonality condition (13.4.24), it is natural to integrate both sides of (13.4.22) against $u_k(x)$. The result is

$$\int_0^\pi f(x) \cdot u_k(x)dx = \int_0^\pi \left(\sum_{j=0}^\infty b_j u_j(x) \right) \cdot u_k(x)dx$$

$$= \sum_{j=0}^\infty b_j \int_\pi^\pi u_j(x)u_k(x)dx$$

$$= \frac{\pi}{2}b_k.$$

The b_k are the Fourier coefficients that we studied earlier in this chapter. Using these coefficients, we have *Bernoulli's solution* (13.4.20) of the wave equation.

Exercises

1. Find the eigenvalues λ_n and the eigenfunctions y_n for the equation $y'' + \lambda y = 0$ in each of the following instances.
   a. $y(0) = 0,$ $y(\pi/2) = 0$
   b. $y(0) = 0,$ $y(2\pi) = 0$
   c. $y(0) = 0,$ $y(1) = 0$
   d. $y(0) = 0,$ $y(L) = 0$ for $L > 0$
   e. $y(-L) = 0,$ $y(L) = 0$ for $L > 0$
   f. $y(a) = 0,$ $y(b) = 0$ for $a < b$
   Solve the following two exercises without worrying about convergence of series or differentiability of functions.

2. If $y = F(x)$ is an arbitrary function, then $y = F(x + at)$ represents a wave of fixed shape that moves to the left along the x-axis with velocity a (Figure 13.12).

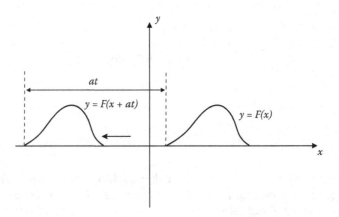

FIGURE 13.12
Wave of fixed shape moving to the left.

Similarly, if $y = G(x)$ is another arbitrary function, then $y = G(x - at)$ is a wave moving to the right, and the most general one-dimensional wave with velocity a is

$$y(x, t) = F(x + at) + G(x - at). \tag{$*$}$$

a. Show that $(*)$ satisfies the wave equation.
b. It is easy to see that the constant a in the wave equation has the dimension of velocity. Also, it is intuitively clear that if a stretched string is disturbed, then the waves will move in both directions away from the source of the disturbance. These considerations suggest introducing the new variables $\alpha = x + at$, $\beta = x - at$. Show that with these independent variables, the wave equation becomes

$$\frac{\partial^2 y}{\partial \alpha \partial \beta} = 0.$$

From this derive $(*)$ by integration. Formula $(*)$ is called *d'Alembert's solution* of the wave equation. It was also obtained, slightly later and independently, by Euler.

3. Solve the vibrating string problem in the text if the initial shape $y(x, 0) = f(x)$ is specified by the given function. In each case, sketch the initial shape of the string on a set of axes.

a. $f(x) = \begin{cases} 2cx/\pi & \text{if } 0 \le x \le \pi/2 \\ 2c(\pi - x)/\pi & \text{if } \pi/2 \le x \le \pi \end{cases}$

b. $f(x) = \frac{1}{\pi}x(\pi - x)$

c. $f(x) = \begin{cases} x & \text{if} & 0 \leq x \leq \pi/4 \\ \pi/4 & \text{if} & \pi/4 < x < 3\pi/4 \\ \pi - x & \text{if} & 3\pi/4 \leq x \leq \pi \end{cases}$

4. Solve the vibrating string problem in the text if the initial shape $y(x, 0) = f(x)$ is that of a single arch of the sine curve $f(x) = c \sin x$. Show that the moving string always has the same general shape, regardless of the value of c. Do the same for functions of the form $f(x) = c \sin nx$. Show in particular that there are $n - 1$ points between $x = 0$ and $x = \pi$ at which the string remains motionless; these points are called *nodes*, and these solutions are called *standing waves*. Draw sketches to illustrate the movement of the standing waves.

5. The problem of the *struck string* is that of solving the wave equation with the boundary conditions

$$y(0, t) = 0, \quad y(\pi, t) = 0$$

and the initial conditions

$$\left.\frac{\partial y}{\partial t}\right|_{t=0} = g(x) \quad \text{and} \quad y(x, 0) = 0.$$

(These initial conditions mean that the string is initially in the equilibrium position, and has an initial velocity $g(x)$ at the point x as a result of being struck.) By separating variables and proceeding formally, obtain the solution

$$y(x, t) = \sum_{j=1}^{\infty} c_j \sin jx \sin jat,$$

where

$$c_j = \frac{2}{\pi j a} \int_0^\pi g(x) \sin jx \; dx \;.$$

6. Consider an infinite string stretched taut on the x-axis from $-\infty$ to $+\infty$. Let the string be drawn aside into a curve $y = f(x)$ and released, and assume that its subsequent motion is described by the wave equation.

a. Use $(*)$ in Exercise **2** to show that the string's displacement is given by *d'Alembert's formula*

$$y(x, t) = \frac{1}{2}[f(x + at) + f(x - at)]. \qquad (**)$$

Hint: Remember the initial conditions.

b. Assume further that the string remains motionless at the points $x = 0$ and $x = \pi$ (such points are called *nodes*), so that $y(0, t) = y(\pi, t) = 0$, and use $(**)$ to show that f is an odd function that is periodic with period 2π (that is, $f(-x) = f(x)$ and $f(x + 2\pi) = f(x)$).

c. Show that since f is odd and periodic with period 2π then f necessarily vanishes at 0 and π.

d. Show that Bernoulli's solution of the wave equation can be written in the form $(**)$.
   Hint: $2 \sin nx \cos nat = \sin[n(x + at)] + \sin[n(x - at)]$.

7. If $y = F(x)$ is an arbitrary function, then $y = F(x + at)$ represents a wave of fixed shape that moves to the left along the x-axis with velocity a (Figure 13.12).
   Similarly, if $y = G(x)$ is another arbitrary function, then $y = G(x - at)$ is a wave moving to the right, and the most general one-dimensional wave with velocity a is

$$y(x, t) = F(x + at) + G(x - at). \qquad (*)$$

a. Show that $(*)$ satisfies the wave equation (13.4.14).

b. It is easy to see that the constant a in equation (13.4.14) has the dimensions of velocity. Also, it is intuitively clear that if a stretched string is disturbed, then the waves will move in both directions away from the source of the disturbance. These considerations suggest introducing the new variables $\alpha = x + at$, $\beta = x - at$. Show that with these independent variables, equation (13.4.14) becomes

$$\frac{\partial^2 y}{\partial \alpha \partial \beta} = 0.$$

From this derive $(*)$ by integration. Formula $(*)$ is called *d'Alembert's solution* of the wave equation. It was also obtained, slightly later and independently, by Euler.

8. Consider an infinite string stretched taut on the x-axis from $-\infty$ to $+\infty$. Let the string be drawn aside into a curve $y = f(x)$ and released, and assume that its subsequent motion is described by the wave equation (13.4.14).

a. Use (∗) in Exercise **2** to show that the string's displacement is given by *d'Alembert's formula*

$$y(x, t) = \frac{1}{2}[f(x + at) + f(x - at)] . \qquad (\ast\ast)$$

Hint: Remember the initial conditions (13.4.15) and (13.4.16).

b. Assume further that the string remains motionless at the points $x = 0$ and $x = \pi$ (such points are called *nodes*), so that $y(0, t) = y(\pi, t) = 0$, and use (∗∗) to show that f is an odd function that is periodic with period 2π (that is, $f(-x) = f(x)$ and $f(x + 2\pi) = f(x)$).

c. Show that since f is odd and periodic with period 2π then f necessarily vanishes at 0 and π.

d. Show that Bernoulli's solution (13.4.20) of the wave equation can be written in the form (∗∗).
Hint: $2 \sin nx \cos nat = \sin[n(x + at)] + \sin[n(x - at)]$.

13.5 The Heat Equation

Fourier's Point of View

In [FOU], Fourier considered variants of the following basic question. Let there be given an insulated, homogeneous rod of length π with initial temperature at each $x \in [0, \pi]$ given by a function $f(x)$ (Figure 13.13). Assume that the endpoints are held at temperatureheated rod 0, and that the temperature of each cross-section is constant. The problem is to describe the temperature $u(x, t)$ of the point x in the rod at time t. Fourier perceived the fundamental importance of this problem as follows:

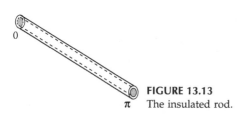

FIGURE 13.13
The insulated rod.

Primary causes are unknown to us; but are subject to simple and constant laws, which may be discovered by observation, the study of them being the object of natural philosophy.

··· ···

Heat, like gravity, penetrates every substance of the universe, its rays occupying all parts of space. The object of our work is to set forth the mathematical laws which this element obeys. The theory of heat will hereafter form one of the most important branches of general physics.

I have deduced these laws from prolonged study and attentive comparison of the facts known up to this time; all these facts I have observed afresh in the course of several years with the most exact instruments that have hitherto been used.

Let us now describe the manner in which Fourier solved his problem. First, it is required to write a differential equation which u satisfies. We shall derive such an equation using three physical principles:

1. The density of heat energy isdensity of heat energy proportional to the temperature u, hence the amount of heat energy in any interval $[a, b]$ of the rod is proportional to $\int_a^b u(x, t)dx$.

2. **(Newton's law of cooling)** The rate at which heat flows from a hot place to a cold one is proportional to the difference in temperature. The infinitesimal version of this statement is that the rate of heat flow across a point x (from left to right) is some negative constant times $\partial_x u(x, t)$.

3. **(Conservation of energy)** Heat has no sources or sinks.

Now **(3)** tells us that the only way that heat can enter or leave any interval portion $[a, b]$ of the rod is through the endpoints. And **(2)** tells us exactly how this happens. Using **(1)**, we may therefore write

$$\frac{d}{dt}\int_a^b u(x, t)dx = \eta^2[\partial_x u(b, t) - \partial_x u(a, t)].$$

We may rewrite this equation as

$$\int_a^b \partial_t u(x, t)dx = \eta^2 \int_a^b \partial_x^2 u(x, t)dx.$$

Differentiating in b, we find that

$$\partial_t u = \eta^2 \partial_x^2 u, \tag{13.5.1}$$

and that is the heat equation.

Suppose for simplicity that the constant of proportionality η^2 equals 1. Fourier guessed that equation (13.5.1) has a solution of the form $u(x, t) = \alpha(x)\beta(t)$. Substituting this guess into the equation yields

$$\alpha(x)\beta'(t) = \alpha''(x)\beta(t)$$

or

$$\frac{\beta'(t)}{\beta(t)} = \frac{\alpha''(x)}{\alpha(x)}.$$

Since the left side is independent of x and the right side is independent of t, it follows that there is a constant K such that

$$\frac{\beta'(t)}{\beta(t)} = K = \frac{\alpha''(x)}{\alpha(x)}$$

or

$$\beta'(t) = K\beta(t)$$

$$\alpha''(x) = K\alpha(x).$$

We conclude that $\beta(t) = Ce^{Kt}$. The nature of β, and hence of α, thus depends on the sign of K. But physical considerations tell us that the temperature will dissipate as time goes on, so we conclude that $K \le 0$. Therefore $\alpha(x) = \cos\sqrt{-K}x$ and $\alpha(x) = \sin\sqrt{-K}x$ are solutions of the differential equation for α. The initial conditions $u(0, t) = u(\pi, t) = 0$ (since the ends of the rod are held at constant temperature 0) eliminate the first of these solutions and force $K = -j^2$, j an integer. Thus Fourier found the solutions

$$u_j(x, t) = e^{-j^2 t}\sin jx, \quad j \in \mathbb{N}$$

of the heat equation. By linearity, any finite linear combination

$$u(x, t) = \sum_j b_j e^{-j^2 t}\sin jx \tag{13.5.2}$$

of these solutions is also a solution. It is plausible to extend this assertion to infinite linear combinations. Using the initial condition $u(x, 0) = f(x)$ again raises the question of whether "any" function $f(x)$ on $[0, \pi]$ can be written as a (infinite) linear combination of the functions $\sin jx$.

Fourier's solution to this last problem (of the sine functions spanning essentially everything) is roughly as follows. Suppose f is a function that is so representable:

$$f(x) = \sum_j b_j \sin jx. \tag{13.5.3}$$

Setting $x = 0$ gives

$$f(0) = 0.$$

Differentiating both sides of (13.5.3) and setting $x = 0$ gives

$$f'(0) = \sum_{j=1}^{\infty} jb_j. \tag{13.5.4}$$

Successive differentiation of (13.5.3), and evaluation at 0, gives

$$f^{(k)}(0) = \sum_{j=1}^{\infty} j^k b_j (-1)^{\lfloor k/2 \rfloor}$$

for k odd (by oddness of f, the even derivatives must be 0 at 0). Here $\lfloor \ \rfloor$ denotes the greatest integer function. Thus Fourier devised a system of infinitely many equations in the infinitely many unknowns $\{b_j\}$. He proceeded to solve this system by truncating it to an $N \times N$ system (the first N equations restricted to the first N unknowns), solving that truncated system, and then letting N tend to ∞. Suffice it to say that Fourier's arguments contained many dubious steps (see [FOU] and [LAN]).

The upshot of Fourier's intricate and lengthy calculations was that

$$b_j = \frac{2}{\pi} \int_0^{\pi} f(x) \sin jx \, dx. \tag{13.5.5}$$

By modern standards, Fourier's reasoning was specious; for he began by assuming that f possessed an expansion in terms of sine functions. The formula (13.5.5) hinges on that supposition, together with steps in which one compensated division by zero with a later division by ∞. Nonetheless, Fourier's methods give an actual *procedure* for endeavoring to expand any given f in a series of sine functions.

Fourier's abstract arguments constitute the first part of his book. The bulk, and remainder, of the book consists of separate chapters in which the expansions for particular functions are computed.

Example 13.23: Suppose that the thin rod in the setup of the heat equation is first immersed in boiling water so that its temperature is uniformly 100°C. Then imagine that it is removed from the water at time $t = 0$ with its ends immediately put into ice so that these ends are kept at temperature 0°C. Let us find the temperature $u = u(x, t)$ under these circumstances.

The initial temperature distribution is given by the constant function

$$f(x) = 100, \quad 0 < x < \pi.$$

The two boundary conditions, and the other initial condition, are as usual. Thus our job is simply this: to find the sine series expansion of this function f. We calculate that

$$b_j = \frac{2}{\pi} \int_0^\pi 100 \, \sin \, jx \, dx$$

$$= -\left. \frac{200}{\pi} \frac{\cos jx}{j} \right|_0^\pi$$

$$= -\frac{200}{\pi} \left[\frac{(-1)^j}{j} - \frac{1}{j} \right]$$

$$= \begin{cases} 0 & \text{if } j = 2\ell \text{ is even} \\ \dfrac{400}{\pi j} & \text{if } j = 2\ell - 1 \text{ is odd.} \end{cases}$$

Thus

$$f(x) = \frac{400}{\pi} \left(\sin x + \frac{\sin 3x}{3} + \frac{\sin 5x}{5} + \cdots \right).$$

Now, referring to formula (13.5.2) from our general discussion of the heat equation, we know that

$$u(x, t) = \frac{400}{\pi} \left(e^{-t} \sin x + \frac{1}{3} e^{-9t} \sin 3x + \frac{1}{5} e^{-25t} \sin 5x + \cdots \right). \qquad \square$$

Example 13.24: Let us find the steady-state temperature of the thin rod from our analysis of the heat equation if the fixed temperatures at the ends $x = 0$ and $x = \pi$ are w_1 and w_2 respectively.

The phrase "steady-state" means that $\partial u / \partial t = 0$, so that the heat equation reduces to $\partial^2 u / \partial x^2 = 0$ or $d^2 u / dx^2 = 0$. The general solution is then $u = Ax + B$. The values of these two constants A and B are forced by the two boundary conditions.

In fact a little high school algebra tells us that

$$u = w_1 + \frac{1}{\pi}(w_2 - w_1)x .$$

□

The steady-state version of the 3-dimensional heat equation

$$a^2 \left(\frac{\partial^2 u}{\partial x^2} + \frac{\partial^2 u}{\partial y^2} + \frac{\partial^2 u}{\partial z^2} \right) = \frac{\partial u}{\partial t}$$

is

$$\frac{\partial^2 u}{\partial x^2} + \frac{\partial^2 u}{\partial y^2} + \frac{\partial^2 u}{\partial z^2} = 0.$$

This last is called *Laplace's equation*. The study of this equation and its so-lutions and subsolutions and their applications is a deep and rich branch of mathematics called *potential theory*. There are applications to heat, to Laplace equation gravitation, to electromagnetics, and to many other parts of phy-sics. The equation plays a central role in the theory of partial differential equations, and is also an integral part of complex variable theory.

Exercises

1. Solve the boundary value problem

$$a^2 \frac{\partial^2 w}{\partial x^2} = \frac{\partial w}{\partial t}$$
$$w(x, 0) = f(x)$$
$$w(0, t) = 0$$
$$w(\pi, t) = 0$$

if the last three conditions—the boundary conditions—are changed to

$$w(x, 0) = f(x)$$
$$w(0, t) = w_1$$
$$w(\pi, t) = w_2 .$$

2. In the solution of the heat equation, suppose that the ends of the rod are insulated instead of being kept fixed at 0°C. What are the new boundary conditions? Find the temperature $w(x, t)$ in this case by using just common sense.

3. Solve the problem of finding $w(x, t)$ for the rod with insulated ends at $x = 0$ and $x = \pi$ (see the preceding exercise) if the initial tem-perature distribution is given by $w(x, 0) = f(x)$.

Solve the following two exercises without worrying about convergence of series or differentiability of functions.

4. Solve the Dirichlet problem for the unit disc when the boundary function $f(\theta)$ is defined by

a. $f(\theta) = \cos\theta/2, \, -\pi \le \theta \le \pi$

b. $f(\theta) = \theta, \, -\pi < \theta < 0$

c. $f(\theta) = \begin{cases} 0 & \text{if } -\pi \le \theta < 0 \\ \sin\theta & \text{if } 0 \le \theta \le \pi \end{cases}$

d. $f(\theta) = \begin{cases} 0 & \text{if } -\pi \le \theta < \pi/2 \\ 1 & \text{if } \pi/2 \le \theta \le \pi \end{cases}$

e. $f(\theta) = \theta^2/4, \, -\pi \le \theta \le \pi$

*5. Suppose that the lateral surface of the thin rod that we analyzed in the text is not insulated, but in fact radiates heat into the surrounding air. If Newton's law of cooling (that a body cools at a rate proportional to the difference of its temperature with the temperature of the surrounding air) is assumed to apply, then show that the 1-dimensional heat equation becomes

$$a^2\frac{\partial^2 w}{\partial x^2} = \frac{\partial w}{\partial t} + c(w - w_0)$$

where c is a positive constant and w_0 is the temperature of the surrounding air.

6. In Exercise 5, find $w(x, t)$ if the ends of the rod are kept at 0°C, $w_0 = 0$°C, and the initial temperature distribution on the rod is $f(x)$.

*7. The 2-dimensional heat equation is

$$a^2\left(\frac{\partial^2 w}{\partial x^2} + \frac{\partial^2 w}{\partial y^2}\right) = \frac{\partial w}{\partial t}.$$

Use the method of separation of variables to find a steady-state solution of this equation in the infinite strip of the x-y plane bounded by the lines $x = 0$, $x = \pi$, and $y = 0$ if the following boundary conditions are satisfied:

$$w(0, y) = 0 \qquad w(\pi, y) = 0$$
$$w(x, 0) = f(x) \qquad \lim_{y \to +\infty} w(x, y) = 0.$$

*8. Show that the Dirichlet problem for the disc $\{(x, y): x^2 + y^2 \le R^2\}$, where $f(\theta)$ is the boundary function, has the solution

$$w(r, \theta) = \frac{1}{2}a_0 + \sum_{j=1}^{\infty} \left(\frac{r}{R}\right)^j \left(a_j \cos j\theta + b_j \sin j\theta\right)$$

where a_j and b_j are the Fourier coefficients of f. Show also that the Poisson integral formula for this more general disc setting is

$$w(r, \theta) = \frac{1}{2\pi} \int_{-\pi}^{\pi} \frac{R^2 - r^2}{R^2 - 2Rr \cos(\theta - \phi) + r^2} f(\phi) \, d\phi \; .$$

*9. Let w be a harmonic function (that is, a function annihilated by the Laplacian) in a planar region, and let C be any circle entirely contained (along with its interior) in this region. Prove that the value of w at the center of C is the average of its values on the circumference.

10. The 2-dimensional heat equation is

$$a^2 \left(\frac{\partial^2 w}{\partial x^2} + \frac{\partial^2 w}{\partial y^2}\right) = \frac{\partial w}{\partial t} \; .$$

Use the method of separation of variables to find a steady-state solution of this equation in the infinite half-strip of the x-y plane bounded by the lines $x = 0$, $x = \pi$, and $y = 0$ if the following boundary conditions are satisfied:

$$w(0, y, t) = 0 \qquad w(\pi, y, t) = 0$$
$$w(x, 0, 0) = f(x) \quad \lim_{y \to +\infty} w(x, y, t) = 0.$$

11. Derive the 3-dimensional heat equation

$$a^2 \left(\frac{\partial^2 w}{\partial x^2} + \frac{\partial^2 w}{\partial y^2} + \frac{\partial^2 w}{\partial z^2}\right) = \frac{\partial w}{\partial t}$$

by adapting the reasoning in the text to the case of a small box with edges Δx, Δy, Δz contained in a region R in x-y-z space where the temperature function $w(x, y, z, t)$ is sought. *Hint:* Consider the flow of heat through two opposite faces of the box, first perpendicular to the x-axis, then perpendicular to the y-axis, and finally perpendicular to the z-axis.

Notes

1 Notice that the result enunciated here is a decisive improvement over what we know about Taylor series. We have asserted that a function that is only continuously differentiable has a Fourier series that converges at every point. But even an infinitely differentiable function can have Taylor series that converges at no point.

2 More explicitly, $\lambda = 0$ gives a linear function for a solution and $\lambda < 0$ gives an exponential function for a solution.

Appendix: Review of Linear Algebra

Section A1. Linear Algebra Basics

When we first learn linear algebra, the subject is difficult because it is not usually presented in the context of applications. In the current text we see one of the most important applications of linear algebra: to provide a language in which to do analysis of several real variables. We now give a quick review of elementary linear algebra.

The principal properties of a vector space are that it have an additive structure and an operation of scalar multiplication. If $\mathbf{u} = (v_1, v_2, \ldots, v_k)$ and $\mathbf{v} = (v_1, v_2, \ldots, v_k)$ are elements of $\mathbb{R}^k$ and $a \in \mathbb{R}$ then define the operations of addition and scalar multiplication as follows:

$$\mathbf{u} + \mathbf{v} = (u_1 + v_1, u_2 + v_2, \ldots, u_k + v_k)$$

and

$$a \cdot \mathbf{u} = (au_1, au_2, \ldots, au_k).$$

Notice that the vector $\mathbf{0} = (0, 0, \ldots, 0)$ is the additive identity: $\mathbf{u} + \mathbf{0} = \mathbf{u}$ for any element $\mathbf{u} \in \mathbb{R}^k$. Also every element $\mathbf{u} = (u_1, u_2, \ldots, u_k) \in \mathbb{R}^k$ has an additive inverse $-\mathbf{u} = (-u_1, -u_2, \ldots, -u_k)$ that satisfies $\mathbf{u} + (-\mathbf{u}) = \mathbf{0}$.

Example A.1: We have

$$(3, -2, 7) + (4, 1, -9) = (7, -1, -2)$$

and

$$5 \cdot (3, -2, 7, 14) = (15, -10, 35, 70). \qquad \square$$

The first major idea in linear algebra is that of linear dependence:

Definition A.2: A collection of elements $\mathbf{u}^1, \mathbf{u}^2, \ldots, \mathbf{u}^m \in \mathbb{R}^k$ is said to be *linearly dependent* if there exist constants $a_1, a_2, \ldots, a_m$, not all zero, such that

$$\sum_{j=1}^{m} a_j \mathbf{u}^j = 0.$$

Example A.3: The vectors $\mathbf{u} = (1, 3, 4)$, $\mathbf{v} = (2, -1, -3)$, and $\mathbf{w} = (5, 1, -2)$ are linearly dependent because $1 \cdot \mathbf{u} + 2 \cdot \mathbf{v} - 1 \cdot \mathbf{w} = 0$.

However, the vectors $\mathbf{u}' = (1, 0, 0)$, $\mathbf{v}' = (0, 1, 1)$, and $\mathbf{w}' = (1, 0, 1)$ are *not* linearly dependent since, if there were constants a, b, c such that

$$a\,\mathbf{u}' + b\,\mathbf{v}' + c\,\mathbf{w}' = 0,$$

then

$$(a + c, b, b + c) = 0.$$

But this means that

$$
\begin{aligned}
a + c &= 0 \\
b &= 0 \\
b + c &= 0.
\end{aligned}
$$

We conclude that a, b, c must all be equal to zero. That is not allowed in the definition of linear dependence. □

A collection of vectors that is not linearly dependent is called *linearly independent*. The vectors linear independence $\mathbf{u}', \mathbf{v}', \mathbf{w}'$ in the last example are linearly independent. Any set of k linearly independent vectors in $\mathbb{R}^k$ is called a **basis** for $\mathbb{R}^k$.

How do we recognize a basis? Notice that k vectors

$$
\begin{aligned}
\mathbf{u}^1 &= (u_1^1, u_2^1, \ldots, u_k^1) \\
\mathbf{u}^2 &= (u_1^2, v_2^2, \ldots, v_k^2) \\
&\cdots \\
\mathbf{u}^k &= (u_1^k, u_2^k, \ldots, u_k^k)
\end{aligned}
$$

are linearly dependent if and only if there are numbers $a_1, a_2, \ldots, a_k$, not all zero, such that

$$a_1\,\mathbf{u}^1 + a_2\,\mathbf{u}^2 + \cdots + a_k\,\mathbf{u}^k = 0.$$

This in turn is true if and only if the system of equations

$$a_1 u_1^1 + a_2 u_1^2 + \cdots + a_k u_1^k = 0$$
$$a_1 u_2^1 + a_2 u_2^2 + \cdots + a_k u_2^k = 0$$
$$\cdots$$
$$a_1 u_k^1 + a_2 u_k^2 + \cdots + a_k u_k^k = 0$$

has a nontrivial solution. But such a system has a nontrivial solution if and only if

$$\det \begin{pmatrix} u_1^1 & u_1^2 & \cdots & u_1^k \\ u_2^1 & u_2^2 & \cdots & u_2^k \\ \cdots & & & \\ u_k^1 & u_k^2 & \cdots & u_k^k \end{pmatrix} = 0.$$

So a basis is a set of k vectors as above such that this determinant is *not* 0.

Bases are important because if $\mathbf{u}^1, \mathbf{u}^2, \ldots, \mathbf{u}^k$ form a basis then every element x of $\mathbb{R}^k$ can be expressed in one and only one way as

$$x = a_1 \mathbf{u}^1 + a_2 \mathbf{u}^2 + \cdots + a_k \mathbf{u}^k,$$

with $a_1, a_2, \ldots, a_k$ scalars. We call this a representation of x as a *linear combination* of $\mathbf{u}^1, \mathbf{u}^2, \ldots, \mathbf{u}^k$. To see that such a representation is always possible, and is unique, let $x = (x_1, x_2, \ldots, x_k)$ be any element of $\mathbb{R}^k$. If $\mathbf{u}^1, \mathbf{u}^2, \ldots, \mathbf{u}^k$ form a basis then we wish to find $a_1, a_2, \ldots, a_k$ such that

$$x = a_1 \mathbf{u}^1 + a_2 \mathbf{u}^2 + \cdots + a_k \mathbf{u}^k.$$

But, as above, this leads to the system of equations

$$a_1 u_1^1 + a_2 u_1^2 + \cdots + a_k u_1^k = x_1$$
$$a_1 u_2^1 + a_2 u_2^2 + \cdots + a_k u_2^k = x_2 \qquad \text{(A.4)}$$
$$\cdots$$
$$a_1 u_k^1 + a_2 u_k^2 + \cdots + a_k^k u_k^k = x_k.$$

Now Cramer's Rule tells us that the unique solution of the system (A.4) is given by Cramer's Rule

$$a_1 = \frac{\det \begin{pmatrix} x_1 & u_1^2 & \cdots & u_1^k \\ x_2 & u_2^2 & \cdots & u_2^k \\ & \cdots & & \\ x_k & u_k^2 & \cdots & u_k^k \end{pmatrix}}{\det \begin{pmatrix} u_1^1 & u_1^2 & \cdots & u_1^k \\ u_2^1 & u_2^2 & \cdots & u_2^k \\ & \cdots & & \\ u_k^1 & u_k^2 & \cdots & u_k^k \end{pmatrix}}, \quad a_2 = \frac{\det \begin{pmatrix} u_1^1 & x_1 & \cdots & u_1^k \\ u_2^1 & x_2 & \cdots & u_2^k \\ & \cdots & & \\ u_k^1 & x_k & \cdots & u_k^k \end{pmatrix}}{\det \begin{pmatrix} u_1^1 & u_1^2 & \cdots & u_1^k \\ u_2^1 & u_2^2 & \cdots & u_2^k \\ & \cdots & & \\ u_k^1 & u_k^2 & \cdots & u_k^k \end{pmatrix}},$$

$$\cdots$$

$$\cdots, a_k = \frac{\det \begin{pmatrix} u_1^1 & u_1^2 & \cdots & x_1 \\ u_2^1 & u_2^2 & \cdots & x_2 \\ & \cdots & & \\ u_k^1 & u_k^2 & \cdots & x_k \end{pmatrix}}{\det \begin{pmatrix} u_1^1 & u_1^2 & \cdots & u_1^k \\ u_2^1 & u_2^2 & \cdots & u_2^k \\ & \cdots & & \\ u_k^1 & u_k^2 & \cdots & u_k^k \end{pmatrix}}.$$

Notice that the nonvanishing of the determinant in the denominator is crucial for this method to work.

In practice we will be given a basis u^1, u^2, ..., u^k for $\mathbb{R}^k$ and a vector x and we wish to express x as a linear combination of u^1, u^2, ..., u^k. We may do so by solving a system of linear equations as above. A more elegant way to do this is to use the concept of the inverse of a matrix.

Definition A.5: If

$$M = (m_{pq})_{\substack{p=1,\ldots,k \\ q=1,\ldots,\ell}}$$

is a $k \times \ell$ matrix (where k is the number of rows, ℓ the number of columns, and m_{pq} is the element in the pth row and qth column) and

$$N = (n_{rs})_{\substack{r=1,\ldots,\ell \\ s=1,\ldots,m}}$$

is an $\ell \times m$ matrix, then the *product* $M \cdot N$ is defined to be the matrix

$$T = (t_{uv})_{\substack{u=1,\ldots,k \\ q=1,\ldots,m}}$$

where

$$t_{uv} = \sum_{q=1}^{\ell} m_{uq} \cdot n_{qv}.$$

Example A.6: Let

$$M = \begin{pmatrix} 2 & 3 & 9 \\ -1 & 4 & 0 \\ 5 & -3 & 6 \\ 4 & 4 & 1 \end{pmatrix}$$

and

$$N = \begin{pmatrix} -3 & 0 \\ 2 & 5 \\ -4 & -1 \end{pmatrix}.$$

Then $T = M \cdot N$ is well defined as a 4×2 matrix. We notice, for example, that

$$t_{11} = 2 \cdot (-3) + 3 \cdot 2 + 9 \cdot (-4) = -36$$

and

$$t_{32} = 5 \cdot 0 + (-3) \cdot 5 + 6 \cdot (-1) = -21.$$

Six other easy calculations of this kind yield that

$$M \cdot N = \begin{pmatrix} -36 & 6 \\ 11 & 20 \\ -45 & -21 \\ -8 & 19 \end{pmatrix}.$$

□

Definition A.7: Let M be a $k \times k$ matrix. A matrix N is called the *inverse* of M if $M \cdot N = N \cdot M = I_k = I$, where

$$I = \begin{pmatrix} 1 & 0 & \cdots & 0 \\ 0 & 1 & \cdots & 0 \\ & & \cdots & \\ 0 & 0 & \cdots & 1 \end{pmatrix}.$$

When M has an inverse then it is called *invertible*. We denote the inverse by M^{-1}.

It follows immediately from the definition that, in order for a matrix to be a candidate for being invertible, it must be square.

Proposition A.8: *Let M be a k × k matrix with nonzero determinant. Then M is invertible and the elements of its inverse are given by*

$$n_{ij} = \frac{(-1)^{i+j} \cdot \det M(i, j)}{\det M}.$$

Here $M(i, j)$ is the $(k-1) \times (k-1)$ matrix obtained by deleting the jth row and ith column from M.

Proof: This is a direct calculation.   □

Definition A.9: If M is either a matrix or a vector, then the *transpose* ^tM of M is defined as follows: If the *ij*th entry M is m_{ij} then the *ij*th entry of ^tM is m_{ij}.

We will find the transpose notion useful primarily as notation. When we want to multiply a vector by a matrix, the multiplication will only make sense (in the language of matrix multiplication) after we have transposed the vector.

Proposition A.10: *If*

$$\begin{aligned}
\mathbf{u}^1 &= (u_1^1, u_2^1, ..., u_k^1) \\
\mathbf{u}^2 &= (u_1^2, u_2^2, ..., u_k^2) \\
&\cdots \\
\mathbf{u}^k &= (u_1^k, u_2^k, ..., u_k^k)
\end{aligned}$$

form a basis for $\mathbb{R}^k$ *then let M be the matrix of the coefficients of these vectors and* M^{-1} *the inverse of M (which we know exists because the determinant of the matrix is nonzero). If* $x = (x_1, x_2, ..., x_k)$ *is any element of* $\mathbb{R}^k$ *then*

$$x = a_1 \cdot \mathbf{u}^1 + a_2 \cdot \mathbf{u}^2 + \cdots + a_k \cdot \mathbf{u}^k,$$

where

$$(a_1, a_2, ..., a_k) = x \cdot M^{-1}.$$

Proof: Let A be the vector of unknown coefficients $(a_1, a_2, ..., a_k)$. The system of equations that we need to solve to find $a_1, a_2, ..., a_k$ can be written in matrix notation as

$$A \cdot M = x.$$

Applying the matrix M^{-1} to both sides of this equation (on the right) gives

$$(A \cdot M) \cdot M^{-1} = x \cdot M^{-1}$$

or

$$A \cdot I = x \cdot M^{-1}$$

or

$$A = x \cdot M^{-1},$$

as desired. □

The *standard basis* for $\mathbb{R}^k$ consists of the vectors standard basis

$$
\begin{aligned}
\mathbf{e}^1 &= (1, 0, \ldots, 0) \\
\mathbf{e}^2 &= (0, 1, \ldots, 0) \\
&\cdots \\
\mathbf{e}^k &= (0, 0, \ldots, 1).
\end{aligned}
\tag{A.11}
$$

If $x = (x_1, x_2, \ldots, x_k)$ is any element of $\mathbb{R}^k$, then we may write

$$x = x_1 \mathbf{e}^1 + x_2 \mathbf{e}^2 + \cdots + x_k \cdot \mathbf{e}^k.$$

In other words, the usual coordinates with which we locate points in k-dimensional space are the coordinates with respect to the special basis (A.11). We write this basis as $\mathbf{e}^1, \mathbf{e}^2, \ldots, \mathbf{e}^k$.

If $x = (x_1, x_2, \ldots, x_k)$ and $\mathbf{y} = (y_1, y_2, \ldots, y_k)$ are elements of $\mathbb{R}^k$ then we define

$$\|x\| = \sqrt{(x_1)^2 + (x_2)^2 + \cdots + (x_k)^2}$$

and

$$x \cdot \mathbf{y} = x_1 y_1 + x_2 y_2 + \cdots + x_k y_k.$$

Proposition A.12: (The Schwarz Inequality)
If x and y are Schwarz inequality elements of $\mathbb{R}^k$ then

$$|x \cdot \mathbf{y}| \le \|x\| \|\mathbf{y}\|.$$

Proof: Write out both sides and square. If all terms are moved to the right then the right side becomes a sum of perfect squares and therefore the inequality is obvious. □

Table of Notation

Notation	Section	Definition
$\mathbb{Q}$	1.1	the rational numbers
$\sup X$	1.1	supremum of X
$\mathrm{lub}\, X$	1.1	least upper bound of X
$\inf X$	1.1	infimum of X
$\mathrm{glb}\, X$	1.1	greatest lower bound of X
$\mathbb{R}$	1.1	the real numbers
$|x|$	1.1	absolute value
$|x + y|\,| \le |x| + |y|$	1.1	triangle inequality
C	1.1AP	a cut
$\mathbb{C}$	1.2	the complex numbers
z	1.2	a complex number
i	1.2	the square root of -1
$\bar{z}$	1.2	complex conjugate
$|z|$	1.2	modulus of z
$e^{i\theta}$	1.2	complex exponential
$\{a_j\}$	2.1	a sequence
a_j	2.1	a sequence
a_{j_k}	2.2	a subsequence
$\liminf a_j$	2.3	limit infimum of a_j
$\limsup a_j$	2.3	limit supremum of a_j
a^j	2.4	a power sequence
e	2.4	Euler's number e
$\sum_{j=1}^{\infty} a_j$	3.1	a series
S_N	3.1	a partial sum
$\sum_{j=1}^{N} a_j$	3.1	a partial sum
$\sum_{j=1}^{\infty} (-1)^j b_j$	3.3	an alternating series
$j!$	3.4	j factorial
$\sum_{n=0}^{\infty} \sum_{j=0}^{n} a_j \cdot b_{n-j}$	3.5	the Cauchy product of series
(a, b)	4.1	open interval
$[a, b]$	4.1	closed interval
$[a, b)$	4.1	half-open interval

(Continued)

Notation	Section	Definition
$(a, b]$	4.1	half-open interval
U	4.1	an open set
F	4.1	a closed set
∂S	4.1	boundary of S
$^c S$	4.1	complement of S
$\bar{S}$	4.2	closure of S
$\overset{\circ}{S}$	4.2	interior of S
$\{O_\alpha\}$	4.3	an open cover
S_j	4.4	step in constructing the Cantor set
C	4.4	the Cantor set
$\lim_{E \ni x \to P} f(x)$	5.1	limit of f at P
ℓ	5.1	a limit
$f + g$	5.1	sum of functions
$f - g$	5.1	difference of functions
$f \cdot g$	5.1	product of functions
f/g	5.1	quotient of functions
$f \circ g$	5.2	composition of functions
f^{-1}	5.2	inverse function
$f^{-1}(W)$	5.2	inverse image of a set
$f(L)$	5.3	image of the set L
m	5.3	minimum for a function f
M	5.3	maximum for a function f
$\lim_{x \to P^-} f(x)$	5.4	left limit of f at P
$\lim_{x \to P^+} f(x)$	5.4	right limit of f at P
$f'(x)$	6.1	derivative of f at x
df/dx	6.1	derivative of f
$Lip_\alpha(I)$	6.3	space of Lipschitz-α functions
$C^{k,\alpha}(I)$	6.3	space of smooth functions of order k, α
p	7.1	a partition
I_j	7.1	interval from the partition
Δ_j	7.1	length of I_j
$m(\mathcal{P})$	7.1	mesh of the partition
$\mathcal{R}(f, \mathcal{P})$	7.1	Riemann sum
$\int_a^b f(x)dx$	7.1	Riemann integral
$\int_b^a f(x)dx$	7.2	integral with reverse orientation
$\mathcal{U}(f, \mathcal{P}, \alpha)$	7.3	upper Riemann sum
$\mathcal{L}(f, \mathcal{P}, \alpha)$	7.3	lower Riemann sum
$I^*(f)$	7.3	upper integral of f

Notation	Section	Definition
$I_*(f)$	7.3	lower integral of f
$\int f\,d\alpha$	7.3	Riemann–Stieltjes integral
Vf	7.4	total variation of f
f_j	8.1	sequence of functions
$\{f_j\}$	8.1	sequence of functions
$\lim_{x \to s} f(x)$	8.2	limit of f as x approaches s
$\sum_{j=1}^{\infty} f_j(x)$	8.3	series of functions
$S_N(x)$	8.3	partial sum of a series of functions
$p(x)$	8.4	a polynomial
$\sum_{j=0}^{\infty} a_j(x-c)^j$	9.1	a power series
R_N	9.1	tail of the power series
ρ	9.2	radius of convergence
$f(x) = \sum_{j=0}^{k} f^{(j)}(a)\frac{(x-a)^j}{j!} + R_{k,a}(x)$	9.2	Taylor expansion
$exp(x)$	9.3	the exponential function
$\sin x$	9.3	the sine function
$\cos x$	9.3	the cosine function
$Sin\ x$	9.3	sine with restricted domain
$Cos\ x$	9.3	cosine with restricted domain
$\ln x$	9.4	the natural logarithm function
$\mathbb{R}^k$	10.1	multidimensional Euclidean space
$x = (x_1, x_2, x_3)$	10.1	a point in multidimensional space
$B(x, r)$	10.1	an (open) ball in multidimensional space
$\bar{B}(x, r)$	10.1	a closed ball in multidimensional space
$\lim_{x \to P} f(x)$	10.1	limit in multidimensional space
M_P	10.1	the derivative of f at P
$\mathcal{R}_P$	10.2	remainder term for the derivative
Jf	10.3	Jacobian matrix
(X, ρ)	11.1	metric space
$B(P, r)$	11.1	open ball
$\bar{B}(P, r)$	11.1	closed ball
$\lim_{x \to P} f(x)$	11.1	limit of f at P
U	11.2	open set
E	11.2	closed set
$\{O_\alpha\}_{\alpha \in A}$	11.2	open covering

(*Continued*)

Notation	Section	Definition
$\bar{S}$	11.3	closure of the set S
$\mathcal{F}$	11.4	family of functions
$dy/dx = F(x, y)$	12.1	first-order differential equation
$y(x) = y_0 + \int_{x_0}^x F(t, y(t))dt$	12.1	integral equation equivalent of first-order ODE
$y_{j+1}(x) = y_0 + \int_{x_0}^x F(t, y_j(t))dt$	12.1	Picard iteration technique
$(j + 1)a_{j+1} + (j - p)a_j = 0$	12.2	a recursion
$- 2m(m - 1) - m + 1 = 0$	12.2	indicial equation
c_n	13.1	Fourier coefficient
$\hat{f}(n)$	13.2	nth Fourier coefficient
Sf	13.2	Fourier series
$S_N f$	13.2	partial sum of Fourier series
D_N	13.2	Dirichlet kernel
$\hat{f}(\xi)$	13.3	Fourier transform of f
$C_0(\mathbb{R})$	13.3	continuous functions that vanish at ∞
a_n	13.4	Fourier cosine coefficient
b_n	13.4	Fourier sine coefficient
Δ	13.4	Laplacian
$w(r, \theta) = \frac{1}{2}a_0$ $+ \sum_{j=1}^\infty r^j(a_j \cos j\theta + b_j \sin j\theta)$	13.4	solving the Dirichlet problem
$\mathbb{N}$	A.1	the natural numbers
$\hat{x}$	A.1	successor
$Q(n)$	A.1	inductive statement
$\begin{pmatrix} n \\ k \end{pmatrix}$	A.1	choose function
$\mathbb{Z}$	A.2	the integers
$[(a, b)]$	A.2	an integer
$\mathbb{Q}$	A.3	the rational numbers
$[(c, d)]$	A.3	a rational number

Glossary

Abel's convergence test	A test for convergence of series that is based on summation by parts.
Absolutely convergent series	A series for which the absolute values of the terms form a convergent series.
Absolute maximum	A number M is the absolute maximum for a function f if $f(x) \leq f(M)$ for every x.
Absolute minimum	A number m is the absolute minimum for a function f if $f(x) \geq f(m)$ for every x.
Absolute value	Given a real number x, its absolute value is the distance of x to 0.
Accumulation point	A point x is an accumulation point of a set S if every neighborhood of x contains infinitely many distinct elements of S.
Algebraic number	A number that is the solution of a polynomial equation with integer coefficients.
Alternating series	A series of real terms that alternate in sign.
Alternating series test	If an alternating series has terms tending to zero, then it converges.
and	The connective that is used for conjunction.
Archimedean Property	If a and b are positive real numbers, then there is a positive integer n so that $na > b$.
Ascoli-Arzela theorem	Let (Y, σ) be a metric space and assume that Y is compact. Let $\mathcal{F}$ be an equibounded, equicontinuous family of functions on Y. Then, there is a sequence $\{f_j\} \subseteq \mathcal{F}$ that converges uniformly to a continuous function on Y.

Atomic sentence	A sentence with a subject and a verb, and sometimes an object, but no connectives.
Baire Category Theorem	The result that a complete metric space is of second category.
Bessel's inequality	An inequality for Fourier coefficients having the form $\sum_{n=-N}^{N} |\hat{f}_n|^2 \leq \int_0^{2\pi} |f(t)|^2 \, dt$.
Bijection	A one-to-one, onto function.
Binomial expansion	The expansion, under multiplication, of the expression$(a + b)^n$.
Bolzano–Weierstrass Theorem	Every bounded sequence of real numbers has a convergent subsequence.
Boundary of a set	The set of boundary points for the set.
Boundary point	The point b is in the boundary of S if each neighborhood of b contains both points of S and points of the complement of S.
Bounded above	A subset $S \subset \mathbb{R}$ is bounded above if there is a real number b such that $s \leq b$ for all $s \in S$.
Bounded below	A subset $S \subset \mathbb{R}$ is bounded below if there is a real number c such that $s \geq c$ for all $s \in S$.
Bounded sequence	A sequence a_j with the property that there is a number M so that $|a_j| \leq M$ for every j.
Bounded set	A set S with the property that there is a number M with $|s| \leq M$ for every $s \in S$.
Bounded variation	A function having bounded total oscillation.
Cantor set	A compact set that is uncountable, has zero length, is perfect, is totally disconnected, and has many other unusual properties.
Cardinality	Two sets have the same cardinality when there is a one-to-one correspondence between them.
Cauchy Condensation Test	A series of decreasing, nonnegative terms converges if and only if its dyadically condensed series converges.
Cauchy criterion	A sequence a_j is said to be Cauchy if, for each $\varepsilon > 0$, there is an $N > 0$ so that, if $j, k > N$, then $|a_j - a_k| < \varepsilon$.
Cauchy criterion for a series	A series satisfies the Cauchy criterion if and only if the sequence of partial sums satisfies the Cauchy criterion for a sequence.

Cauchy product	A means for taking the product of two series.
Cauchy's Mean Value Theorem	A generalization of the Mean Value Theorem that allows the comparison of two functions.
Chain Rule	A rule for differentiating the composition of functions.
Change of variable	A method for transforming an integral by subjecting the domain of integration to a one-to-one function.
Closed ball	The set of points at distance less than or equal to some $r > 0$ from a fixed point P.
Closed set	The complement of an open set.
Closure of a set	The set together with its boundary points.
Common refinement of two partitions	The union of the two partitions.
Compact set	A set E is compact if every sequence in E contains a subsequence that converges to an element of E.
Comparison test for convergence	A series converges if it is majorized in absolute value by a convergent series.
Comparison test for divergence	A series diverges if it majorizes a positive divergent series.
Complement of a set	The set of points not in the set.
Complete space	A space in which every Cauchy sequence has a limit.
Complex conjugate	Given a complex number $z = x + iy$, the conjugate is the number $\bar{z} = x - iy$.
Complex numbers	The set $\mathbb{C}$ of ordered pairs of real numbers equipped with certain operations of addition and multiplication.
Composition	The composition of two functions is the succession of one function by the other.
Conditionally convergent series	A series that converges, but not absolutely.
Connected set	A set that cannot be separated by two disjoint open sets.
Connectives	The words that are used to connect atomic sentences. These are "and," "or," "not," "if-then," and "if and only if."
Continuity at a point	The function f is continuous at P if the limit of f at P equals the value of f at P. Equivalently, given $\varepsilon > 0$, there is a $\delta > 0$ so that $|x - P| < \delta$ implies $|f(x) - f(P)| < \varepsilon$.

Continuous function	A function for which the inverse image of an open set is open.
Continuously differentiable function	A function that has a derivative at every point, and so that the derivative function is continuous.
Contrapositive	For a statement **"A implies B"**, the contrapositive statement is " **~B implies ~A"**.
Convergence of a sequence (of scalars)	A sequence a_j with the property that there is a limiting element ℓ so that, for any $\varepsilon > 0$, there is a positive integer N so that, if $j > N$, then $|a_j - \ell| < \varepsilon$.
Convergence of a series	A series converges if and only if its sequence of partial sums converges.
Converse	For a statement **"A implies B"**, the converse statement is **"B implies A"**.
Cosine function	The function $\cos x = \sum_{j=0}^{\infty} (-1)^j x^{2j} / (2j)!$.
Countable set	A set that has the same cardinality as the natural numbers.
Cramer's Rule	A device in linear algebra for solving systems of linear equations.
Decreasing sequence	The sequence $\{a_j\}$ of real numbers is decreasing if $a_1 \geq a_2 \geq a_3 \geq \cdots$.
Dedekind cut	A rational halfline. Used to construct the real numbers.
De Morgan's Laws	The identities $^c(A \cup B) = {}^cA \cap {}^cB$ and $^c(A \cap B) = {}^cA \cup {}^cB$.
Density Property	If $c < d$ are real numbers, then there is a rational number q with $c < q < d$.
Denumerable set	A set that is either empty, finite, or countable.
Derivative	The limit $\lim_{t \to x}(f(x) - f(t))/(t - x)$ for a function f on an open interval.
Derived power series	The series obtained by differentiating a power series term by term.
Determinant	The signed sum of products of elements of a matrix.
Difference quotient	The quotient $(f(t) - f(x))/(t - x)$ for a function f on an open interval.
Differentiable	A function that possesses the derivative at a point. This will be written differently in one variable and in several variables.
Dirichlet function	A function, taking only the values 0 and 1, which is highly discontinuous.

Dirichlet kernel	A kernel that represents the partial sum of a Fourier series. The kernel has the form $D_N(t) = (\sin(N + \frac{1}{2})t)/(\sin\frac{1}{2}t)$.
Dirichlet problem on the disc	The problem of finding a harmonic function on the disc with specified boundary values.
Disconnected set	A set that can be separated by two disjoint open sets.
Discontinuity of the first kind	A point at which a function f is discontinuous because the left and right limits at the point disagree.
Discontinuity of the second kind	A point at which a function f is discontinuous because either the left limit or the right limit at the point does not exist.
Diverge to infinity	A sequence with elements that become arbitrarily large.
Domain of a function	See *function*.
Domain of integration	The interval over which the integration is performed.
Dummy variable	A variable whose role in an argument or expression is formal. A dummy variable can be replaced by any other variable with no logical consequences.
Element of	A member of a given set.
Empty set	The set with no elements.
Equibounded	A family of functions $\mathcal{F} = \{f_\alpha\}$ of functions is equibounded if there is a number $M > 0$ such that $|f(x)| \leq M$ for all $x \in X$ and all $f \in \mathcal{F}$.
Equicontinuous family of functions	A family $\mathcal{F} = \{f_\alpha\}$ of functions is equicontinuous if for every $\varepsilon > 0$ there is a $\delta > 0$ such that when $\rho(s, t) < \delta$, then $|f_\alpha(s) - f_\alpha(t)| < \varepsilon$.
Equivalence classes	The pairwise disjoint sets into which an equivalence relation partitions a set.
Equivalence relation	A relation that partitions the set in question into pairwise disjoint sets, called *equivalence classes*.
Euler's formula	The identity $e^{iy} = \cos y + i \sin y$.
Euler's number	This is the number $e = 2.71828\ldots$ that is known to be irrational, indeed transcendental.
Exponential function	The function $exp(z) = \sum_{j=0}^{\infty} z^j/j!$.

Field	A system of numbers equipped with operations of addition and multiplication and satisfying eleven natural axioms.
Finite-dimensional space	A linear space with a finite basis.
Finite set	A set that can be put in one-to-one correspondence with a set of the form $\{1, 2, \ldots, n\}$ for some positive integer n.
Finite subcovering	An open covering $\mathcal{U} = \{U_j\}_{j=1}^{k}$ is a finite subcovering of E if each element of $\mathcal{U}$ is an element of a larger covering $\mathcal{V}$.
First category	A space that can be written as the countable union of nowhere dense sets.
First-order differential equation (ODE)	An equation of the form $dy/dx = F(x, y)$.
For all	The quantifier $\forall$ for making a statement about all objects of a certain kind.
Fourier coefficient	The coefficient $\hat{f}(n) = (1/[2\pi]) \int_{0}^{2\pi} f(t)e^{-int}dt$ of the Fourier series for the function f.
Fourier series	A series of the form $f(t) \sim \Sigma_j c_j e^{ijt}$ that decomposes the function f as a sum of sines and cosines. We sometimes write $Sf \sim \Sigma_{j=-\infty}^{\infty} \hat{f}(j)e^{ijt}$.
Fourier transform	Given a function f on the real line, its Fourier transform is $\hat{f}(\xi) = \int_{\mathbb{R}} f(t)e^{it\xi}dt$.
Function	A *function* from a set A to a set B is a relation f on A and B such that for each $a \in A$ there is one and only one pair $(a, b) \in f$. We call A the *domain* and B the *range* of the function.
Fundamental Theorem of Calculus	A result relating the values of a function to the integral of its derivative: $f(x) - f(a) = \int_{a}^{x} f'(t)dt$.
Geometric series	This is a series of powers of a fixed base.
Greatest lower bound	The real number c is the greatest lower bound for the set $S \subset \mathbb{R}$ if b is a lower bound and if there is no lower bound that is greater than c.
Green's function	A function $G(x, y)$ that is manufactured from the fundamental solution for the Laplacian and is useful in solving partial differential equations.
Harmonic function	A function that is annihilated by the Laplacian.

Hausdorff space	A topological space in which distinct points p, q are separated by disjoint neighborhoods.
Higher derivative	The derivative of a derivative.
Hilbert transform	The singular integral operator $f \mapsto P.\,V.\ \int f(t)/(x - t)dt$ that governs convergence of Fourier series and many other important phenomena in analysis.
i	The square root of -1 in the complex number system.
Identity matrix	The square matrix with 1s on the diagonal and 0s in the other entries.
If and only if	The connective that is used for logical equivalence.
If–then	The connective that is used for implication.
Image of a function	See *function*. The image of the function f is Image $f = \{b \in B : \exists\, a \in A$ such that $f(a) = b\}$.
Image of a set	If f is a function, then the image of E under f is the set $\{f(e) : e \in E\}$.
Imaginary part	Given a complex number $z = x + iy$, its imaginary part is y.
Implicit function theorem	A result that gives sufficient conditions, in terms of the derivative, on an equation of several variables to be able to solve for one variable in terms of the others.
Increasing sequence	The sequence of real numbers a_j is increasing if $a_1 \leq a_2 \leq a_3 \leq \cdots$.
Infimum	See *greatest lower bound*.
Infinite set	A set is infinite if it is not finite.
Initial condition	For a first-order differential equation, this is a side condition of the form $y(x_0) = y_0$.
Integers	The natural numbers, the negatives of the natural numbers, and zero.
Integral equation equivalent of a first-order ODE	An equation of the form $y(x) = y_0 + \int_{x_0}^{x} F(t, y(t))dt$.
Integration by parts	A device for integrating a product.
Interior of a set	The collection of interior points of the set.
Interior point	A point of the set S that has a neighborhood lying in S.
Intermediate value theorem	The result that says that a continuous function does not skip values.

Intersection of sets	The set of elements common to two or more given sets.
Interval	A subset of the reals that contains all its intermediate points.
Interval of convergence of a power series	An interval of the form $(c - \rho, c + \rho)$ on which the power series converges (uniformly on compact subsets of the interval).
Inverse function theorem	A result that gives sufficient conditions, in terms of the derivative, for a function to be locally invertible.
Inverse of a matrix	Given a square matrix A, we say that B is its inverse if $A{\cdot}B = B{\cdot}A = I$, where I is the identity matrix.
Invertible matrix	A matrix that has an *inverse*.
Irrational number	A real number that is not rational.
Isolated point of a set	A point of the set with a neighborhood containing no other point of the set.
Jacobian matrix	The matrix of partial derivatives of a mapping from $\mathbb{R}^k$ to $\mathbb{R}^k$.
k times continuously differentiable	A function that has k derivatives, each of which is continuous.
Lambert W function	A transcendental function W with the property that any of the standard transcendental functions (sine, cosine, exponential, logarithm) can be expressed in terms of W.
Laplacian	The partial differential operator given by $\Delta = \partial^2/\partial x_1^2 + \partial^2/\partial x_2^2 + \cdots + \partial^2/\partial x_k^2$.
Least upper bound	The real number b is the least upper bound for the set $S \subset \mathbb{R}$ if b is an upper bound and if there is no other upper bound that is less than b.
Least Upper Bound Property	The important defining property of the real numbers.
Left limit	A limit of a function at a point P that is calculated with values of the function that are to the left of P.
Legendre's equation	The ODE $(1 - x^2)y'' - 2xy' + p(p + 1)y = 0$.
l'Hôpital's Rule	A rule for calculating the limit of the quotient of two functions in terms of the quotient of the derivatives.
Limit	The value ℓ that a function approaches at a point of or an accumulation point P of the domain. Equivalently, given $\varepsilon > 0$,

	there is a $\delta > 0$ so that $|f(x) - \ell| < \varepsilon$ whenever $|x - P| < \delta$.
Limit infimum	The least limit of any subsequence of a given sequence.
Limit supremum	The greatest limit of any subsequence of a given sequence.
Linear combination	If $v_1, v_2, , v_k$ are vectors, then a linear combination is an expression of the form $c_1v_1 + c_2v_2 + \cdots c_kv_k$ for scalar coefficients c_j.
Linearly dependent set	In a linear space, a set that is not linearly independent.
Linearly independent set	In a linear space, a set that has no nontrivial linear combination giving 0.
Linear operator	A function between linear spaces that satisfies the linearity condition $T(cx + dy) = cT(x) + dT(y)$.
Lipschitz function	A function that satisfies a condition of the form $|f(s) - f(t)| \leq C|s - t|$ or $|f(s) - f(t)| \leq |s - t|^\alpha$ for $0 < \alpha \leq 1$.
Local extrema	Either a local maximum or a local minimum.
Local maximum	The point x is a local maximum for the function f if $f(x) \geq f(t)$ for all t in a neighborhood of x.
Local minimum	The point x is a local minimum for the function f if $f(x) \leq f(t)$ for all t in a neighborhood of x.
Logically equivalent	Two statements are logically equivalent if they have the same truth table.
Logically independent	Two statements are logically independent if neither one implies the other.
Lower bound	A real number c is a lower bound for a subset $S \subset \mathbb{R}$ if $s \geq c$ for all $s \in S$.
Lower Riemann sum	A Riemann sum devised for defining the Riemann–Stieltjes integral.
Mean Value Theorem	If f is a continuous function on $[a, b]$, differentiable on the interior, then the slope of the segment connecting $(a, f(a))$ and $(b, f(b))$ equals the derivative of f at some interior point.
Mesh of a partition	The maximum length of any interval in the partition.
Metric	The distance function on a metric space.

Metric space	A space X equipped with a distance function ρ.
$m \times n$ Matrix	A matrix with m rows and n columns.
Modulus	The modulus of a complex number $z = x + iy$ is $|z| = \sqrt{x^2 + y^2}$.
Monotone sequence	A sequence that is either increasing or decreasing.
Monotonically decreasing function	A function whose graph goes downhill when moving from left to right: $f(s) \geq f(t)$ when $s < t$.
Monotonically increasing function	A function whose graph goes uphill when moving from left to right: $f(s) \leq f(t)$ when $s < t$.
Monotonic function	A function that is either monotonically increasing or monotonically decreasing.
Natural logarithm function	The inverse function to the exponential function.
Natural numbers	The counting numbers 1, 2, 3, ...
Necessary for	An alternative phrase for converse implication.
Neighborhood of a point	An open set containing the point.
Neumann series	A series of the form $1/(1 - \alpha) = \sum_{j=0}^{\infty} \alpha^j$ for $|\alpha| < 1$.
Newton quotient	The quotient $(f(t) - f(x))/(t - x)$ for a function f on an open interval.
Non-terminating decimal expansion	A decimal expansion for a real number that has infinitely many nonzero digits.
Norm	The notion of distance on a normed linear space.
Normed linear space	A linear space equipped with a norm that is compatible with the linear structure.
Not	The connective that is used for negation.
Nowhere dense	A space is nowhere dense if its closure contains no balls.
One-to-one function	A function that takes different values at different points of the domain.
Only if	An alternative phrase for implication.
Onto	A function whose image equals its range.
Open ball	The set of points at distance less than some $r > 0$ from a fixed point P.
Open covering	A collection $\{U_\alpha\}_{\alpha \in A}$ of open sets is an open covering of a set S if $\cup_\alpha U_\alpha \supset S$.
Open Mapping Principle	The result that says that a bounded, surjective linear mapping is open.

Open set	A set that contains a neighborhood of each of its points.
Or	The connective that is used for disjunction.
Ordered field	A field equipped with an order relation that is compatible with the field structure.
Ordinary differential equation (ODE)	An equation involving a function of one variable and some of its derivatives.
Partial derivative	For a function of several variables, this is the derivative calculated in just one variable, with the other variables held fixed.
Partial sum of a Fourier series	The sum of the terms of a Fourier series having index between $-N$ and N.
Partial sum of functions	The sum of the first N terms of a series of functions.
Partial sum (of scalars)	The sum of the first N terms of a series of scalars.
Partition of the interval $[a, b]$	A finite, ordered set of points $\mathcal{P} = \{x_0, x_1, x_2, \ldots, x_{k-1}, x_k\}$, such that $a = x_0 \le x_1 \le x_2 \le \ldots \le x_{k-1} \le x_k = b$.
Peano axioms	An axiom system for the natural numbers.
Perfect set	A set that is closed and in which every point is an accumulation point.
Picard iteration technique	An iteration scheme for solving a first-order ODE using the steps $y_{j+1}(x) = y_0 + \int_{x_0}^{x} F(t, y_j(t))dt$.
Pinching Principle	A criterion for convergence of a sequence that involves bounding it below by a convergent sequence and bounding it above by another convergent sequence with the same limit.
Pointwise convergence of a sequence of functions	A sequence f_j of functions converges pointwise if $f_j(x)$ convergence for each x in the common domain.
Poisson kernel	The reproducing kernel for harmonic functions.
Polar form of a complex number	The polar form of a complex number z is $re^{i\theta}$, where r is the modulus of z and θ is the angle that the vector from 0 to z subtends with the positive x-axis.
Power series expanded about the point c	A series of the form $\sum_{j=0}^{\infty} a_j(x - c)^j$.

Power set	The collection of all subsets of a given set.
Principle of Induction	A proof technique for establishing a statement $Q(n)$ about the natural numbers.
Quantifier	A logical device for making a quantitative statement. Our standard quantifiers are "for all" and "there exists."
Radius of convergence of a power series	Half the length ρ of the interval of convergence.
Range of a functon	See *function*.
Rational numbers	Numbers that may be represented as quotients of integers.
Ratio Test for Convergence	A series converges if the limit of the sequence of quotients of summands is less than 1.
Ratio Test for Divergence	A series diverges if the limit of the sequence of quotients of summands is greater than 1.
Real analytic function	A function with a convergent power series expansion about each point of its domain.
Real numbers	An ordered field $\mathbb{R}$ containing the rationals $\mathbb{Q}$ so that every nonempty subset with an upper bound has a least upper bound.
Real part	Given a complex number $z = x + iy$, its real part is x.
Rearrangement of a series	A new series obtained by permuting the summands of the original series.
Relation	A relation on sets A and B is a subset of $A \times B$.
Remainder term for the Taylor expansion	The term $R_{k,a}(x)$ in the Taylor expansion.
Riemann integrable	A function for which the Riemann integral exists.
Riemann integral	The limit of the Riemann sums.
Riemann–Lebesgue Lemma	The result that says that the Fourier transform of an integrable function vanishes at infinity.
Riemann's lemma	A result guaranteeing the existence of the Riemann–Stieltjes integral in terms of the proximity of the upper and lower Riemann sums.

Riemann–Stieltjes integral	A generalization of the Riemann integral, which allows measure of the length of the interval in the partition by a function α.
Riemann sum	The approximate integral based on a partition.
Right limit	A limit of a function at a point P that is calculated with values of the function to the right of P.
Rolle's Theorem	The special case of the Mean Value Theorem when $f(a) = f(b) = 0$.
Root Test for Convergence	A series converges if the limit of the nth roots of the nth terms is less than one.
Root Test for Divergence	A series is divergent if the limit of the nth roots of the nth terms is greater than one.
Scalar	An element of either $\mathbb{R}$ or $\mathbb{C}$.
Schroeder–Bernstein Theorem	The result that says that if there is a one-to-one function from the set A to the set B and a one-to-one function from the set B to the set A, then A and B have the same cardinality.
Schwarz inequality	The inequality $|v \cdot w| \leq \|v\|\|w\|$.
Second category	A space that is not of first category.
Sequence of functions	A function from $\mathbb{N}$ into the set of functions on some space.
Sequence (of scalars)	A function from $\mathbb{N}$ into $\mathbb{R}$ or $\mathbb{C}$ or a metric space. We often denote the sequence by a_j.
Series of functions	An infinite sum of functions.
Series (of scalars)	An infinite sum of scalars.
Set	A collection of objects.
Setbuilder notation	The notation $\{x : P(x)\}$ for specifying a set.
Set-theoretic difference	The set-theoretic difference $A \setminus B$ consists of those elements that lie in A but not in B.
Set-theoretic isomorphism	A one-to-one, onto function.
Set-theoretic product	If A and B are sets, then their set-theoretic product is the set of ordered pairs (a, b) with $a \in A$ and $b \in B$.
Sine function	The function $\sin x = \sum_{j=0}^{\infty} (-1)^j x^{2j+1} / (2j+1)!$.
Smaller cardinality	The set A has smaller cardinality than the set B if there is a one-to-one mapping of A to B but none from B to A.
Strictly monotonically decreasing function	A function whose graph goes strictly downhill when moving from left to right: $f(s) > f(t)$ when $s < t$.

Strictly monotonically increasing function	A function whose graph goes strictly uphill when moving from left to right: $f(s) > f(t)$ when $s < t$.
Subcovering	A covering that is a subcollection of a larger covering.
Subfield	Given a field k, a subfield m is a subset of k that is also a field with the induced field structure.
Subsequence	A sequence that is a subset of a given sequence with the elements occurring in the same order.
Subset of	A subcollection of the members of a given set.
Successor	The natural number that follows a given natural number.
Suffices for	An alternative phrase for implication.
Summation by parts	A discrete analogue of integration by parts.
Supremum	See *least upper bound*.
Taylor's expansion	The expansion $f(x) = \sum_{j=0}^{k} f^{(j)}(a)\frac{(x-a)^j}{j!} + R_{k,a}(x)$ for a given function f.
Terminating decimal	A decimal expansion for a real number that has only finitely many nonzero digits.
There exists	The quantifier $\exists$ for making a statement about some objects of a certain kind.
Totally disconnected set	A set in which any two points can be separated by two disjoint open sets.
Transcendental number	A real number that is not algebraic.
Transpose of a matrix	Given a matrix $A = \{a_{ij}\}$, the transpose is the matrix obtained by replacing a_{ij} with a_{ji}.
Triangle inequality	The inequality $|a + b| \le |a| + |b|$ for real numbers.
Truth table	An array that shows the possible truth values of a statement.
Uncountable set	An infinite set that does not have the same cardinality as the natural numbers.
Uniform convergence of a sequence of functions	The sequence f_j of functions converges uniformly to a function f if, given $\varepsilon > 0$, there is an $N > 0$ so that, if $j > N$, then $|f_j(x) - f(x)| < \varepsilon$ for all x.
Uniform convergence of a series of functions	A series of functions such that the sequence of partial sums converges uniformly.

Uniformly Cauchy sequence of functions	A sequence of functions f_j with the property that, for $\varepsilon > 0$, there is an $N > 0$ so that, if $j, k > N$, then $|f_j(x) - f_k(x)| < \varepsilon$ for all x in the common domain.
Uniformly continuous	A function f is uniformly continuous if, for each $\varepsilon > 0$, there is a $\delta > 0$ so that $|f(s) - f(t)| < \varepsilon$ whenever $|s - t| < \delta$.
Union of sets	The collection of objects that lie in any one of a given collection of sets.
Universal set	The set of which all other sets are a subset.
Upper bound	A real number b is an upper bound for a subset $S \subset \mathbb{R}$ if $s \le b$ for all $s \in S$.
Upper Riemann sum	A Riemann sum devised for defining the Riemann–Stieltjes integral.
Venn diagram	A pictorial device for showing relation ships among sets.
Weierstrass Approximation Theorem	The result that any continuous function on $[0, 1]$ can be uniformly approximated by polynomials.
Weierstrass M-Test	A simple scalar test that guarantees the uniform convergence of a series of functions.
Weierstrass nowhere differentiable function	A function that is continuous on $[0, 1]$ that is not differentiable at any point of $[0, 1]$.
Well defined	An operation on equivalence classes is well defined if the result is independent of the representatives chosen from the equivalence classes.
Zero Test	If a series converges, then its summands tend to zero.

Bibliography

1. R. P. Boas, *A Primer of Real Functions*, Carus Mathematical Monograph No. 13, John Wiley & Sons, Inc., New York, 1960.
2. G. Birkhoff and G.-C. Rota, *Ordinary Differential Equations*, John Wiley & Sons, New York, 1978.
3. R. C. Buck, *Advanced Calculus*, 2d ed., McGraw-Hill Book Company, New York, 1965.
4. E. Coddington and N. Levinson, *Theory of Ordinary Differential Equations*, McGraw-Hill, New York, 1955.
5. N. Dunford and J. Schwartz, *Linear Operators*, Interscience Publishers, New York, 1958–1971.
6. H. Federer, *Geometric Measure Theory*, Springer-Verlag, New York, 1969.
7. J. Fourier, *The Analytical Theory of Heat*, G. E. Stechert & Co., New York, 1878.
8. K. Hoffman, *Analysis in Euclidean Space*, Prentice Hall, Inc., Englewood Cliffs, NJ, 1962.
9. S. Kakutani, Some Characterizations of Euclidean Space, *Japanese Journal of Mathematics.* 16(1939), 93–97.
10. Y. Katznelson, *Introduction to Harmonic Analysis*, John Wiley and Sons, New York, 1968.
11. J. L. Kelley, Banach Spaces with the Extension Property, *Transactions of the American Mathematical Society* 72(1952), 323–326.
12. A. N. Kolmogorov, *Grundbegriffe der Wahrscheinlichkeitsrechnung*, Springer-Verlag, Berlin, 1933.
13. S. G. Krantz, *The Elements of Advanced Mathematics,* 3rd ed., CRC Press, Boca Raton, FL, 2012.
14. S. G. Krantz, *A Panorama of Harmonic Analysis*, Mathematical Association of America, Washington, DC, 1999.
15. S. G. Krantz, *Partial Differential Equations and Complex Analysis*, CRC Press, Boca Raton, FL, 1992.
16. S. G. Krantz, *Handbook of Logic and Proof Techniques for Computer Scientists*, Birkhäuser, Boston, 2002.
17. S. G. Krantz, *Real Analysis and Foundations*, 3rd ed., CRC Press, Boca Raton, FL, 2013.
18. S. G. Krantz, *Function Theory of Several Complex Variables*, 2nd ed., American Mathematical Society, Providence, RI, 2001.
19. S. G. Krantz *Differential Equations: Theory, Technique, and Practice*, 2nd ed., Taylor & Francis/CRC Press, 2015.
20. S. G. Krantz, *Convex Analysis*, Taylor & Francis, Boca Raton, FL, 2015.

21. S. G. Krantz and H. R. Parks, *A Primer of Real Analytic Functions*, 2nd ed., Birkhäuser Publishing, Boston, 2002.
22. R. E. Langer, *Fourier Series: The Genesis and Evolution of a Theory*, Herbert Ellsworth Slaught Memorial Paper I, *The American Mathematical Monthly* 54(1947).
23. P. D. Lax, On the Existence of Green's Functions, *Proceedings of the American Mathematical Society* 3(1952), 526–531.
24. L. Loomis and S. Sternberg, *Advanced Calculus*, Addison-Wesley, Reading, MA, 1968.
25. N. Luzin, The Evolution of "Function", Part I, Abe Shenitzer, ed., *The American Mathematical Monthly* 105(1998), 59–67.
26. I. Niven, *Irrational Numbers*, Carus Mathematical Monograph No. 11, John Wiley & Sons, Inc., New York, 1956.
27. M. Reed and B. Simon, *Methods of Modern Mathematical Physics*, Academic Press, New York, 1972.
28. H. Royden, *Real Analysis*, Macmillan, New York, 1963.
29. W. Rudin, *Principles of Mathematical Analysis*, 3rd ed., McGraw-Hill Book Company, New York, 1976.
30. W. Rudin, *Real and Complex Analysis*, McGraw-Hill Book Company, New York, 1966.
31. W. Rudin, *Functional Analysis*, McGraw-Hill, New York, 1973.
32. A. Sobczyk, On the Extension of Linear Transformations, *Transactions of the American Mathematical Society* 55(1944), 153–169.
33. E. M. Stein and G. Weiss, *Introduction to Fourier Analysis on Euclidean Spaces*, Princeton University Press, Princeton, NJ, 1971.
34. K. Stromberg, *An Introduction to Classical Real Analysis*, Wadsworth Publishing, Inc., Belmont, CA, 1981.
35. J. Walker, *A Primer on Wavelets and their Scientific Applications*, 2nd ed., CRC Press, Boca Raton, FL, 2008.
36. K. Yosida, *Functional Analysis*, 6th ed., Springer-Verlag, New York, 1980.

Index

Printed in the United States
by Baker & Taylor Publisher Services